Animal Biodiversity: Patterns and Processes

Animal Biodiversity: Patterns and Processes

Gaurav Tyagi

RANDOM PUBLICATIONS
NEW DELHI (INDIA)

Animal Biodiversity: Patterns and Processes

ISBN 978-93-5111-785-8

Published in 2016 in India by

RANDOM PUBLICATIONS

Reprint, 2017

4376-A/4B, Gali Murari Lal, Ansari Road

New Delhi-110 002

Phone : +9111-43580356, 011-23289044, 011-43142548

e-mail: sales@randompublications.com,

info@randompublications.com, randomexports@gmail.com

Type Setting by : Friends Media, Delhi-110089

Digitally Printed at : Replika Press Pvt. Ltd.

Preface

This book contains a series of scientific articles regarding Animal Biodiversity. It addresses the economic, institutional and social challenges confronting scientists and policy makers in conserving biodiversity and ecosystem services that are critical for sustaining human well being and development.

The number and variety of plants, animals and other organisms that exist is known as biodiversity. It is an essential component of nature and it ensures the survival of human species by providing food, fuel, shelter, medicines and other resources to mankind. The richness of biodiversity depends on the climatic conditions and area of the region. All species of plants taken together are known as flora and about 70,000 species of plants are known till date. All species of animals taken together are known as fauna which includes birds, mammals, fish, reptiles, insects, crustaceans, molluscs, etc.

In an ecosystem, there may exist different landforms, each of which supports different and specific vegetation. Ecosystem diversity is difficult to measure since the boundaries of the communities, which constitute the various sub ecosystems, are elusive. Ecosystem diversity could best be understood if one studies the communities in various ecological niches within the given ecosystem; each community is associated with definite species complexes. These complexes are related to composition and structure of biodiversity.

During the last century, human impacts on our planet have led to an increasing and alarming loss of biodiversity. Scientists estimate that current extinction rates exceed those of prehistoric mass extinctions. Loss of biodiversity also means loss of genetic diversity and loss of ecosystems. Loss of suitable habitat is the major cause for declines in species populations. Development, wetland filling, and other activities not only reduce the total amount of habitat, but also fragment remaining forest, grassland, and wetland habitat into patches too small and too isolated to support some animal species.

The book therefore, attempts to provide an overall emphasis of diverse aspects of animal biodiversity, including soil, vectors of animal and plant diseases, agroecosystem diversity, forest biodiversity, marine, fresh water and island biodiversity.

– ***Author***

Contents

1

Biodiversity: Dimensions and Implications

BIODIVERSITY

The word 'biodiversity' is a combination of two words: biological and diversity. It refers to the variety of life on Earth. Biodiversity encompasses all the living things that exist in a certain area, in the air, on land or in water: plants, animals, microorganisms, fungi. The area considered may be as small as your backyard compost heap — or as big as our whole planet.

Animals and plants don't exist in isolation. All living things are connected to other living things and to their non-living environment (earth forms, rocks and rivers). If one tiny species in an ecosystem becomes extinct, we may not notice, or think it's important. But the biodiversity of that ecosystem will be altered, and all the ecosystems that the species belonged to will be affected.

IMPORTANT OF BIODIVERSITY

The natural environment is the source of all our resources for life. Environmental processes provide a wealth of services to the living world — providing us with air to breathe, water to drink and food to eat, as well as materials to use in our daily lives and natural beauty to enjoy.

Complex ecosystems with a wide variety of plants and animals tend to be more stable. A highly diverse ecosystem is a sign of a healthy system. Since all the living world relies on the natural environment, especially us, it is in our best interests and the interests of future generations to conserve biodiversity and our resources.

Some might argue that some species have become extinct, with no obvious effect on the environment. But the Earth's systems are so complex that we are still learning about environmental processes and resources and the roles they play. The careless loss of any part of the natural environment means that we may never know what use it was or could have been in terms of future technologies, say, or for medical science, or indeed for the health of the planet itself. It's important to understand that environments are constantly changing. A healthy, robust environment evolves and adapts to naturally changing conditions. It is fascinating to observe the far-reaching effects even small changes can make and the importance of genetic diversity for species to adapt, survive and evolve. Preservation of biodiversity is not necessarily about preserving everything currently in existence. It's more a question of 'walking lightly' on the Earth — a balance of respecting the natural changes that occur and of protecting species and environments from wanton extinction and destruction. Life on Earth would not be the same if our planet's biodiversity were to be radically affected.

THREE TYPES OF BIODIVERSITY

There are three aspects to biodiversity: species diversity, genetic diversity and ecosystem diversity. All three interact and change over time and from place to place. Species diversity refers to the variety of different living things. Genetic diversity refers to the variations between individuals of a species — characteristics passed down from parents to their offspring. Ecosystem diversity refers to the great variety of environments produced by the interplay of the living (animals and plants) and non-living world (earth forms, soil, rocks and water).

THREATS TO BIODIVERSITY

There are many threats to biodiversity. One of the problems is simply the number of people on Earth. As the Earth's population grows, and people encroach on more and more land, habitats can become fragmented and degraded. Another threat to biodiversity is the issue of climate change. There is world-wide concern over global warming. Pollution of our air and waterways also affects biodiversity. Overharvesting depletes the Earth's stocks of animal and plant resources and the ill-considered introduction of exotic species also has a negative effect on biodiversity. Many of these threats to biodiversity are the result of human activity. But it is also humans who can make positive changes — adopting attitudes to the Earth's rich resources that will respect and maintain this diversity.

TYPES OF BIODIVERSITY

The term biodiversity includes three different aspects, which are closely related to each other. Following are the types of biodiversity:

GENETIC DIVERSITY

It refers to the variation of genes within the species. This constitutes distinct population of the same species or genetic variation within population or varieties within a species.

Species Diversity

It refers to the variety of species within a region. Such diversity could be measured on the basis of number of species in a region.

Ecosystem Diversity

In an ecosystem, there may exist different landforms, each of which supports different and specific vegetation. Ecosystem diversity is difficult to measure since the boundaries of the communities, which constitute the various sub ecosystems, are elusive.

Ecosystem diversity could best be understood if one studies the communities in various ecological niches within the given ecosystem; each community is associated with definite species complexes. These complexes are related to composition and structure of biodiversity.

Agro-Biodiversity

The agricultural biological diversity more commonly referred to as the agro-biodiversity has been fast emerging as a strong, evolutionary divergent line from the biodiversity, which deals with the life forms at large.

It has been specifically recognized to differentiate between concern for ecosystems versus agro-ecosystems, wild forest flora and fauna versus agriculture related plants, reptiles, insects, avian and microbes; in situ conservation of wild forms versus on farm conservation of landgraves and traditional/ primitive cultivars or ex-situ conservation of plant genetic resources, etc.

Agro-biodiversity in a traditional farming system is as follows:

- Rich in plant and animal species
- A wide diversity of niches in the local environment utilized
- Reuse of organic residues, consuming biomass enabled
- Ecosystem functions, such as pest, weed and disease management enhanced
- Locally available resources consumed to an advantage
- Reduction of risk and optimization of resources use
- Associated with farmers time tested local knowledge about resources.

BIODIVERSITY AT GLOBAL LEVELS

It is estimated that there exists 5-30 million species of living forms on our earth and of these, only 1.5 million have been identified and include 3,00,000 species of green plants and fungi, 8,00,000 species of insects, 40,000 species of vertebrates and 3,60,000 species of micro-organisms. Recently it has been estimated that the number of insects alone may be as high as 10 million, but many believe it to be around 5 million.

The tropical forests are regarded as the riches in biodiversity. According to the opinion of the scientists more than half of the species on the earth live in moist tropical forests, which is only 7% of the total land surface. Insects (80%) and primates (90%) make up most of the species.

Table. Estimated Number of Species Worldwide.

Taxonomic Group	Number of Species
Bacteria	3600
Blue Green Algae	1700
Fungi	46983
Bryophytes	17000
Gymnosperms	750
Angiosperms	250000

STATUS IN INDIA

During the last few years, the subject of conservation of biological diversity has attracted considerable attention at the national and global levels. India is a rich centre of biodiversity and has contributed many economic plants to the world and many useful genes for genetic upgrading of cultivated plants and domesticated animals. India has a land mass of 329 million hactares with a diversified eco-geographical regions. Almost all types of habitats available in the world are found in India.There are two biogeographical realms in India and it is the confluence of floras and faunas of Africa, Mediterranean, European, Sino-Japanese and Malayan regions. As a result, we have a rich biological diversity.

Plant Species Diversity

At present 1.7 million species have been recorded so far in the world. India's contribution to this record stands at 7 %. Surveys conducted so far have inventorised over 47,000 species of plants and over 89,000 species of animals. Survey and inventorisation of India's biodiversity is still far from complete especially the lower groups of plants and invertibrate animals.

Table. India's Biological Wealth

Plant Taxa	Species	Animal Taxa	Species
Bacteria	850	Protista	2577
Viruses	Unknown	Mollusca	5070
Algae	6500	Arthropoda (Insecta, Crustacea etc.)	68389
Fungi	14500	Other Invertebrates	8329
Lichens	2000	Protochordata	119
Bryophytes	2850	Pisces	2546
Pteridophytes	1100	Amphibian	209
Gymnosperms	64	Reptilian	456
Angiosperms	17500	Aves	1232
		Mammalian	390
Total	**45364**	**Total**	**89317**

Based on the available data, India ranks 10th in the world and 4th in Asia in plant diversity and ranks 11th in the number of Angiosperm species. India ranks 10th in the number of mammalian species and 11th in the number of endemic species of higher vertebrates in the world.

Table. Number of Angiosperm Species in Different Countries.

Country	Angiospermic species
Brazil	55000
Colombia	45000
Ecuador	29000
China	27000
Mexico	25000
Australia	23000
South Africa	21000
Indonesia	20000
Venezuela	20000
Peru	20000
India	17000

Floristic Status

As noted earlier 47,000 species of plants representing about 12 % of the recorded world's flora have already been identified. Comparative statement of recorded number of plant species in India and the world is given in Table.

Table. Comparative Statement of Recorded Number of Plant Species in India.

Plant Taxa	Species		Percentage of India
	India	World	to the World
Bacteria	850	4000	21.25
Viruses	Unknown	4000	-
Algae	6500	40000	16.25
Fungi	14500	72000	20.14
Lichens	2000	17000	11.80
Bryophytes	2850	16000	17.80
Pteridophytes	1100	13000	8.64
Gymnosperms	64	750	8.53
Angiosperms	17500	25000	7.00

Flowering plants accounts nearly 17,500 among 45,000 species of plants. The important economic species includes rice, sugar cane, coix, beans, cowpeas, banana, Citrus, mango, coconut, cardamoms, nutmeg, tea, cotton, jute, colocasia, pepper, ginger, Rhododendron, Jasmines, bamboos, Orchids, betel leaf etc.

Two regions of our country harbours maximum diversity, they are North-East and Sourth-West India. The North- East region is a very active centre of evolution and has diversity for a number of plants like Rhododndron, Camelia, Magnolia, Buddleia, etc. On the basis of distribution pattern of plants, Good, divided plant wealth into 37 floristic zones. With in these zones, pockets of diversity of plants species arose and they were domesticated by human kind during the past 10,000 years. During these years enormous variability was generated because of mutation, recombination and selection process.The result being complex variation pattern in plants. These evolved plants bear little or no resemblance with their ancestors.

Endemic Species

Every major habitat, from areas of heavy rainfall to the dry desert, from coldest to the hottest climatic conditions, from highest elevation down to the sea level is found in the country..India has a rich endemic flora. Endemism of Indian biodiversity is significant. About 4950 species of flowering plants or 33% of this recorded flora are endemic to the country.These are distributed over 141 genera belonging to 47 families.

These are concentrated in the floristically rich areas of North-East India, the Western Ghats, North-West Himalayas and the Andaman & Nicobar Islands. The Western Ghats and Eastern Himalayas are reported to have 16000 and 3500 endemic species of flowering plants, respectively. These areas constitute two of the 34 hot spots identified in the world.

Cultivated Plants

Indian region alone has given to the world nearly 167 economical plants whose centre of origin/ diversity lie in India along with their 320 species of wild relatives and land races.

India is considered to be the centre of origin of rice, sugar cane, minor millets, pigeon pea, brassicas, rice-bean, Asiatic vignas, egg plant, banana, citrus, mango, cardamom, jack fruit, jute, edible diascorea, black pepper, seed amaranths, turmeric, ginger, several umbellifers and cucurbits, bitter gourd, colacasia, okra, coconut, bamboo, taro, indigo, sun hemp, gooseberries and many herbal drugs, rhododendron, jasmine, some orchids and betel nut.

This remarkable diversity of life-forms in a single country is because of the great diversity of ecosystems. The gene bank of National Bureau of Plant Genetic Resources (NBPGR) has a collection of over 1,59,080 varieties. The details of the active germplasm holding and base collections of NBPGR are given in Table below.

Wild relatives of Crops

There are several hundred species of wild crop relatives distributed all over the country. A major centre for wild rice is the eastern peninsular India, i.e., West Bengal, Orissa and Andhra Pradesh.

The North-Eastern hills and Tamil Nadu hills are rich in wild relatives of millets. Wild relatives of wheat and barley have been located in the western and North-Eastern Himalaya. Table gives the statement of wild relatives of crops recorded so far.

Table. Active Germplasm Holding and Base Collections at NBPGR.

Crop groups	Active Germplasm	Base Collection Holdings
Cereals	12086	43409
Pulses	38695	22269
Millets & Minor Millets	10349	14488
Oilseeds	19808	14278
Vegetables	12146	5681
Medicinal & Aromatic Plants	870	942
Pseudocereals	4739	736
Tuber Crop/ Spices	2053	-
Forage Crop	4060	-
Horticultural/ Ornamentals	22212	-
Fibre Crops	-	3212
Released crop Varieties	-	904
Reference Samples (Medium Term)	-	53161
Total	**107018**	**159080**

Table. Wild Relatives of Crop.

Crop Wild relatives	No. of
Millets	51
Fruits	104
Spices and condiments	27
Vegetables and pulses	55
Fibre crop	24
Oil seeds, tea, coffee, Tobacco, & sugarcane	12
Medicinal plants	3000

DECLINES IN BIODIVERSITY

The major causes of biodiversity decline are land use changes, pollution, changes in atmospheric CO_2 concentrations, changes in the nitrogen cycle and acid rain, climate alterations, and the introduction of exotic species, all coincident to human population growth. For rainforests, the primary factor is land conversion. Climate will probably change least in tropical regions, and nitrogen problems are not as important because growth in rainforests is usually limited more by low phosphorus levels than by nitrogen insufficiency. The introduction of exotic species is also less of a problem than in temperate areas because there is so much diversity in tropical forests that newcomers have difficulty becoming established.

HUMAN POPULATION GROWTH

The geometric rise in human population levels during the twentieth century is the fundamental cause of the loss of biodiversity. It exacerbates every other factor having an impact on rainforests (not to mention other ecosystems). It has led to an unceasing search for more arable land for food production and livestock grazing, and for wood for fuel, construction, and energy. Previously undisturbed areas (which may or may not be suitable for the purposes to which they are constrained) are being transformed into agricultural or pasture land, stripped of wood, or mined for resources to support the energy needs of an ever-growing human population. Humans also tend to settle in areas of high biodiversity, which often have relatively rich soils and other attractions for human activities.

This leads to great threats to biodiversity, especially since many of these areas have numerous endemic species. Balmford, et al., have demonstrated that human population size in a given tropical area correlates with the number of endangered species, and that this pattern holds for every taxonomic group. Most of the other effects mentioned below are either consequent to the human population expansion or related to it. The human population was approximately 600,000 million in 1700, and one billion in 1800. Just now it exceeds six billion, and low estimates are that it may reach 10 billion by the mid-21st century and 12 billion by 2100. The question is whether many ecological aspects of biological systems can be sustained under the pressure of such numbers. Can birds continue to migrate, can larger organisms have space (habitat) to forage, can ecosystems survive in anything like their present form, or are they doomed to impoverishment and degradation?

HABITAT DESTRUCTION

Habitat destruction is the single most important cause of the loss of rainforest biodiversity and is directly related to human population growth. As rainforest land is converted to ranches, agricultural land (and then, frequently, to degraded woodlands, scrubland, or desert), urban areas and other human usages, habitat is lost for forest organisms. Many species are widely distributed and thus, initially, habitat destruction may only reduce local population numbers. Species which are local, endemic, or which have specialized habitats are much more vulnerable to extinction, since once their particular habitat is degraded or converted for human activity, they will disappear. Most of the habitats being destroyed are those which contain the highest levels of biodiversity, such as lowland tropical wet forests. In this case, habitat loss is caused by clearing, selective logging, and burning.

POLLUTION

Industrial, agricultural and waste-based pollutants can have catastrophic effects on many species. Those species which are more tolerant of pollution will survive; those requiring pristine environments (water, air, food) will not. Thus, pollution can act as a selective agent. Pollution of water in lakes and rivers has degraded waters so that many freshwater ecosystems are dying. Since almost 12% of animals species live in these ecosystems, and most others depend on them to some degree, this is a very serious matter. In developing countries approximately 90% of wastewater is discharged, untreated, directly into waterways.

AGRICULTURE

The dramatic increase in the number of humans during the twentieth century has instigated a concomitant growth in agriculture, and has led to conversion of wildlands to croplands, massive diversions of water from lakes, rivers and underground aquifers, and, at the same time, has polluted water and land resources with pesticides, fertilizers, and animal wastes. The result has been the destruction, disturbance or disabling of terrestrial ecosystems, and polluted, oxygen-depleted and atrophied water resources. Formerly, agriculture in different regions of the world was relatively independent and local. Now, however, much of it has become part of the global exchange economy and has caused significant changes in social organization.

Earlier agricultural systems were integrated with and co-evolved with technologies, beliefs, myths and traditions as part of an integrated social system. Generally, people planted a variety of crops in different areas, in the hope of obtaining a reasonably stable food supply. These systems could only be maintained at low population levels, and were relatively nondestructive (but not always).

More recently, agriculture has in many places lost its local character, and has become incorporated into the global economy. This has led to increased pressure on agricultural land for exchange commodities and export goods. More land is being diverted from local food production to "cash crops" for export and exchange; fewer types of crops are raised, and each crop is raised in much greater quantities than before. Thus, ever more land is converted from forest (and other natural systems) for agriculture for export, rather than using land for subsistence crops.

The introduction of monocropping and the use of relatively few plants for food and other uses – at the expense of the wide variety of plants and animals utilized by earlier peoples and indigenous peoples – is responsible for a loss of diversity and genetic variability. The native plants and animals adapted to the local conditions are now being replaced with "foreign" (or "exotic") species which require special inputs of food and nutrients, large quantities of water.

Such exotic species frequently drive out native species. There is pressure to conform to crop selection and agricultural techniques – all is driven by global markets and technologies.

GLOBAL WARMING

There is recent evidence that climate changes are having effects on tropical forest ecology. Warming in general (as distinct from the effects of increasing concentrations of CO_2 and other greenhouse gases) can increase primary productivity, yielding new plant biomass, increased organic litter, and increased food supplies for animals and soil flora (decomposers). Temperature changes can also alter the water cycle and the availability of nitrogen and other nutrients. Basically, the temperature variations which are now occurring affect all parts of forest ecosystems, some more than others. These interactions are unimaginably complex. While warming may at first increase net primary productivity (NPP), in the longer run, because plant biomass is increasing, more nitrogen is taken up from the soil and sequestered in the plant bodies. This leaves less nitrogen for the growth of additional plants, so the increase in NPP over time (due to a rise in temperature or CO_2 levels) will be limited by nitrogen availability. The same is probably true of other mineral nutrients.

The consequences of warming-induced shifts in the distribution of nutrients will not be seen rapidly, but perhaps only over many years. These events may effect changes in species distribution and other ecosystem processes in complex ways. We know little about the reactions of tropical forests, but they may differ from those of temperate forests. In tropical forests, warming may be more important because of its effects on evapotranspiration and soil moisture levels than because of nutrient redistribution or NPP (which is already very high because tropical temperatures are close to the optimum range for photosynthesis and there is so much available light energy). And warming will obviously act in concert with other global or local changes – increases in atmospheric CO_2 (which may modify plant chemistry and the water balance of the forest) and land clearing (which changes rainfall and local temperatures), for examples.

Root, et al. have determined that more than 80% of plant and animal species on which they gathered data had undergone temperature-related shifts in physiology. Highland forests in Costa Rica have suffered losses of amphibian and reptile populations which appear to be due to increased warming of montane forests. The golden toad *Bufo periglenes* of Costa Rica has become extinct, at least partly because of the decrease in mist frequency in its cloud forest habitat. The changes in mists appear to be a consequence of warming trends. Other suspected causes are alterations in juvenile growth or maturation rates or sex ratios due to temperature shifts. Parmesan and Yohe, in a statistical analysis, determined that climate change had biological effects on the 279 species which they examined. The migratory patterns of some birds which live in both tropical and temperate regions during the year seem to be shifting, which is dangerous for these species, as they may arrive at their breeding or wintering grounds at an inappropriate time. Or they may lose their essential interactions with plants which they pollinate or their insect or plant food supplies. Perhaps for these reasons, many migratory species are in decline, and their inability to co-ordinate migratory clues with climatic actualities may be partly to blame.

The great tit, which still breeds at the same time as previously, now misses much of its food supply because its plant food develops at an earlier time of year, before the birds have arrived from their wintering grounds. Also, as temperatures rise, some bird populations have shifted, with lowland and foothill species moving into higher areas. The consequences for highland bird populations are not yet clear. And many other organisms, both plant and animal, are being affected by warming. An increase in infectious diseases is another consequence of climate change, since the causative agents are affected by humidity, temperature change, and rainfall. Many species of frogs and lizards have declined or disappeared, perhaps because of the increase in parasites occasioned by higher temperatures. As warming continues, accelerating plant growth, pathogens may spread more quickly because of the increased availability of vegetation (a "density" effect) and because of increased humidity under heavier plant cover.

As mentioned above, the fungus *Phytophtora cinnamoni* has demolished many *Eucalyptus* forests in Australia. In addition, the geographical range of pathogens can expand when the climate moderates, allowing pathogens to find new, nonresistant hosts. On the other hand, a number of instances of amphibian decline seem to be due to infections with chrytid fungi, which flourish at cooler temperatures. An excellent review of this complex issue may be found in Harvell, *et al.* There may be a link between augmented carbon dioxide levels and marked increase in the density of lianas in Amazonian forests. This relationship is suggested by the fact that growth rates of lianas are highly sensitive to CO_2 levels. As lianas become more dense, tree mortality rises, but mortality is not equal among species because lianas preferentially grow on certain species. Because of this biodiversity may be reduced by increased mortality in some species but not others.

FOREST FRAGMENTATION

The fragmentation of forests is a general consequence of the haphazard logging and agricultural land conversion which is occurring everywhere, but especially in tropical forests. When forests are cut into smaller and smaller pieces, there are many consequences, some of which may be unanticipated. Fragmentation decreases habitat simply through loss of land area, reducing the probability of maintaining effective reproductive units of plant and animal populations. Most tropical trees are pollinated by animals, and therefore the maintenance of adequate pollinator population levels is essential for forest health. When a forest becomes fragmented, trees of many species are isolated because their pollinators cannot cross the unforested areas. Under these conditions, the trees in the fragments will then become inbred and lose genetic variability and vigour.

Other species, which have more wide-ranging pollinators, may suffer less from fragmentation. For instance, the pollen of several species of strangler figs (the fruit of which is an essential element in the diets of many animals) is dispersed by wasps over distances as great as 14.2 km. Thus "breeding units" of these figs are extremely large, comprising hundreds of plants located in huge areas of forest. Isolated fig populations seem to survive and help to maintain frugivore numbers (if not diversity), so long as the number of trees within the range of the wasps does not fall below a critical minimum. Most species are not so tolerant, however. Animals, particularly large ones, cannot maintain themselves in small fragmented forests. Many large mammals have huge ranges and require extensive areas of intact forest to obtain sufficient food, or to find suitable nesting sites. Additionally, their migrations may be interrupted by fragmentation. These animals are also much more susceptible to hunting in forest fragments, which accounts for much of the decline in animal populations in rainforests.

Species extinctions occur more rapidly in fragments, for these reasons, and also because species depend upon each other. The dissection of forests into fragments in certain parts of the Amazon has led to extreme hunting pressures on peccaries, for instance, and in some places where they are locally extinct, three species of frogs have also disappeared, since they depended upon peccary wallows for breeding ponds. The absence of large predator species leads to imbalances in prey populations, and, since many of the prey species are seed-eaters, to declines in the population levels of many plant species. The prey, now at high population levels, consume most available seeds, leaving few to germinate. On small islands created after dam construction on the Chagres River in Panama, even large seed predators could not survive, and after 70 years, the former mixed tropical forest has become a forest of large-seeded plants only. As Terborgh states, and we should attend to this lesson, "Distortions in any link of the interaction chain will induce changes in the remaining links." When forests are cut down or burned, the resulting gaps are too large to be filled in by the normal regeneration processes. This permits the ascendancy of rapid-growing, light-tolerant species and grasses. Large gaps may then be converted to scrub or grassland.

The "edge" effect: The cutting of forest into fragments creates many "edges" where previously there was deep forest. Many effects are consequent upon this. Edges are lighter, warmer and windier than the forest interior. These changes in microclimate alter plant reproduction, animal distribution, the biological structure and many other features of the forest. Tree mortality is much greater near edges, and climax species will be replaced by pioneer species. These effects can be seen as far as one kilometre into the forest.

The drier and warmer conditions also make the fragment more flammable, with a concomitant increase in the frequency of fires. Without further stress, the forest may regenerate. However, if the fragment is surrounded by a human-dominated landscape, it may be inhibited from regeneration. This has occurred in certain areas of Brazil, where forest fragments are surrounded by sugar cane and Eucalyptus plantations. Thus, species requiring large areas of undisturbed primary forest are sacrificed to the benefit of those species which can exist on forest margins.

Fire is particularly frequent in fragments. Recently, many forests have been subjected to deliberately-set and accidental fires, to which they have little resistance, and to which they are rarely naturally subjected. People often set fire to cut-over areas adjacent to forests to clear them of debris. These fires often get out of control and burn large areas, extend into the forest interior, and inhibit edge regeneration by killing pioneer forest vegetation. More than 90% of forest fires in certain eastern Amazon forest areas were associated with the edges of forest fragments. If conditions remain severe, the forest will recede and be replaced by scrub.

The use of herbicides and the introduction of exotic species into areas surrounding forest fragments are detrimental to forest health. Herbicides blow from cleared agricultural areas into forests, and exotic species introduced by farmers and ranchers spread, often displacing native species. These exotic organisms interrupt the forest ecosystem and, since they have few or no natural enemies in their new environment, they are difficult to eradicate. According to Vitousek, there are many islands where fewer than half of the species are native, and in many other terrestrial environments, more than 20% of species are foreign. These invasions drive the loss of indigenous species. For unknown reasons, fragmentation leads to the death of large canopy trees, even in the interior of fragments. Canopy trees dominate the forest structure, and they provide fruits and shelter for many animals. The mortality of trees in fragmented patches in Brazil has been found to be twice that of similar trees in the forest interior. Not only that, but tree mortality is confined disproportionately to large trees (an almost 40% increase in mortality). Large trees may be more vulnerable in fragmented forests because they are not as well buffered from wind and natural forces, because there are more tree parasites (lianas), and because they are more subject to desiccation at forest edges.

Loss of these largest trees has several corollary effects – the alteration of biogeochemical cycles (transpiration, carbon cycles), the reduction of species complexity, and the reduction of fecundity. As mentioned above, large trees are essential habitats and food sources for many other organisms, both plant and animal; they are the source of much of the primary productivity of the forest;

and they are responsible for many effects on the water and nutrient cycles. They are irreplaceable in the forest ecosystem. The fragmentation of forests by logging and agricultural conversion also exaggerates the probability of major epidemics. Pathogens introduced through human activities by land use practices in areas surrounding the forest can be lethal to forest plants and animals.

Rainforests are losing species, not only because of the disappearance of their habitat, but also because essential ecological processes are being interrupted by fragmentation. Fragments are much more easily accessible to human incursions than are intact forests. This leads to a variety of extractive activities within the forest interior. Intensive hunting, by depleting animal populations, inhibits plant reproduction, since many seeds can neither be dispersed, nor flowers be pollinated without them. Where these seed dispersers have been eliminated, are at low population densities, or cannot move between forest fragments, seed dispersal will be very limited, and as a result tree species dependent upon animal dispersers may become locally extinct. In the remnants of the Atlantic forest of Brazil, the seeds of 71% of tree species are dispersed by vertebrates (birds and mammals), and about 48% of these dispersers are birds which are deep-forest dwellers.

As this forest becomes more and more fragmented, these birds are disappearing, so eventually the trees dependent upon them will be unable to replace themselves. In some fragments, all large vertebrates (including seed-eaters) have been hunted to extinction, and in some places the fragments are so distant from each other that these animals cannot pass from one to another. The Alagoas curassow, a large fruit-eating bird of this area, is now extinct; many others (toucans, aracaris, guans) are endangered by hunting pressures. Other species are sensitive to disturbance of their environment and they become locally extinct.

The tiny bits of Atlantic forest remaining are becoming dominated by trees which are wind or water-pollinated or whose seeds are dispersed by animal species which can tolerate disturbed habitats or edges ("edge species"). In addition, in fragmented forests, seeds will frequently land in deforested areas (where they are in the open, and exposed to heat, light and desiccation) in which they cannot germinate, and the seedlings cannot survive. In Brazil, three to seven times as many Heliconia acuminata seedlings planted in continuous areas of forest germinated as compared to those planted in fragmented areas.

Whatever the explanation for the lower rate of seedling germination in fragmented forests, whether due to inbreeding or other causes, fewer and fewer individuals in fragments grow to adulthood. Those which do will breed, but since populations are small, inbreeding occurs and the downward spiral continues until the population becomes locally extinct. This effect is seen frequently in forest fragments.

HUNTING, FISHING, AND GATHERING

Many forests which appear intact are in fact "empty forests," since most large animals have been hunted to unsustainable levels. These animals are mainly hunted for meat, but also for skins (jaguar, ocelot) or medicinal/chemical properties (poison-arrow frogs, collected to provide poisons for arrow tips, and the midwife toad, which in the Amazon is thought to have medicinal value). Turtles are heavily harvested for meat and their eggs are collected for food almost everywhere in the tropics and subtropics.

Asian tropical freshwater turtles are in serious decline because they are extensively hunted for food or for use in traditional Chinese medicines. Thousands of tons of live turtles are caught or sent to China annually, a completely unsustainable level of collection. There are apparently no turtles left in the wild in Vietnam for this reason. More than 80 species of Asian turtles are at such low population numbers that they will become extinct unless emergency measures – restrictions on international trade, increased habitat protection, captive breeding programs – are taken immediately.

Some of the hunting is done for subsistence purposes by villagers; some by farmers, miners and loggers, who live in the forest and use forest animals as a major food source; some by commercial hunters to supply urban markets. This is a major source of income in many rural tropical areas. In Gabon alone, a tiny country, 3,600 tons of bushmeat are consumed annually. The popularity of "bushmeat" in cities and towns located in or close to rainforests is rising. Surveys of bushmeat consumption in Bolivia and Honduras showed that people will eat more bushmeat as their income rises, but when they become more affluent, consumption declines. Consumption declines as well when bushmeat prices rise and the price of alternative sources of protein declines. Part of the remedy for overexploitation of wildlife resources, then, lies with improving the income levels of local residents (so bushmeat becomes less attractive as a protein source), in increasing the costs of hunting, and in lowering the prices of alternative protein sources.

Many animals are trapped for the pet trade (tropical fish, birds, reptiles, monkeys) or for zoos or medical research. A 25-week survey of the Bangkok weekend market found specimens of 225 species of birds (most "protected" by government decree) for sale. Other animals are trapped for their hides or furs, and some are killed because they live too close to human habitation and impinge on human activities. For instance, ocelots and other small carnivores may be shot when they attempt to prey on chickens or other domestic animals. Many tropical animals are hunted mercilessly for their value in traditional Asian medicines. Tigers, bears, deer, snakes, and many other animals are near extinction in many places because of this trade.

Tigers in India are almost gone and only 3-5000 tigers still exist in the wild anywhere. Many of these animals, or their parts, are smuggled illegally from Southeast Asian (and other) countries to China and other countries with large Chinese populations for these uses.

The effects of hunting are not just on the animals "taken." Many animals which are human prey eat fruits and seeds, and are major seed dispersers in tropical forests, and the seeds of certain species of trees must pass through the gut of an animal in order to germinate. In these ways many tropical plants and trees depend upon animals, for, without them, they will not be able to reproduce. For instance, the seeds of Inga ingoides, a South American tree, are dispersed widely by the spider monkey *Ateles paniscus*.

Where this monkey is locally extinct (due to hunting pressures), the trees do not "outbreed"; the seeds fall to the forest floor and patches of seedlings of low genetic diversity surround the parent trees. This can be very detrimental to forests, which generally have high genetic diversity, because more homogeneous plants are generally less fit. The loss of elephants in African countries due to hunting has led to a loss of reproductive ability in many valuable tree species.

Fish and aquatic animals are killed indiscriminately by fishing techniques which employ insecticides and/or dynamite. These techniques not only catch the few desired specimens, but kill all of the other animals in the area. Commercial fishing operations are not sensitive to issues of sustainability. They catch as many marketable fish as possible, and intensify their efforts when fish populations drop (declines due in the first place to overfishing). Such unsustainable fishing operations have led and are leading to severe declines in fish in major river systems within tropical rainforests. Fragmentation may be more serious than previously imagined because the consequences of fragmentation are not static, but progressive. The edges of cut areas do not remain "in place" but gradually recede, further reducing the size of the fragments. Eventually fragments may disappear altogether or undergo ecological collapse.

Why do people heedlessly decimate the precious biodiversity of their planet? Some of them feel they have no economic alternative, while others are driven by the desire for short-term profit. Still others are uncomprehending. Unfortunately, so much of the depredation which is being inflicted upon areas of great biodiversity is, in the long run, and often in the short run, in vain. While tropical forests now occupy less than half of their former range, and much of what remains is damaged or fragmented, the net profit to humanity is slight. Clearing of tropical forests has provided only a relatively small percentage of total agricultural land, since much of the land converted for farms becomes rapidly degraded and is abandoned.

Logging results in a one-time profit, mainly to large companies. Ranching is an activity which, on former rainforest land, is uneconomical, requires subsidizing, and is eventually abandoned. But the damage is permanent and the forest irreplaceable, so forest destruction has dire consequences. It degrades aquatic fisheries, causes floods and has many other consequences – so much harm for so little benefit.

BIODIVERSITY: VALUES, GENERATION AND MAINTENANCE

Biodiversity, short for biological diversity, is the term used to describe the variety of life found on Earth and all of the natural processes. This includes ecosystem, genetic and cultural diversity, and the connections between these and all species. The different aspects of biodiversity all have a very strong influence on each other. We have only just started to understand the relationships between living things and their environments. It is helpful to think of an ecosystem as a woven carpet; if you pull on a loose thread it might only affect the thread and those closest to it or it might unravel the whole carpet.

Biodiversity also helps us in our day-to-day lives. Unfortunately, the greenhouse gases produced by human activities are building up in the atmosphere and causing climate change. Climate change is a major threat to biodiversity. Climate change affects air and ocean temperatures, the length of seasons, sea levels, the pattern of ocean and wind currents, levels of precipitation, as well as other things. These changes affect the habitats and behaviour of many different species. Many will not be able to adapt fast enough and may become extinct.

There are many things that you can do to combat climate change. Planting deciduous trees on the south side of your house or school will keep your house/school cooler in the summer. They will help produce oxygen and remove carbon dioxide. You can also encourage your friends and family to use public transit, carpool and walk or bike when they can.

"Biodiversity" is most commonly used to replace the more clearly defined and long established terms, species diversity and species richness. Biologists most often define biodiversity as the "totality of genes, species, and ecosystems of a region". An advantage of this definition is that it seems to describe most circumstances and presents a unified view of the traditional three levels at which biological variety has been identified:

- Species diversity
- Ecosystem diversity
- Genetic diversity

In 2003 Professor Anthony Campbell at Cardiff University, UK and the Darwin Centre, Pembrokeshire, defined a fourth level: Molecular Diversity. This multilevel construct is consistent with Dasmann and Lovejoy. An explicit definition consistent with this interpretation was first given in a paper by Bruce A. Wilcox commissioned by the International Union for the Conservation of Nature and Natural Resources (IUCN) for the 1982 World National Parks Conference. Wilcox's definition was "Biological diversity is the variety of life forms...at all levels of biological systems (*i.e.*, molecular, organismic, population, species and ecosystem)...". The 1992 United Nations Earth Summit defined "biological diversity" as "the variability among living organisms from all sources, including, 'inter alia', terrestrial, marine, and other aquatic ecosystems, and the ecological complexes of which they are part: this includes diversity within species, between species and of ecosystems". This definition is used in the United Nations Convention on Biological Diversity.

One textbook's definition is "variation of life at all levels of biological organisation". Genetically biodiversity can be defined as the diversity of alleles, genes, and organisms. They study processes such as mutationand gene transfer that drive evolution. Measuring diversity at one level in a group of organisms may not precisely correspond to diversity at other levels. However, tetrapod (terrestrial vertebrates) taxonomic and ecological diversity shows a very close correlation.

TYPES OF DIVERSITY AND SIGNIFICANCE

TYPES OF BIODIVERSITY

Biodiversity is a generic term that can be related to many environments and species, for example, forests, freshwater, marine and temperate environments, the soil, crop plants, domestic animals, wild species and micro-organisms.

Basically it can be classified according to three types of diversity:

- Ecosystems and landscapes (habitat diversity)
- Animal, plant, bacterial species (species diversity)
- All genes (genetic diversity)

Of particular importance are the taxonomically isolated species, as they have little similarity to other species and therefore are unique with respect to their genetic structure. These species are often endemic meaning limited to one specific area. Their extinction would mean a greater loss for global biodiversity rather than just the extinction of a species.

Why is Biodiversity Important?

Biodiversity increases ecosystem productivity; all of the species in that ecosystem, no matter their size, have a big role. A diverse ecosystem can prevent and recover from lots of disasters. Humans depend on plants and animals. For example, one quarter of all prescription medicines in the U.S. have ingredients from plants. If a diverse ecosystem is more productive, it's easier to get these plants. Humans also directly benefit from a diverse ecosystem: plants, clean water and air, provide oxygen, and control erosion.

Here are some of the major ways biodiversity helps humans:

- Plants absorb greenhouse gases and help stop global warming.
- It is easier for biodiverse ecosystems to recover from natural disasters.

- Healthy biodiversity of species can provide a variety of food (like meat and produce).
- Many of our medicinal drugs come from plants.
- All of our wood products come from nature.
- We can learn more about our earth by observing a diverse ecosystem.
- Many recreational areas benefit from a healthy ecosystem, which promotes tourism.
- Biodiversity is beautiful and should be enjoyed.

These values are free to us, but as we lose biodiversity the cost of replacing these (if even possible) would be very high!

THE SIGNIFICANCE OF BIODIVERSITY

"The fundamental property of ecological systems is a certain mixture, or diversity of living things..."(1). Biodiversity, or the variety of living things that exist, is fundamental to the existence of life on Earth, and the importance of it cannot be underestimated. In the past few centuries, humans have had an especially negative affect on biodiversity, although, in general, are becoming more aware of its role. However, due to the damage we have caused, and the value that biodiversity has to us as humans, protection of the natural environment is necessary. Biodiversity is an extremely important part of life on Earth. It is not only the variety of living organisms on our planet, but also the interdependence of all these living things, including humans. It thus creates and maintains ecological systems; the most recognisable of which are Earth's biomes, which can be divided into the broad categories of Forests, Tundra, Aquatic, Grasslands, and Deserts. Life is, in fact, one of the major features that distinguishes biomes from one another. "Biomes are defined as 'the world's major communities, classified according to the predominant vegetation and characterised by adaptations of organisms to that particular environment'. Without vegetation or organisms, these landscapes would be virtually indistinguishable from one another. Clearly life plays a major role in the function of ecosystems, and the variety, or diversity, of this life has played a major role in the evolution of the world.

In evolutionary theory, it has become clear that the greater the diversity that exists within a family or genus, the more likely it is to survive environmental change. Thus, evolution depends on biodiversity. However, humans have been the main cause of recent rapid "evolutionary" change. Ecosystems are being destroyed, animals and plants becoming extinct, and biodiversity is being lost due to increased human activity. Although environments would be shifting and evolving regardless of human influence, it is necessary to understand that humans are causing the rate of change to become particularly dangerous. Environmental conditions are changing so quickly that individual species as well as entire ecosystems are struggling, and often failing, to adapt. For these reasons, it is very important that we protect biodiversity and the natural environment.

Currently, there are many manners in which ecosystems and species are being negatively affected. The first is land use, which is most responsible for the contemporary decrease in biodiversity. About one quarter of the Earth's surface is covered by farm land, a problem that is often overlooked. The most fertile soil is usually found in the best climates, which also usually happens to be where the largest amount of biodiversity is. The best example of this is in the tropics, where both tropical rainforests and cloud forests are being cut down and turned into "patchwork" farms. Furthermore, intensive agriculture is a growing concern. Fertilizers and pesticides used to treat crops harm land and drive animals away. Eventually, a given plot of farmland will contain too many chemicals to continue farming on, and the farmer will have to move to a new one, creating a vicious cycle of destruction of natural land.

Another threat to biodiversity is the loss and extinction of species. This topic is better known and publicised than the farmland issue, and many organisations are working towards the preservation of wild animals. However, it is important to understand that we need to pay as much, if not more, attention to reductions in species as extinction's. We often wait until a species is highly endangered before helping, at which point it is often too late. Endangerment occurs both directly by humans, such as fishing and hunting to excess, and indirectly, such as reducing habitats to the point where animals can no longer live.

The dramatic disappearance of many species is often referred to as gene erosion, which is now happening very quickly. It must also be noted that introducing alien species in new habitats can greatly affect the natural environment. Many ecosystems have little immunity to new species, especially when the "intruder" has different traits than the original species. For example, the introduction of the house cat can be dangerous to an otherwise safe bird and small mammal population.

While many developed countries are now regulating pollution and toxification, the degrees to which this is done vary, and it is still a major concern. Many pollutants travel incredibly quickly and cover a broad area. Long-term pollution is a great concern as well, even at low levels, because it can affect entire ecosystems through the chain of life. Furthermore, pollutants in soil and ground water cannot travel quickly, and thus do not filter out well. Reproductive anomalies in animals, especially frogs, are being attributed to pollution and toxification, and some scientists fear that these could eventually affect humans. Climate change is a growing concern as well, though it is somewhat debatable as to whether or not humans caused it. Natural changes in weather have had perhaps the greatest affect on biodiversity and ecological systems. The threat of humans shifting the climate is therefore extremely threatening to the natural environment. "Were the average temperature to rise by several degrees Celsius, that warming would probably be followed by potentially large reorganisations of some ecological communities."

One last issue concerning the affects that humans have on biodiversity is that of overpopulation. Recent advances in science and medicine have allowed for much greater life span and a very small infant mortality rate. We are increasing in population more rapidly than ever before. The growing population causes displacement of natural environments, not only because we need more living space, but also because the demand for agriculture and industry becomes higher as a result. It is painfully clear that in many ways humans have had a significantly negative affect on biodiversity and Earth's natural environment as a whole. It is essential to realise that as rational beings, humans have the ability to not only understand the problems we have created and what needs to be done to amend them, but also the capability of accomplishing these tasks. There are two basic venues of thought as to why we should protect biodiversity and our natural environment, one being intrinsic reasoning, and the other being anthropocentric.

Many believe that there are intrinsic reasons to protect biodiversity, separate from all human needs and desires. These arguments are based on the idea that humans are part of nature, not separate from it. Evolution, for example, is what allowed us to come into being originally, and humans are now destroying the same biodiversity that allowed evolution to happen. A similar, but slightly different principle behind the intrinsic theory, is that people did not create nature, and therefore should not have the right to destroy it. Every species has a right to not be eliminated by humans. Furthermore, because humans destroy natural habitats consciously, we should be responsible for fixing any unnecessary damage that we have done.

A somewhat contradictory view is the anthropocentric theory. This is based on the idea that biodiversity has value for us as humans. The first, most direct example of this lies in goods obtained from nature. The most important, and often overlooked, is food. It is natural and necessary for us to consume a variety of living things, from vegetables to animals, in order to remain healthy. Cloth is another such example; we need the diversity of life in order to make clothes for ourselves, whether they be cotton, as many are now, or animal skin, as used in the past. Other goods include pharmaceuticals and medicines that are derived from naturally existing sources. These have proven to be incredibly valuable to us, and millions of plants have never been chemically tested, which leaves many open opportunities for discovery of new organic remedies.

The natural environment provides other services which benefit the economy as well. For instance, biodiversity helps keep water clean and naturally manages waterflow and watershed. Trees and plants keep air clean through the constant transfer of carbon dioxide and oxygen, and overall biodiversity helps regulate climate. It is estimated that it would cost over three trillion dollars to replace these natural services with man-made ones. The recreational and aesthetic benefits of nature are also considered anthropocentric. A growing number of people are participating in activities such as hiking, camping, and birding. In addition, ecotourism is becoming increasingly popular, which has not only raised awareness about biodiversity, but helps the natural environment economically as well.

Biodiversity is clearly a fundamental component of life on Earth. It creates complex ecosystems that could never be reproduced by humans. The value of that biodiversity, both intrinsically and to humans, is immeasurable, and thus must be protected. In the end, we both want and need biodiversity. Although we continue to harm the natural environment, often without realising the impact that we have, an increasing number of people are becoming aware of the need to protect biodiversity. Hopefully humans will continue to pursue the issue so we can eventually live entirely with nature, not harm the very system that allows us to exist.

2

Animal Biodiversity and Soil Nutrient

SOIL ANIMALS

The most important group of larger soil animals are the earthworms, of which there may be a dozen species in a healthy Canadian soil. Earthworms perform the final task of humification -- the conversion of decomposed organic matter to stable humus colloids -- and mix the humus with material from the lower soil horizons. The digestive tract of the earthworm has a remarkable capacity to literally alter the chemical and physical nature of soil. Earthworms are major agents in the process of soil creation through the formation of clay-humus complexes and they play a key role in the management of calcium. By inoculating their castings with intestinal flora, earthworms distribute microbial populations throughout the soil. Earthworms can increase the availability of phosphorus from rock phosphate by 15-39 per cent. They act as mini-subsoilers, their burrows increasing soil aeration, drainage and porosity. In the process of burrowing, earthworms mix the subsoil with the topsoil and deposit their nutrient-rich castings on or near the soil surface. The presence of a large earthworm population indicates good soil fertility. They can be encouraged by adding lime when needed to correct soil acidity and organic matter to provide the worms with food. Note that the red wriggler, or manure worm, prefers an environment higher in organic matter and cannot survive in most soils; inoculating a field with these worms will not improve soil fertility.

Mites are the most abundant of the soil arthropods. Most mites are beneficial, feeding on micro-organisms and other small animals. They assist with decomposition by browsing on preferred fungi, thus preventing any one species from becoming dominant, and by transporting the spores through the soil. Springtails perform similar functions. Larger arthropods, slugs and snails burrow through the soil and feed on dead plant material. By maintaining a suitable environment for the hundreds of species of soil creatures, large and small, organic farmers provide their crops with an abundant supply of plant nutrients.

NITROGEN CYCLE

The vegetative growth of plants (leaves, stems, and roots) is especially dependent on nitrogen. The atmosphere contains 78 per cent nitrogen by volume, yet it is the element that most often limits plant growth. Plants cannot use gaseous nitrogen, but require nitrogen in the form of nitrate (NO3-) or ammonium (NH4+). Atmospheric nitrogen is converted into NO3- and NH4+ in the soil by nitrogen fixation, which is performed by certain soil micro-organisms. These include the symbiotic Rhizobia bacteria associated with legumes, and the non-symbiotic bacteria Clostridium and Azoterbacter which are free-living in the soil.

Once gaseous nitrogen is incorporated into plant material as proteins and amino acids, it may be recycled many times through the activity of the soil decomposers. Young plants are especially rich in nitrogen and, when they are incorporated in the surface layers of the soil as green manure, this nitrogen is released by biological activity. The ammonium (NH^{4+}) ions can be stored on the clay-humus complex for long periods. The nitrate ions (NO^{3-}) are subject to leaching if not taken up by the crop.

Deficiencies of nitrogen may occur not because there is not enough entering the system but because of the way it cycles round the system. Cycling is increased by maximizing biological activity which is determined by the way different components of the system, such as residues, manure, weeds and drainage, are managed.

Carbon cycle

Carbon is the building block of life. Plants obtain carbon from atmospheric carbon dioxide (CO_2) through photosynthesis, during which the chloroplasts in the plant cells convert CO_2 to carbohydrates. It is the cycling of carbon from the atmosphere through plants and algae, to animals and micro-organisms and back to the atmosphere, that maintains earth's atmosphere and climate in its current balance. The greenhouse effect, or warming of the planet, is a consequence of an excess of atmospheric CO_2 caused by deforestation (reduced CO2 consumption) and compounded by excessive fossil fuel energy use (increased CO2 production). Keeping the soil covered with growing plants can make a contribution to reducing global warming.

Carbon is a critical element in the formation of stable humus. The carbon:nitrogen (C:N) ratio of the organic matter supplied to the soil is a controlling factor in this process. A ratio of about 20:1 is considered ideal. If greater amounts of carbon are present, decomposition slows as micro-organisms become nitrogen-starved and compete with the plants for available nitrogen. Nitrate nitrogen practically disappears from the soil because microbes need nitrogen to build their tissues. If there is too much soil nitrogen, the decomposers produce soluble nutrients in the form of effective humus, but little stable humus. These conditions can give the advantage to weeds rather than the crop. A good C:N ratio will result in the formation of both effective humus and stable humus. As decay occurs, the C:N ratio of the plant material decreases since carbon is being lost as CO2, and nitrogen is conserved. This process continues until the micro-organisms run out of easily-oxidized carbon. The exuded, undecomposed carbon persists as stable humus.

Phosphorus cycle

Phosphorus (P) is important in plant-cell division and growth. It is a difficult nutrient to manage because, although abundant in the soil, it is often in a form unavailable to plants. In acidic soils (pH below 5) the phosphorus gets tied up with iron and aluminum, and in alkaline soils (pH above 7) it gets tied up with calcium. Even with a favorable pH, phosphorus readily becomes immobilized by other soil minerals. Phosphorus anions may also be physically trapped in the clay-humus complex. Phosphorus is lost from soils through soil erosion, often at a greater rate than it can be replaced from the underlying subsoils. It accumulates in lakes and slow-flowing rivers, causing eutrophication. The elimination of soil erosion is the first step in phosphorus conservation. The addition of powdered rock phosphate or colloidal phosphate is a precautionary measure which, used in conjunction with the biological measures described below, can avoid phosphorus deficiency.

The release of P to plants depends on soil biological activity, particularly that of certain bacteria and mycorrhizal fungi. Soil acids, produced by these micro-organisms and by OM decomposition, release phosphates. Phosphorus availability is therefore dependent on the maintenance of high levels of biological activity and stable humus in the soil. Under these conditions, phosphorus is continually recycled through the processes of OM decay. Some plants produce acidity around their roots which assists in the uptake of P; examples of these are legumes actively fixing nitrogen, rapeseed, oilradish and buckwheat.

Potassium Cycle

Potassium (K) is important as an enzyme activator in plants. It is involved in facilitating membrane permeability and translocation of sugars. Potassium is also needed for photosynthesis, fruit formation, winter hardiness, disease resistance, and amino acid and protein formation. Potassium builds plant stalk strength. It does not, however, form a permanent part of plant tissues, but is translocated to the stems and roots during ripening. Thus, potassium is readily available in crop residues -- roots, straw and corn stalks. Very little potassium is removed with a grain crop at harvest if the straw is left on the field. Repeated cutting for hay or silage without returning potash in the form of manure or crop residues will quickly induce K deficiency. Soil potassium is present in minerals that dissolve slowly, thereby limiting its availability. Potassium availability is regulated by cation exchange. Potassium leaching increases as the amounts of clay and humus decrease and therefore may be a problem in sandy soils. A deep-rooting green manure will help prevent losses. Increased biological activity and colloidal humus formation will increase potassium availability by enhancing the CEC in the soil. The addition of powdered basalt, green sand and clay minerals has been found to correct potassium deficiencies in a biologically active soil. Manure is a good source of K if care has been taken to minimize leaching during storage.

It has been reported that in some organic systems, low available potash levels, according to soil analyses, are not necessarily associated with plant deficiencies or lower yields. This may be because available K is immediately taken up by the growing plant.

Micronutrients

About one hundred elements have been found in living plants. Carbon, hydrogen, and oxygen are the most abundant and are derived from water, oxygen and carbon dioxide. The nutrients N, P, K, calcium and magnesium have been discussed above. Of the other elements, we know that sulfur, iron, copper, manganese, zinc, molybdenum, boron and chlorine are required by plants in trace amounts. They are not constituents of the plant structure, but contribute to plant growth and development. Other elements, such as iodine, are essential to the animals that eat the plants. Deficiencies occur in soils that lack an inherent source of an element, or they can be caused by an imbalance in soil pH. Conversely, if certain micronutrients exceed trace levels, they can be toxic to plants.

The range between deficiency and excess is very small. Therefore, micronutrients should not be applied unless a deficiency is shown by leaf analysis or by visible plant symptoms. Micronutrients are best applied via compost, or by a foliar spray. Either of these methods is preferable to applying a trace mineral directly to the soil. In a biologically-active soil with good CEC and balanced pH, micronutrient deficiencies are rare. Products based on seaweed (kelp) contain more than 80 elements, and organic farmers feed kelp meal as mineral supplement to their livestock, or incorporate small amounts of kelp products into compost as a precautionary measure against micronutrient deficiency.

Water, Air and Drainage

Fundamental to soil ecology is the cycling of water to the soil through precipitation and its return to the air through evaporation and transpiration. Biological activity is dependent upon the balance of air and water in the soil. Too much water causes aerobic decomposition to cease and anaerobic bacteria to take over, with damaging effects. For example, nitrification, or the breakdown of nitrate nitrogen to gaseous nitrogen, occurs as a result of anaerobic biological activity in the soil. Too little water also causes biological activity to slow down and hence reduces the availability of nutrients.

The water available to plants is the moisture held mostly by capillarity in small soil pores. A soil with a large number of small pores, such as a clay-loam, will withstand drought much better than a sandy soil, which has few capillary pores. Large pores allow drainage and air flow that supplies oxygen and nitrogen for root and microbial growth. Both types of pore space are important for soil fertility, and both can be maintained and enhanced by the addition of organic matter and humus to the soil.

An ideal soil has a high infiltration rate, and fairly slow hydraulic conductivity. The infiltration rate is the rate at which water soaks into the ground; if the infiltration rate is slower than the rate of precipitation, the excess water will become surface run-off, with attendant erosion and pollution hazards. Hydraulic conductivity is the rate at which water drains through a saturated soil. This action transports nutrients from the surface layers to the rhizosphere. If the hydraulic conductivity is too fast, nutrients will be leached out of the soil and groundwater may become polluted. Organic matter in the form of cover crops or mulch improves the infiltration rate. When converted into humus through biological activity, organic matter can lower the hydraulic conductivity of sandy soils.

Wet soils, if caused by high groundwater levels, tend to be unsuitable for organic field crops and are often better left as permanent pasture, or allowed to revert to natural habitat. If the water problem is caused by compaction or hardpan, chisel plowing or subsoiling may correct the situation. Earthworms, and crops with long tap roots such as alfalfa, can then help to maintain the field in improved condition. Solutions such as ditching or tile drainage should be very carefully assessed for their environmental implications.

MAJOR SOIL PROCESSES

SOIL AND SOIL PROCESSES

Soil is the naturally occurring unconsolidated mineral or organic material at the surface of the earth which is capable of supporting plant growth. The particular type of soil at any position on the earth is a result of the geologic or parent material, climate, biota, topography and time. These factors of soil formation do not act singly but together in various ways to produce a particular soil. The soil must not be regarded as a passive, inert body on the earth's surface. It is a continually changing system within the total environment. Soils possess properties that have been inherited from the parent material, for example the amount of stones, mineralogy and thickness. They also have properties and features that are the result of soil processes, including such things as the kind and amount of organic matter, the formation of horizons (layers), new minerals and changes in porosity. For instance the features of a valley bottom soil will be largely inherited from the conditions of sediment accumulation in that part of the valley, whereas the humus content of the topsoil will be the result of the amount of organic litter added to the surface and the balance of the constructive and destructive processes acting on this litter.

Soils have both internal and external properties. The internal properties are often shown as various layers or horizons that are exposed in a vertical cross-section of the soil body. External properties include such attributes as slope and aspect. Soils often grade imperceptibly from one kind to another and occur as a relatively continuous blanket on the land surface of the earth. This continuous blanket is divided into soil classes defined upon the basis of soil properties, many of which reflect processes of soil genesis. Local changes in slope, elevation or aspect are dominant factors in determining changes in soil properties. It is on the basis of the soil's internal and external properties that soil resource inventories are prepared.

Soil Components

Soils are multiphase systems, composed of a mixture of numerous kinds and sizes of mineral grains, organic fragments, air and water. The proportions of these components vary from point to point on the surface of the earth, and from time to time, depending upon the factors of soil formation and the influence of man's activity. Soil air plays an important role in plant growth and the activity of soil organisms. Its main constituents, as in the atmosphere, are nitrogen, oxygen and carbon dioxide.

Various organisms use the oxygen and give off carbon dioxide. The supply of oxygen has to be constantly replenished and this is the importance of adequate pore spaces or aeration within the soil. For adequate aeration approximately 10% by volume of the soil should be occupied by air.

The importance of soil water lies in its roles in plant growth and in the chemical and physical reactions that occur in soils. Water provides the medium by which plants take up their nutrients in solution through their root systems; it may also be considered a plant nutrient itself. Thus water is crucial to plant growth. Molecules of water also play a central role in the chemical weathering processes of hydration, hydrolysis, oxidation, reduction and solution.

In physical weathering soil particles and rock fragments are disrupted by the expansion of ice in cycles of freezing and thawing and during wetting and drying. Wetting and drying cycles play an important role in the production of soil structure and in the physical translocation of material such as clay within a soil.

Air and water in the soil have a reciprocal relationship since both compete for the same pore spaces. For example, after a rain or if the soil is poorly drained, the pores are filled with water and air is excluded. Conversely, as water moves out of a moist soil, the pore space is filled with air. Thus the relationship between air and water in soils is continually changing.

The mineral portion of soils is composed of a wide variety of inorganic compounds. The minerals are commonly referred to as primary or secondary. Primary minerals originate from the geologic substrata. They include quartz, feldspar, mica, amphibole and pyroxene. They respond to the environmental factors and in the presence of water and air are altered or weathered. As a result of chemical weathering processes the primary minerals become secondary minerals such as illite, vermiculite, chlorite, montmorillonite and kaolinite (often called the clay minerals) and the hydrous oxides of iron and aluminum.

Particle size of the mineral fraction varies widely, from stones (over 250 mm in diameter), through cobbles (250-75 mm), gravel (75-2 mm), sand (2-0.05 mm), silt (0.05-0.002 mm) to clay (less than 0.002 mm). All combinations of these particle sizes can occur in soils, and the expression of the amounts of the various size fractions is called soil texture. Although coarse fragments and sand sized particles play but a small role in chemical processes occurring in soils, they are very important in soil drainage and in the engineering uses of soils. Silts and clays play the most important role in water and nutrient retention as well as the swelling and shrinking properties of soils.

The organic portion of soils results from the accumulation of animal and plant residues added to the mineral soil. It is this organic portion that differentiates soil from geological material occurring below the earth's surface which otherwise may have many of the properties of a soil. Organic compounds undergo decomposition by soil flora and fauna. This process produces humus, the most active and important form of organic matter for crop growth and soil formation. Organic matter is an extremely important constituent of soils, as it is the source of many nutrients necessary for plant growth, it enhances the soil's water holding capacity and its presence favours good soil structure especially in the topsoil.

Soil Processes

There are four basic processes that occur in the formation of soils, namely additions, losses, translocations and transformations. The two driving forces for these processes are climate (temperature and precipitation) and organisms, (plants and animals). Parent material is usually a rather passive factor in affecting soil processes because parent materials are inherited from the geologic world. Topography (or relief) is also rather passive in affecting soil processes, mainly by modifying the climatic influences of temperature and precipitation.

Losses occur both from the surface and from the deep subsoil. For instance, water is lost by evapotranspiration and carbon dioxide by diffusion at the surface and, on a more catastrophic level, large masses of soil can be stripped by erosion. Materials suspended or dissolved in water are the main forms of losses from the subsoil.

Translocation refers to the physical movement of material within soil. The material can be in the solid, liquid or gaseous form, the movement can be in any direction from and to any horizon. For instance clay, organic matter and iron and aluminum hydrous oxides are commonly moved from the surface horizon to a subsurface horizon. Conversely, in very dry climates salts are moved upwards in solution by capillarity, and in very cold climates solid mineral fragments are moved upwards by frost action.

Additions, losses and translocations all involve movement. Figure 2.2.1 shows this movement in two dimensions. Since a soil is three dimensional and most soils occur on slopes, it must be remembered that these movements can occur laterally as well as vertically. Water can carry clay in suspension laterally through a subsurface horizon, or topsoil can move slowly en mass from a surface horizon upslope to form a new surface downslope.

Transformations involve the change of some soil constituent without any physical displacement. Chemical and physical weathering and the decomposition of organic matter are included here. All these processes occur to a greater or lesser extent in all soils. The properties that characterize one soil are the result of a particular balance among all the processes. Other soils will be different because they have been formed by groups of processes having different balances.

SOIL FORMING FACTORS

Soils are a product of the environment. Often this is illustrated in the form of a mathematical expression, namely:

soil = f (parent material, climate, biota, topography, time)

Although the above expression shows soils to be the function of a number of individual factors, the factors are not mutually exclusive but interdependent. For example, the kind of vegetation found at any one location on the earth's surface is dependent on climate, parent material, topography, time and, in fact, soil. It is obvious that numerous combinations of the factors are possible. This leads to many different kinds of soils, each representing a certain combination of the factors of soil formation. Parent material, topography and time are sometimes referred to as the passive factors of soil formation, as they do not form soils without the active factors of climate and biota.

Parent material may be of two general kinds, organic or mineral. Organic parent materials are formed by biological action where plant and animal tissue is produced faster than it is decomposed. This occurs in cold and wet regions where biological decomposition is very slow. Soil inherits many properties from the parent material from which it forms, for example, the kinds of minerals, particle size and the chemical elements. Thus, parent materials are the building blocks upon which the other factors of soil formation manifest their effects.

Climate is an active factor of soil formation because it governs the amount, distribution and kind of precipitation as well as the amount and distribution of solar energy available at any point on the earth's surface. Biological, physical and chemical processes take place in water, thus the amount of water available leaches the soil, allows nutrients to be obtained by plants and governs the kind of vegetative cover. The form of precipitation, whether rain or snow, is also important in relation to processes which take place in soils. Solar energy, usually expressed as temperature, is an important part of climate because it controls the form of water falling onto the soil surface as well as in the soil. Also as temperature, it increases the rate of reactions, such as chemical reactions, evapotranspiration and biological processes. Wide fluctuations in temperature, especially in the presence of water cause shrinking and swelling, frost action and general weathering in soils. It can be seen that both water and temperature complement each other in changing parent material and affecting the biota on any soil landscape.

Biota is another active factor in soil formation. The kind and amount of plants and animals that exist bring organic matter into the soil system as well as nutrient elements. This has a great effect on the kind of soil that will form. For example forest vegetation tends to accumulate organic matter on top of the mineral soil where it decomposes and adds organic acids to the soil below. This aids in weathering mineral soil particles. Soils under grassland, however, accumulate organic matter within the mineral soil as a result of the extensive root system which dies back periodically. Animals play an important part in soils; in some cases animals ingest soil particles and mix the mineral and organic portions, e.g. earthworms. Other animals tunnel through soils mixing the various components. Similarly, microorganisms have great effects on soils in decomposing organic matter, fixing nitrogen from the atmosphere, and making other soil nutrients available for plant growth. Biota, in conjunction with climate, modify parent material to produce soil.

Topography is often considered a passive factor modifying the effects of climate. Everyone is aware of the effect that aspect, or orientation of the soil surface to the sun, has on the type of vegetation. South-facing slopes tend to be drier than north-facing slopes as the result of higher temperatures. Similarly, as one goes from the valley bottom to the tops of mountains or plateaus the climate becomes cooler. Topography also redistributes the water reaching the soil surface. Runoff from uplands creates wetter conditions on the lowlands, in some cases saline sloughs or organic soils. Thus as a redistributor of the climate features, topography affects soil processes, soil distribution and the type of vegetation at the site.

Finally it must also be remembered that soils develop over long periods and the balance between processes is liable to change in that time. For instance, chemical weathering is likely to be a very important form of transformation when a river sediment is first exposed. But as vegetation slowly becomes established the addition and decomposition of organic matter will become more important and chemical weathering less so.

Soil Horizons

The effect of the processes described acting within the soil is to form different horizontal layers or horizons down a vertical section. Soil horizons run roughly parallel to the surface of the earth and may be composed of mineral or organic material. They differ from adjacent horizons in such properties as colour, structure, texture and consistence. A soil profile is a vertical section down through these horizons to the parent material. Soil horizons are labelled with capital letters to denote the major horizon followed by small letters to denote its particular characteristics.

In mineral soils there are generally three major horizons; A, B and C. The A horizon is uppermost, where the maximum accumulation of organic matter in situ occurs and where maximum removal of materials occurs by solution, suspension or erosion. It is also where the largest transfers of energy take place, both solar and biological. This is truly where the action is! The B horizon is where the materials accumulate that were freed in the upper portion of the soil body. There is a close relationship between the A and B horizons. Translocations as well as many biological and chemical reactions take place between them. The B horizon, however, tends to be more stable than the A for short term differences. The C horizon is often termed the parent material. The effects of soil processes have not manifested themselves appreciably and thus the material is little modified.

Small letters are used to modify the major mineral soil horizon letters A, B and C. These small letters denote the predominant properties of the major horizon. Thus h indicates a horizon enriched with organic matter and t indicates a horizon where clay has accumulated. Small letters can be combined; for example, a horizon labeled Bhf indicates that it is a layer within the soil where organic matter, iron and aluminum have accumulated.

Major organic soil horizons are found in two distinctive kinds of environments, well aerated or poorly aerated. In well aerated environments leaves, needles, twigs and branches fall to the earth, accumulate and begin to decompose on top of the mineral soil. The degree of decomposition is important in understanding soil processes, and it is the criterion for distinguishing three horizons L, F and H. L has the least decomposition and H the most with F denoting moderate decomposition.

In poorly aerated environments (wet areas) the organic materials also accumulate but their character is different because of the saturated condition. In these areas the horizons are denoted O. They will be composed mainly of mosses, rushes and woody material.

NITROGEN: THE SURFACE LAYER OF SOILS

Over 90% of the nitrogen N in the surface layer of most soils occurs in organic forms, with most of the remainder being present as NH_4^- whichis held within the lattice structures of clayminerals. The surface layerof most cultivated soils contains between0.06 and 0.3% N. Peat soils have high N contents to 3.5%.Plant remains and other debris contribute nitrogen N in the form of:

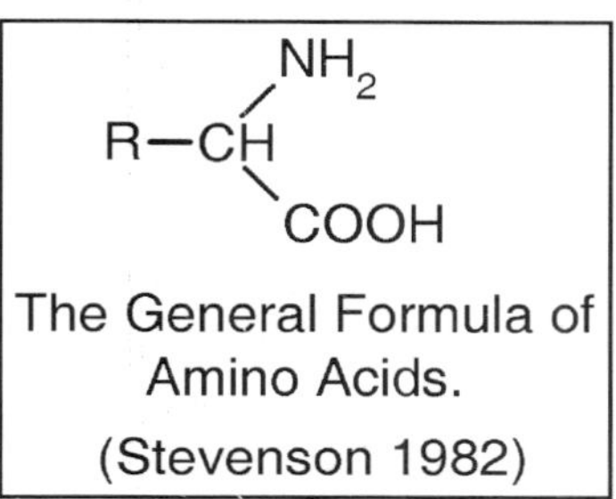

Fig. Amino Acids

Amino acids exist in soil in several different forms, like:

- As free amino acids
  - In the soil solution
  - In soil micropores
- As amino acids, peptides or proteins bound to clay minerals
  - On external surfaces
  - On internal surfaces
- As amino acids, peptides or proteins bound to humic colloids
  - H-bonding and van der Waals' forces
  - In covalent linkage as quinoid-amino acid complexes
- As mucoproteins
- As a muramic acid

Amino acids, being readily decomposed by microor-ganisms, have only anephemeral existence in soil. Thus the amounts present in the soil solutionat any one time represent a balance between synthesis and destruction by microorganisms. The free amino acids content of the soil is strongly influenced by weatherconditions, moisture status of the soil, type of plant and stage of growth,additions of organic residues, and cultural conditions.

AMINO SUGARS

Amino sugars occur as structal components of a broad group of substances, the mucopolysaccharides and they have been found in combination with mucopeptides and mucoproteins. Some of the amino sugar material in soil may exist in the form of an alkali-insoluble polysaccharide referred to as chitin. Generally the amino sugars in soil are of microbial origin.From 5 to 10%of the N in the surface layer of most soils can be accounted for in N-containing carbohydrates or amino sugars.

D - Glucosamine
(Stevenson 1982)

NUCLEIC ACIDS

Nucleic acids, which occur in the cells of all living organisms, consist of individual mononucleotide units (base-sugar-phosphate) joined by a phosphoricacid ester linkage through the sugar.Two types: ribonucleic acid (RNA) anddeoxyribonucleic acid (DNA).They have pentose sugar (ribose or deoxyribose),the purine: adenine, guanine and the pyrimidine: cytosine, thymine.RNA contains also the uracil.

The N in purine and pyrimidine bases is usually considered to account forless than 1% of the total soil N. Small amounts of N are extrcted from soil in the form of glycerophosphatides, amines, vitamins, pesticide and pesticide degradation products.

NITROGEN TRANSFORMATION

A key feature of the internal cycle is the biological turnover of N betweenmineral and organic forms through the opposing processes of mineralizationand immobilization. The latter leads to incorporation of N into microbial tissues. Whereas much of this newly immobilized N is recycled through mineralization, some is converted to stable humus forms. The overall reaction leading to incorporation of inorganic forms of N intostable humus forms is depicted on the picture.

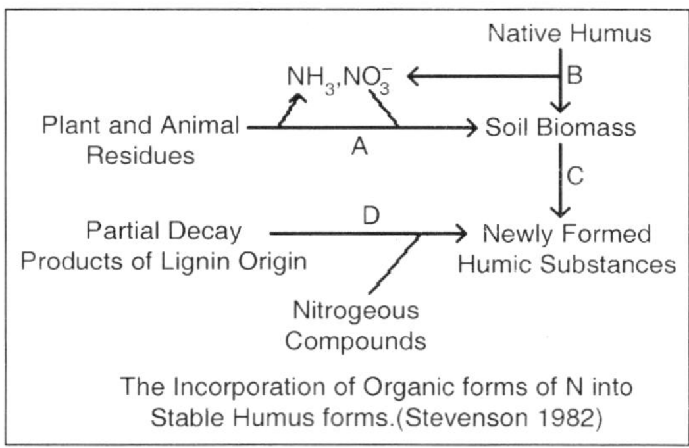

The Incorporation of Organic forms of N into Stable Humus forms.(Stevenson 1982)

Thus the decay of plant and animal residues by microorganisms results in theformation of mineral forms of N (NH_4^+ and NO_3^-) and assimilation of part ofthe C into microbial tissue (reaction A). Part of the native humus undergoes a similar fate (reaction B). Subsequent turnover through mineralization-immobilization leads to incorporation of N into stable humus forms (reaction C). Stabilization of N may also occur through the reaction of partial decay products of lignin with nitrogenous constituents (raection D). Except under unusual circumstances, both mineralization and immobilizationalways function in soil, but in opposite direction.

CHEMICAL REACTION OF AMMONIA AND NITRITE WITH ORGANIC MATTER

The fate of mineral forms of N in soil is determined to some extent bynonbiological reactions involving NH_4^+, $NH_3$and NO_2^- as depicted in fig.

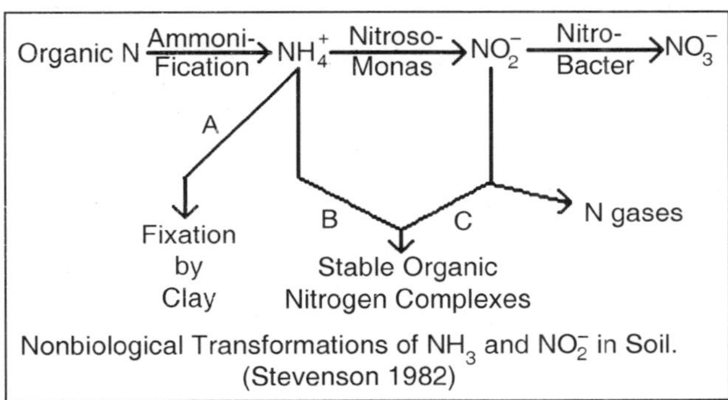

Nonbiological Transformations of NH_3 and NO_2^- in Soil. (Stevenson 1982)

In addition to NH_4^+ fixation by clay minerals (reaction A), NH_3 and NO_2^- react chemically with organic matter to form stable organic N complexes(reaction B and C). The chemical interaction of NO_2^- with organic matter may lead to the generation of N gases. Although both types of reactions can proceed over a wide pH range, fixation of NH_3^+ is favoured by a high pH (>7.0). In contrast, NO_2^- - organic matter interactions occur most readily under highly acidic conditions (pH of 5.0 to 5.5 or below).

STABILITY OF SOIL ORGANIC NITROGEN

- Proteinaceous constituents are stabilized through their reaction withother organic constituents, such as lignins, tanins, quinons.
- Biologically resistant complexes are formed in soil by chemical reactionsinvolving NH_3^+ or NO_2^- with lignins or humic substances.
- Adsorption of organic N compounds by clay minerals (pariculary montmorillinitic types) protects the molecule from decomposition.
- Complexes formed between organic N compounds and polyvalent cations, such as Fe, are biologically stable.

- Some of the organic N occurs in small pores or voids and is physicallyinaccessible to microorganisms.

C/N RATIO

For surface soils, and for the top layer of lake and marine sediments, the ratio generally falls within well-defined limits, usually from about10 to 12. In most soils, the C/N ratio decreases with increasing depth, often attaining values less than 5.0.Native humus would be expected to have a lower C/N ratio than most undecayedplant residues for following reasons. The decay of organic residues by soilorganisms leads to incorporation of part of the C into microbial tissue with the remainder being liberated as CO_2.

As a general rule, about one-third of the applied C in fresh residues will remain in the soil after the first few months of decomposition. The decay process is accompanied by conversion of organic form of N to NH_3 and NO_3^- and soil microorganisms utilize partof this N for synthesis of new cells. The gradual transformation of plantraw material into stable organic matter (humus) leads to the establishmentof reasonably consistent relationship between C and N. Other factors which may be involved in narrowing of the C/N ratio include chemical fixation of NH_3 or amines by ligninlike substances.

The C/N ratio of virigin soils formed under grass vegetation is normally lower than for soils formed under forest vegetation, and for the latter,the C/N ratio of the humus layers is usually higher than for the mineral soil proper. Also the C/N ratio of a well-decomposed muck soil is lower than for a fibrous peat. As a general rule it can be said that conditions which encourage decompositionof organic matter result in narrowing of the C/N ratio. The ratio nearly alwaysnarrows sharply with depth in the profile; for certain subsurface soils C/Nratios lower than 5 are not uncommon.

ORGANIC AND INORGANIC MATTER OF SOIL

Inorganic material composes the rest of the soil and the subsoil, and comes initially from the particular type of subsoil underneath the topsoil area. It's basically ground down rocky material, which may appear as clay, sand, gravel, stones or or chalky particles. Plants need the air spaces between both the organic and inorganic matter in soil, in order for their roots to get oxygen and grow. So a range of sizes of inorganic particles to provide drainage, mixed with organic matter for nourishment, is good for them. A very heavy clay soil may be hard to cultivate because of its lack of drainage - the clay particles stick closely together and plants find it hard to penetrate it, or for their roots to breathe. It needs mixing with organic material, or a lighter inorganic type such as sand, to make it workable. A very light sandy soil may drain too fast to retain water, but again, the answer is more organic material, which in this case will increase the soil's ability to retain moisture.

Soil is a complex mix of organic and inorganic matter that includes thousands of different species, the vast majority of which are still undescribed. Some of the organisms are pests which cause significant crop losses while others perform 'environmental services' such as biological control of pests, aeration, drainage, and nutrient and water cycling. As a dynamic living resource, soil is the basis of sustainable agriculture, as well as the physical support for most other human activities.

In many areas of the world, soil fertility is declining and erosion is getting worse. Marginal lands are especially susceptible. Worldwide, an estimated 2 billion hectares of land are considered degraded, that is, less productive due to deterioration of essential soil processes. The deterioration usually results from interactions among three types of processes: physical, such as erosion, crusting, and sealing; chemical, such as nutrient depletion, acidification, and pollution; and biological, such as organic matter depletion and loss of soil flora and fauna. Asia, Africa, and Latin America together account for an estimated 75% of the global area of degraded land, with 750, 490 and 240 million hectares, respectively. North America, Europe, and Australia each have an estimated 100–200 million hectares of degraded land.

Loss of soil structure and fertility, together with the increasing incidence of pests, weeds, and diseases, are often responsible for the migration of small-scale farmers practicing shifting or semishifting cultivation, as in the forest and savanna areas of Africa, South America, and Asia. Attempts to reverse this global trend by means of more sustainable agricultural practices depend on a thorough understanding of soil structure and function. However, while the physical and chemical characteristics of soils have been extensively studied, the great diversity of soil organisms and the complex interactions among them remain poorly documented and understood. It is only comparatively recently that the importance of the soil biota in maintaining soil quality, plant health, and soil resilience (the ability to recover from natural or anthropogenic disturbance) has been recognized. Despite growing interest in developing more sustainable agricultural systems, soil biota management remains a largely neglected area of agricultural research-for-development.

SOIL HEALTH AND PLANT HEALTH

Soil ecosystems are among the most complex of all terrestrial communities, and the role of the soil biota in maintaining plant health is not fully understood. The composition of the soil biota is strongly influenced not only by the nature of the underlying organic matter and mineral components, but also by environmental variables such as temperature, pH, and moisture. Numbers and types of soil organisms thus vary widely both in time and space, and are greatly influenced by agricultural activities such as tillage and cropping practices. There is a strong relationship between soil fertility and plant health, in the sense of the plant's ability to resist pests and diseases. Poor land management and declining soil fertility often result in a negative feedback cycle characterized in part by an increase in soil-borne pests and diseases. Agricultural practices such as adding lime, inorganic fertilizers, and pesticides can change the physical and chemical nature of the soil environment, thereby altering the number of organisms and the ratio of different groups of organisms. Since plant health is intimately linked to soil health, managing the soil in ways that conserve and enhance the soil biota can improve crop yields and quality. A diverse soil community will not only help prevent losses due to soil-borne pests and diseases but also speed up decomposition of organic matter and toxic compounds, and improve nutrient cycling and soil structure. Colonization of roots by mycorrhizae or other endophytic fungi, for example, can confer resistance to root-feeding insects. Conversely, a plant's ability to compensate for root damage can be compromised by nutrient deficiency if soil fertility is low.

Elusive Targets: Soil-borne Pests and Diseases

From a crop management perspective, soil pests and diseases pose special problems. They are hidden from view and hard to detect until the sudden appearance of damage creates an urgent need for rapid, curative treatments. Even then, damage due to soil pests and pathogens may be misdiagnosed and ascribed to other causes, such as nutrient deficiencies. In general, farmers are less aware of soil-borne pests and diseases (and less able to recognize them) than those attacking the above-ground parts of the plant. In the case of some insect pests, for example, farmers may fail to make the connection between the damaging, soil-based larval stage of the insect and its above-ground adult stage during which little or no damage occurs.

At the research level, the complexity and diversity of subterranean ecosystems pose unique challenges to those seeking to quantify the effects of individual taxa or species assemblages. This work is made even more difficult by the complex nature of many tropical cropping systems. Population densities of soil organisms can vary widely within a few meters, with their distribution being greatly influenced by the physical, chemical, and biological characteristics of the soil. Furthermore, different types of soil organisms require different methods for their extraction, identification, and quantification, and for the estimation of crop losses caused. The diversity of available techniques may make it difficult to compare results from different research groups.

PRINCIPAL SOIL ORGANISMS

Scientists have devised various schemes for characterizing and classifying soil organisms in order to be able to cope with their great diversity. Apart from the conventional taxonomic approach, there are classification schemes based on body size, function (decomposition, etc.), and the role of organisms from an anthropocentric point of view. The last-mentioned of these systems recognizes 'productive' biota (such as crop plants), 'destructive' biota (pests, pathogens, and weeds) and 'resource' biota (species that contribute to soil processes such as decomposition but that do not produce a harvestable product). Under body size, there are three main groups: the microbiota (<100 μm diameter), the mesobiota (100 μm to 2 mm diameter), and the macrobiota (>2 mm diameter). Microorganisms are the most abundant members of the soil biota. They include species responsible for nutrient mineralization and cycling, antagonists (biological control agents against plant pests and diseases), species that produce substances capable of modifying plant growth, and species that form mutually beneficial (symbiotic) relationships with plant roots. This last group includes mycorrhizal fungi, various actinomycetes, and some bacteria. Within the soil biota, the most important groups of both destructive and resource organisms are the bacteria, fungi, nematodes, arthropods (such as mites and insects), earthworms (mostly beneficial), and weeds.

Bacteria

The most abundant members of the vast community of soil organisms are bacteria. Per gram of soil, they can reach densities of one billion (one thousand million) individuals—and an estimated 20 000 to 40 000 species. The magnitude of bacterial biodiversity has only recently been revealed through molecular techniques that can differentiate between hard-to-culture taxa. Bacteria play important roles in many soil processes, including the cycling of nitrogen, carbon, and phosphorus, and the degradation of pesticides and other potential pollutants. They can multiply rapidly under favorable conditions, although very high growth rates are generally limited by the availability of nutrients. Bacterial populations are typically greater and more diverse close to plant roots. It has been estimated that 5 to 10% of root surfaces may be occupied by bacteria.

Various endophytic bacteria (e.g., species of *Azotobacter*, *Acetobacter*, and *Azospirillium*) not only fix nitrogen but also stimulate the production of root hairs. This increases the host plant's capacity to take up water and nutrients and compensate to some extent for grazing by root-feeding pests. Various bacteria that colonize seed coats and plant roots produce compounds capable of affecting plant growth. These growth-promoting rhizobacteria (including species of *Pseudomonas*, *Bacillus*, *Serratia* and *Arthrobacter*) can also increase plant resistance to pests and diseases through a variety of mechanisms, including the production of structural materials that strengthen root tissues and antibiotic metabolites that directly affect other elements of the soil biota. In some cases, these beneficial bacteria displace potential pathogens by competing for nutrients. Or, they may stimulate increased synthesis of defensive compounds by the plant itself, resulting in so-called induced resistance or systemic acquired resistance. *Rhizobium etli*, for example, induces systemic resistance in potato roots to the potato cyst nematode (*Globodera pallida*). Some rhizobacteria are able to suppress weed growth while leaving crops unaffected. The relationship between bacterial endophytes and their host plants is a subject of increasing interest to those seeking better ways to manage the soil biota. Plant roots themselves produce a number of compounds that affect microbial populations in their vicinity (the so-called rhizosphere). These compounds may act as attractants, repellents, biocides, or biostats (compounds that inhibit microbial growth), and as such offer new possibilities for the manipulation of both pathogens and beneficial species.

Fungi

As with the bacteria, the great diversity of fungi remains poorly documented. It has been estimated that only about 5% of fungi have so far been described. Fungi play as important a role in soil processes as do bacteria, but tend to be more abundant in slightly acid soils. They vary widely in size, preferred habitat, and mode of life. Fungal plant pathogens (e.g., some species of *Fusarium* and *Verticillium*) can cause diseases such as root rots and vascular wilts that are significant problems in many parts of the world. In the early 1990s, for example, the increasing prevalence of fungal root rots severely restricted the viability of bean crops in parts of Kenya.

Many soil-borne fungal pathogens (e.g., *Rhizoctonia solani* and *Pythium* spp.) are capable of infecting a range of plant genera. Furthermore, plants infected with fungal diseases may be more vulnerable to attack by soil-dwelling insects. In Malawi, for example, it was found that groundnuts were more vulnerable to attack by termites if they were infected by fungal pathogens such as *F. solani*. However, some fungi form symbiotic relationships with plant roots, enhancing the plant's ability to take up nutrients. There is some evidence that such relationships can help less competitive plants to become established in pasture systems. Many species of soil fungi are saprophytic (i.e., grow on dead organic matter), while others are parasitic on animals or plants. Some are important antagonists or biological control agents of soil-borne pests or diseases (e.g., species of *Dactylaria* and *Arthobotrys* for nematodes, *Beauveria* and *Metarhizium* for insect pests, and *Trichoderma* and *Coniothyrium* for plantpathogenic fungi).

Nematodes

Nematodes (roundworms or eelworms) are the most abundant microfauna in the soil and are particularly numerous in the top 5 cm. In the top 2 cm of soil, their numbers may exceed two billion per hectare. The number of nematode species worldwide has been estimated at 80 000 to 100 000 species. The majority of soil-dwelling nematodes feed on bacteria and fungi, but many prey on other nematodes, protozoa and rotifers, and some species are parasitic on insect pests.

More than 2000 species are parasites of higher plants. These include cyst and root-knot nematodes, such as certain species of *Heterodera*, *Globodera*, *Cactodera*, and *Meloidogyne*), and 'migratory' species such as *Pratylenchus* spp., *Ditylenchus destructor*, and *Scutellonema bradys*. Their feeding lowers crop yields by disrupting water and nutrient uptake or by decreasing fruit or tuber quality or size; it can also allow fungal and bacterial pathogens to gain access to damaged roots, causing secondary infections.

The presence of root-knot nematodes, for example, is known to increase the incidence of root rots and fusarium wilts in a wide range of crops. Plant parasitic nematodes tend to be more damaging pests in the tropics than in temperate zones. This is because their population growth rate is favored by warm, humid climates and the longer growing seasons in these regions, which allows for more reproductive cycles per year. Nematode pests are associated with almost all crops grown in the tropics and cause losses of millions of dollars each year. It has been estimated that, worldwide, plant-parasitic nematodes annually reduce agricultural production by about 12%. Losses are particularly severe in developing countries due to insufficient expertise for species identification, inadequate quantification of the pest problem, and a limited range of management options. Worldwide, root-knot and cyst nematodes are especially important pests and infestations by them continue to undermine national efforts to improve food security and alleviate poverty.

Losses to individual crops often go unnoticed or are attributed to other causes. This is because most symptoms of nematode damage such as chlorosis (yellowing), patchy growth stunting, and wilting in hot weather, are easily confused with nutrient deficiencies or sometimes with bacterial or fungal diseases. Compared with the plant-parasitic nematodes, much less is known of the biology and ecology of other nematodes. This is particularly true of the bacteria-feeding nematodes, which are generally considered to be beneficial or harmless. These species, often concentrated in the root zone of higher plants, play an important role in soil nutrient cycling and can help distribute bacteria that promote plant growth. Although nematodes are not capable of widespread dispersal on their own, in agricultural systems they may be spread via contaminated machinery, land levelling, irrigation, and soil erosion.

Mites

Although these are among the most common soil-dwelling mesofauna, they tend to be restricted to the leaf litter and surface layers. They are more diverse than any other single group of soil arthropods (including the insects), and this diversity is reflected in their feeding habits and life history. In general, they are more important in their role as resource biota than as destructive biota. Some species feed on fungal spores, while numerous predatory species attack nematodes, other mites, insect eggs, and larvae. Other species feed on plant debris, dung, or carrion and are important members of the decomposer community. Although some mite species feed on living plants and others are parasitic on livestock, in general these potential pests are only a minor component of the mite fauna.

Weeds

In the tropics, crop losses due to competition with weeds are on the order of 25%, with weed removal often the most laborintensive task in smallholder systems. In most areas, the weed population centers on about 20 particularly troublesome species. The individual taxa may be indigenous or introduced, but they usually share a few key traits, especially rapid vegetative growth, high fecundity, and persistence in the soil seed bank. Weed species that resemble the crop in their early stages (e.g., grass weeds in cereal systems) are particularly difficult to deal with when hand weeding is the main control technique. Besides competing with the crop for water and nutrients, weeds can also act as alternative hosts for pest nematodes and plant pathogens. And by boosting the humidity around the base of crop plants, weeds also increase the likelihood of infections by pathogens such as *R. solani* and *Sclerotium rolfsii*.

Some weed species are parasitic on certain staple food crops, in some cases causing total crop failure. The role of weeds in relation to soil-dwelling insect pests has not received a lot of scientific attention, although in some cases weeds are known to act as alternative hosts, maintaining the insect pest between crop cycles. Wild *Ipomoea* species, for example, can support the sweet potato weevil (*Cylas formicarius*) between crops, and in Zimbabwe, the groundnut plant hopper (*Hilda patruelis*) can survive the dry season on a variety of different weeds, allowing it to invade groundnuts as soon as they emerge. But weeds may also play a constructive role by supporting higher numbers of soil-dwelling predators that serve as pest control agents.

Insects

The number of taxa of soil-dwelling insects is relatively small compared with the diversity that marks other major groups of soil organisms. With many species, only part of the life cycle (egg, larva and/or pupa) is spent in the soil, although some taxa are associated with specific soil types. Root-feeding species can reduce crop yields by reducing a plant's ability to absorb water and nutrients and may cause further losses by facilitating the entry of soil-borne pathogens. In contrast, predatory and parasitic insects can contribute to the biological control of both invertebrate pests and plant pathogens. Some *Collembola*, for example, show a marked preference for feeding on the spores of fungal pathogens, including *Rhizoctonia solani* and *Fusarium oxysporum*. Species that feed on organic matter (detritivores) may be important members of the decomposer community. Some insects have ambivalent roles. For example, although termites are usually viewed as pests, in arid tropical soils with low earthworm populations the burrowing activities of termites can also help decompose organic matter and improve soil structure and porosity.

Termites and ants are usually the dominant components of the soil insect biomass. As such, they are probably more important than all other insects in their effects on soil structure. They can also be significant pests. In parts of Malawi, for example, it has been estimated that the two most destructive termites (*Pseudacathotermes militaris* and *Macrotermes michaelseni*) damage crops on 72% and 49% of all smallholdings, respectively. In Africa as a whole, the subfamily *Macrotermitinae* is considered the most damaging group, affecting a wide range of crops, including tree and pasture species. Their success as pests has been attributed to their ability to survive cultivation, feed on both living and dead plant material, and cultivate saprophytic fungi that serve as food in the dry season.

Other important root-feeding insects include the 'white grub' larvae of various scarab beetles (*Scarabeidae*), weevils (*Curculionidae*), wireworms (*Elateridae*), and false wireworms (*Tenebrionidae*). Some *Hemiptera*, *Orthoptera*, and larval *Lepidoptera* (e.g., cutworms) are likewise important in parts of the world. The significance of different species as pests is related to their host range. Those capable of feeding on a wide variety of plants, so called polyphagous species, are generally the most difficult to control. The distribution and damage potential of many soil-dwelling insects is also strongly influenced by soil moisture and rainfall patterns, with the more successful species being able to survive periodic drought.

As noted earlier, damage by root-feeding insects can result in secondary infections by plant pathogens. Root crops are particularly susceptible to this type of damage, since the harvested part of the plant is directly affected. Potatoes, for example, soon become infected with bacteria and fungi if damaged by larvae of the potato tuber moth (*Phthorimaea operculella*), resulting in the rapid rotting of the tuber. The same is true of sweet potato tubers damaged by larvae of the sweet potato weevil (*Cylas formicarius*). In groundnuts, the level of aflatoxin (caused by the fungus *Aspergillus flavus*) can increase if the pods are damaged by termites, since the latter can transmit fungal spores.

Earthworms

The Greek philosopher Aristotle referred to earthworms as the "intestines of the earth". The description is apt not only because these invertebrates physically resemble digestive organs but also because they perform something of a digestive function when they break down plant residues in the soil and recycle nutrients. Experiments with maize, for example, have shown that plant residues can degrade 30% faster in surface soil containing earthworms than in soil without them. Earthworms perform a variety of other ecosystem services as well. For instance, they enhance soil porosity thereby reducing rainwater runoff and allowing for infiltration of moisture to plant roots. They also play a vital role in water purification, agrochemical detoxification, and maintenance of soil stability. When it comes to sheer biomass, earthworms usually account for the bulk of invertebrates that make their home in the soil.

Along with ants and termites, they have a special distinction as 'ecological engineers'. This is due to their ability to excavate remarkably large amounts of soil and to create organomineral structures through their casts (excrement). To date about 3700 species of earthworms have been scientifically described. However, the actual number is estimated to be at least twice that, with the overall knowledge gap being significantly greater for tropical earthworm species than species in temperate zones. Although many aspects of earthworms' soil engineering role have been documented, interactions with other members of the soil biota, and how those affect plant growth and health, positively or negatively, are much less well understood.

ORGANIC MATTER MANAGEMENT AND THE SOIL ECOSYSTEM

Primarily at a functional group level, soil biota regulate vital ecosystem processes such as decomposition (the breakdown of complex organic compounds into nutrients available for plant growth), C sequestration, and nutrient cycling. The rate of decomposition is dependent on the interaction of climate, biota and the quality and quantity of organic matter. Agricultural practices that provide good soil protection and maintain high levels of soil organic matter favour higher biodiversity. Examples include agroforestry systems, intercropping, rotational farming, conservation agriculture, green-cover cropping and integrated arable-livestock systems (Plate 6). Actions that target the joint conservation of both above- and below-ground components of biological diversity directly will have environmental benefits at ecosystem, landscape and global scales.

CROPPING SYSTEM AND SOIL ECOSYSTEM

The successful functioning of most ecosystem processes requires a balance of biotic interactions in a complex soil biota community (detritus food web). Availability of C is one of the important regulating factors of biological activity in soils, which affects the composition of the microbial community and the food-web structure. In addition, the number of trophic levels in a terrestrial food-web community and its stability depend upon the amount and quality of C input and the level and type of disturbance (*e.g.* tillage, genetically modified (GM) crops and use of agrochemicals).

Plants are the main drivers of the dynamics of soil microbial communities via their input of various C sources into the system. Plant residues are the primary source of C in soils, with the majority of biota populations concentrated near residues and in the rhizosphere of plants. Therefore, any changes to the quality of crop residues and rhizosphere inputs will modify the dynamics of soil biota. Hence, a change in vegetation as a result of changes in land use is a major factor affecting the diversity of the microbial community. Moreover, changes in agricultural practice including the intensity of the use of fertilizers and pesticides and crop cover, *e.g.* grass versus arable crops in rotation, may lead to shifts among and within groups of the microbial community.

Diverse habitats support complex mixes of soil organisms. Diversity can be achieved with crop rotations, vegetated field borders, buffer strips, strip cropping, and small fields. Crop rotations provide different food sources into the soil each year and encourage a wider variety of organisms and prevent the buildup of a single pest species.

SOIL RESILIENCE AND RISK ALLEVIATION

Because of time constraints during the workshop, the working group on innovation and risk management agreed to exclude the important issue of genetically modified organisms (GMOs), and to concentrate on other organisms, technologies and methods, including peri-urban and waste management issues (*e.g.* vermicompost).

The knowledge and experience among participants was reviewed, taking into account the different biophysical conditions and range of functional groups: the producers, consumers and decomposers (N fixers, P solubilizers, C and N mineralizers, predators, pests and pathogens, soil aggregation engineers, antibiotics). It was agreed to keep a focus on food security, environmental quality and economic sustainability goals. A holistic systems view is needed to address extensively managed systems, such as shifting cultivation, intensive diverse systems and monocultures.

The first role of biodiversity is to ensure the multiplicity of functions that soil organisms perform. A secondary but important role of biodiversity is to ensure the maintaining of these functions in the face of perturbations. Genetic variability within and between species confers the potential for resistance to perturbations, whether they be short or long term. Understanding the relationship between biodiversity and more complex functions requires the combined study of taxonomically distant groups of organisms that can perform specific functions, and thus belong to the same functional group.

Soil microbial communities represent the largest source of biodiversity on earth. Given the extremely high species diversity in soil, it is estimated that microbial communities contain such high levels of redundancy as to make small changes in soil microbial diversity insignificant. Rather, shifts among groups or species within the microbial community are considered to be of much more relevance for the functioning of terrestrial ecosystems. Shifts that might be relevant for sustainable land use include those in the relative abundance of bacteria and fungi and within groups with specific functions, such as nitrifying bacteria. These shifts could affect vital functions of the soil ecosystems, such as nutrient retention and antagonism against plant diseases.

A greater degree of biodiversity between or within a given species or functional group should logically increase the inherent variability in tolerance or resistance to stress or disturbance. Implicit in these arguments is the assumption that a multiplicity of organisms can perform a particular function, and that the replication of the ability to perform a particular function implies a degree of functional redundancy. Whether organisms are ever truly redundant is a matter of debate. Though redundancy in a single function may be common among many soil biota, the suite of functions attributable to any one species is unlikely to be redundant. Furthermore, functionally similar organisms have different environmental tolerances, physiological requirements and microhabitat preferences. As such, they are likely to play quite different roles in the soil system.

Table. Soil Biological Solutions for Soil Fertility and land Degradation Problems

Physical problems	Chemical problems	Biological problems
Compaction Low water content Poor drainage Erosion Loss of silt or clay	Nutrient depletion Excessive acidity or alkalinity Low phosphate levels Heavy-metal contamination High salinity Pesticide contamination	Low biodiversity Low microbiological activity Low humus content High pest or pathogen levels Lack of natural enemies Low organic matter

POSSIBLE SOIL BIOLOGICAL SOLUTIONS

Aggregation, porosity, regulation of soil hydrological processes - these are improved by bioturbating organisms, plant root, fungal hyphae, microbial secretions.

- Bioremediation.
- Nutrient cycling, decomposition of organic matter, nutrient mineralization, N fixation.
- Crop diversity over space and time (intercropping, diverse rooting depths, rotations).
- P solubilizing bacteria and plant nutrition and plant growth promoters.
- Suppression of pests, parasites and diseases.

Problems	Bioremedial	N fixing	Compost	Manure	Rotation	Extracts	Inoculants
Degraded soils-low aggregation	-	+	+	+	– +	?	+
Degraded soils-low organic matter	-	– +	+	+	– +	–	-
Low saprobes	–	–	+	+	–	+	+
High pesticide levels	+	–	– +	–	+	–	–
High salinity	+	–	+	– +	+	–	–
High pollutants	+	+	– +	–	– +	?	–

Note:

– Unlikely to have beneficial impact

+ Positive impact expected

However, given the estimates for the vast numbers of species present in soils, and the rather limited number of functions that can be ascribed to the soil biota as a whole, a degree of functional redundancy seems inevitable even allowing for the fact that decomposition of plant material may require hundreds of enzymes. The greater the degree of functional redundancy, the greater will be the ability of a particular function to withstand stresses or disturbances, *i.e.* the greater the resilience.

Any novel method for manipulating and managing soil biodiversity and biotic products in situ requires an analysis of risk. There has been considerable interest in evaluation of risks associated with GM crop varieties. However, apart from these cases, little serious attention has been paid to environmental impact. This is the case especially where microbial treatments or manipulations are carried out. For example, little is known of the effect of Rhizobium inoculation on natural microbial populations. Similarly, the effects of herbicide-resistant plants and the use of herbicides on the soil ecosystem are not well known. It is not acceptable to assume that biosafety assessments can be made using external measurements, such as plant health or productivity, without doing the basic research to establish the necessary links.

Risk analysis is more complex than the simple establishment of safety in isolation from the environment in which the new product or process is to be employed.

The following issues need to be addressed:

- *Toxicity*: Is the product safe to eat for consumers (humans or animals) or can it produce toxic products or by-products?
- *Environmental impact*: What are the effects on non-target organisms? Assessments should be made of effects on a range of organisms, including providers of key ecosystem services and prominent species such as birds and butterflies.
- *Genetic drift*: What is the risk of genes from novel crops flowing into the environment? This may happen through hybridization between new varieties of traditional crops and their wild relatives as well as from GM varieties. Terminator gene technology can eliminate this risk for GM crops, but some consider it unacceptable.
- Agronomic merit: do the new varieties perform better than those currently in use; and will pesticide needs be smaller or greater?
- *Socio-economic issues*: Will the crops be acceptable to farmers, and do the farmers have access to any specialized handling equipment needed? For example, a crop designed for mechanical harvesting in the American prairies may not perform well in the small plots of subsistence farmers in Africa.
- *Financial*: Can farmers afford the product, and will increases in production lead to greater income, or a consequent fall in crop price?

Will consumers buy the new product? For example, even if Rhizobium inoculants give proven yield increases, farmers will not adopt them unless their availability is accompanied by information and training on their use and demonstration of the potential benefits of investing in such a product.

SOIL MINERALS

Soil minerals play a vital role in soil fertility since mineral surfaces serve as potential sites for nutrient storage. However, different types of soil minerals hold and retain differing amounts of nutrients. Therefore, it is helpful to know the types of minerals that make up your soil so that you can predict the degree to which the soil can retain and supply nutrients to plants. There are numerous types of minerals found in the soil. These minerals vary greatly in size and chemical composition.

SOIL MINERAL PARTICLE SIZE

Particle size is an important property that allows us to make distinctions among the different soil minerals. Soils contain particles that range from very large boulders to minute particles which are invisible to the naked eye. To further distinguish particles based upon size, particles are separated into the two categories: the coarse fraction and the fine earth fraction.

Fine Earth Fraction

When we refer to most soils, we are generally referring to the second category of particle size: the fine earth fraction. This is because the soils are almost exclusively finely textured.

Table. Description of Sand, Silt, and Clay Classes.

The Fine Earth Fraction			
	Size	**Texture**	**Characteristics**
Sand	2.0 mm - 0.05 mm	gritty	Sand is visible to the naked eye, consists of particles with low surface area, and permits excessive drainage.
Silt	0.05 mm - 0.002 mm	buttery	Silt is not visible to the naked eye and increases the water holding capacity of soil.
Clay	< 0.002 mm	sticky	Clay has a high surface area, high water holding capacity, many small pores, and possesses charged surfaces to attract and hold nutrients.

The fine earth fraction includes any particle less than 2.0 mm (.078 inches) and is divided into three classes of size: sand, silt, or clay. To put this into perspective, the width of the lead in a No. 2 pencil is approximately 2.0 mm. Table provides descriptions of each class in the fine earth fraction.

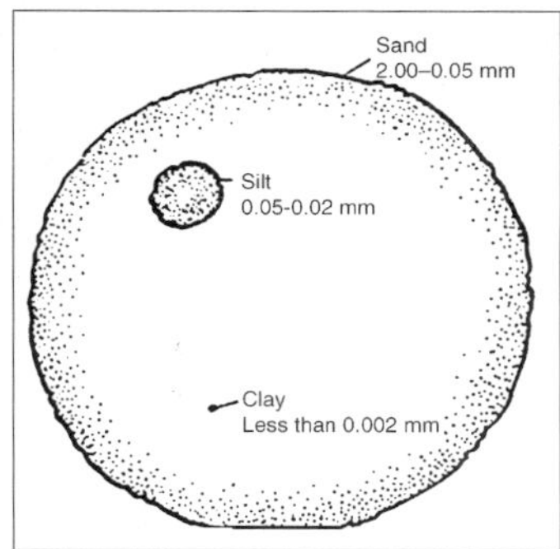

Fig. Relative Size Comparison between Sand, Silt, and Clay of the Fine Earth Fraction.

Coarse Fraction

The coarse fraction of soil includes any soil particles greater than 2mm. The coarse fraction includes boulders, stones, gravels, and coarse sands. These are rocky fragments and are generally a combination of more than one type of mineral. We are not very concerned with the coarse fraction in soil since the soils primarily fall into the fine earth fraction.

MINERAL FRACTION OF SOIL

The mineral fraction of soils is derived largely from weathering of the underlying parent material, which may consist of consolidated bedrock or unconsolidated superficial deposits. Consolidated bedrock can be classified into one of three categories—igneous, sedimentary or metamorphic. Igneous rocks are derived from the consolidation of molten magma, either within the Earth's crust or at its surface, whereas sedimentary rocks are the product of cycles of weathering, erosion and deposition operating at the surface, and metamorphic rocks originate from the alteration of any other rock type by high temperature and/or pressure but without melting.

Unconsolidated superficial deposits are just as variable as solid bedrock materials and are often classified according to the nature of their depositional environment, for example riverine alluvium, glacial till and lacustrine clays. These deposits are often indurated or cemented to some extent by calcareous, siliceous or iron-rich components. In addition to the parent material origin of mineral material, it can also be added to soils by movement from upslope, by aeolian transport or by atmospheric fall-out of materials such as volcanic ash.

Soil minerals occur in a wide variety of forms, but these can be arranged into groups on the basis of their chemical composition and the structural arrangement of their constituent elements.

By far the most abundant group in soils and their parent materials is the *silicate* group. The fundamental building block of all silicate minerals is the silicon-oxygen (Si-O) tetrahedron, which consists of a central silicon ion (Si^{4+}) surrounded by four closely spaced oxygen ions (O^{2-}).

Because the silicon ion has four units of positive charge and each oxygen ion has two units of negative charge, each discrete tetrahedron possesses four units of negative charge. This allows the tetrahedra to link together in a variety of characteristic structural arrangements which forms the basis of their classification. The potential range of silicate minerals is further widened by substitution of one ion for another within the mineral structure, a process known as *isomorphous substitution*.

One of the most common substitutions involves replacement of some Si^{4+}, in Si-O tetrahedra, by aluminium ions (Al^{3+}). Other examples include the substitution of Al^{3+} by magnesium ions (Mg^{2+}) or iron (Fe^{3+} or Fe^{2+}). When ions with the same charge or valency are exchanged, the mineral structure remains electrically neutral. If, however, ions with different valencies are exchanged, there will be a charge imbalance; frequently this results in an excess negative charge, as in the examples above. In this situation, electrical neutrality is achieved either by incorporation of additional cations such as calcium, magnesium, potassium or sodium (Ca^{2+}, Mg^{2+}, K^{+}, Na^{+}) into the crystal lattice, or by structural rearrangements that allow internal compensation of charge. The most common silicate minerals and their associated structures.

Framework silicates or tectosilicates consist of a three-dimensional lattice of Si-O tetrahedra linked through their corners. The two main mineral groups within this category are quartz and feldspars, which are common minerals in many soils. Quartz consists simply of Si-O tetrahedra linked through the O^{2-} ions, and consequently there are twice as many O^{2-} ions as Si^{4+} ions in the mineral structure, which has the general formula $(SiO_2)n$. Unlike quartz, feldspars contain significant amounts of Al^{3+} ions, originating from isomorphous substitution of Si^{4+} ions, and base cations, especially Ca^{2+}, Na^{+} and K^{+}, which satisfy the residual negative charges. There are two main groups of feldspar minerals—potassium feldspars such as orthoclase and microcline, and plagioclase feldspars which form a continuous series of minerals between the sodium end member, albite, and the calcium end member, anorthite.

Chain silicates or *inosilicates* comprise Si-O tetrahedra linked together to form a continuous chain structure of which there are two types. First there is the single chain where the Si-O tetrahedra are linked by sharing two out of the three basal O^{2-} ions, and second there is the double chain where tetrahedra are linked by sharing all three basal O^{2-} ions to form a hexagonal arrangement. Common minerals in these two categories are the pyroxene family and the amphibole family respectively. In chain silicates the chains themselves are linked together by a variety of cations including Ca^{2+}, Mg^{2+}, Fe^{2+}, Na^{+} and Al^{3+}. Minerals in this group are very reactive and more easily weathered than the framework silicates. *Orthosilicates* and *ring-silicates* display a wide range of structures but, in comparison with other groups, their occurrence in soils is relatively minor.

Some minerals in this group are very reactive and therefore particularly susceptible to weathering, whereas others are very resistant. This category can be divided into two groups—*neso-silicates* and *soro-silicates*. In the first group the Si-O tetrahedra occur as separate units, with no shared O^{2-} ions, and are linked by metallic cations.

Examples include olivine and garnet. In soro-silicates the tetrahedra form separate groups in which they share one or more of their O^{2-} ions; when the group is formed by sharing O^{2-} ions by more than two Si-O tetrahedra, a ring structure results. Examples of soro-silicates are beryl and cordierite. *Sheet silicates* or *phyllosilicates* are perhaps the most important group of minerals in soils, playing an important role in the development of many physical and chemical soil characteristics. Unlike the other mineral groups, many sheet silicate minerals occur in the smallest grain-size categories; consequently they are often known as the *clay minerals*. Minerals in this group are made up of various combinations of three fundamental sheet structures:

- Si-O (siloxane) sheet in which the Si-O tetrahedra are linked in a hexagonal arrangement,
- Al-OH (gibbsite) sheet in which the Al^{3+} ions are surrounded by six closely packed hydroxyl (OH^-) ions to form an octahedral structure,
- Mg-OH (brucite) sheet whose structure is similar to that of the gibbsite sheet but contains Mg^{2+} ions rather than Al^{3+}.

Classification of sheet silicate minerals is based on three criteria—the ratio of the above sheets within a unit layer of the mineral structure, the interlayer or basal spacing between unit layers, and the interlayer components or species. The sheet silicate minerals most commonly found in soils include the micas, illite, chlorite, kaolinite, vermiculite and the smectites.

The most frequently occurring micaceous minerals are muscovite and biotite. Both of these are 2:1 layer silicates with an interlayer spacing of approximately 10 Å (1.0 nm) and K^+ as the dominant interlayer component. Illite is very similar in composition and structure to micaceous minerals and is sometimes known as hydrous mica. Chlorite has a 2:1:1 layer structure (two siloxane sheets to one brucite sheet, forming a biotite mica layer, to one further brucite sheet which occupies the interlayer position) with an interlayer spacing of about 14 Å (1.4 nm). Kaolinite has a 1:1 layer structure and an interlayer spacing of around 7 Å (0.7 nm). Layer units of kaolinite are attracted by hydrogen bonding and the crystals have a characteristic hexagonal appearance. Vermiculite is a 2:1 layer silicate with a basal spacing of about 14 Å (1.4nm) and interlayer positions occupied by Mg^{2+} ions and water molecules.

The most common smectitic mineral in soils is montmorillonite. This is a 2:1 layer silicate with interlayer positions dominated by Na^+ and Ca^{2+} ions, and water molecules. Although its basal spacing is approximately 14 Å (1.4 nm), this varies somewhat due to its shrink-swell characteristics which occur on wetting and drying.

Other sheet silicate minerals sometimes found in soils include halloysite, imogolite and allophane, and mixed layer or interstratified minerals. Halloysite is similar in many ways to kaolinite, but its crystal layers are curved to form a tubular structure. Imogolite and allophane are gel-like hydrated alumino-silicate minerals with a poorly crystalline tube- or thread-like structure. Mixed layer minerals result from the interlayering of more than one sheet silicate mineral. Interlayering may be regular, where there is a regular repetition of the different mineral layers, or random where there is no clear pattern of repetition. These minerals may exist as intergrades which represent stages of transition between one mineral and another during weathering. Examples of mixed layer minerals are mica- or illite-vermiculite (hydrobiotite), illite-montmorillonite and kaolinite-montmorillonite.

In addition to the silicate minerals are the *non-silicate minerals*. These are often of only minor occurrence in soils and are usually referred to as *accessory minerals,* although they play a significant role in the development of some soil characteristics. Among the most common non-silicate minerals in soils are the free oxides and hydroxides of iron, aluminium and manganese; these may exist as crystalline or amorphous forms. Iron minerals include goethite, ferrihydrite, hematite, lepidocrocite, maghemite and magnetite. Aluminium minerals include gibbsite and boehmite, while examples of manganese minerals are birnessite and pyrolusite. Other non-silicate minerals include calcite ($CaCO_3$), which occurs in soils developed from calcareous parent materials, and anatase (TiO_2) and amorphous silica, which are often found in soils developed on recent volcanic deposits.

ORGANIC COMPONENTS

Soil organic matter is derived from a number of sources, of which the most important is usually plant litter. This consists of a variety of plant debris including leaves, stems, flowers, twigs, bark, and the larger branches and trunks of trees. Other organic components include plant roots, root exudates, soil organisms together with their faecal remains and metabolites, and organic substances washed into the soil from vegetation. Soil organisms are responsible for the physical comminution and bio-chemical decomposition, and incorporation of organic matter. They exhibit tremendous diversity in terms of their numbers, size and morphology, and also vary dramatically in their function, mode of nutrition and environmental tolerance. Organisms can be referred to as *producers, consumers* or *decomposers*.

Producers, such as plants, fix carbon from atmospheric carbon dioxide (CO_2) during photosynthesis, consumers feed on plants and other organisms, and decomposers utilise carbon from organic material, returning it to the atmosphere as CO_2 and other gaseous by-products and mineralising nutrients to their original ionic form. Soil organisms can also be classified according to their size *micro-organisms*, which include both microflora and microfauna, *mesofauna* and *macrofauna*. The classification of soil organisms according to their mode of nutrition is based on their sources of carbon and energy.

Essentially, they obtain carbon from either organic material or inorganic sources (largely CO_2); these two groups are known as *heterotrophs* and *autotrophs* respectively. They can also obtain energy either from light (photoheterotrophs or photoautotrophs) or from chemical oxidation (chemoheterotrophs or chemoautotrophs). In addition to their carbon and energy sources, soil organisms, especially bacteria, vary in their oxygen requirements. *Aerobes* have a direct requirement for oxygen, *facultative anaerobes* normally require oxygen but may adapt to oxygen deficit by utilising nitrate and other inorganic compounds as electron receptors, and *obligate anaerobes* require an absence of oxygen as it is toxic to them. Before looking at the composition of organic material, the various groups of micro-organisms, mesofauna and macrofauna will be briefly considered.

Four main groups of micro-organisms can be recognised—bacteria and actinomycetes, fungi, algae and protozoa. Bacteria are very small organisms (1-5 μm) and are often round or rod-like in shape. They tend to exist in thin films of water surrounding soil particles, often reproducing with tremendous rapidity, with numbers in the range of 10^6-10^9g having been widely reported. Many species secrete polysaccharide gums, which facilitate aggregation. Bacteria demonstrate great biochemical versatility in their ability to decompose a wide range of materials under a variety of conditions. For example, *Pseudomonas* sp. can metabolise a range of chemicals including pesticides, *Nitrobacter* sp. derives its energy from the oxidation of nitrite to nitrate, *Thiobacillus ferrooxidans* acquires energy from oxidation of reduced sulphur compounds and Fe^{2+}, and *Rhizobium* sp. forms nitrogen-fixing nodules on the roots of leguminous plants.

Actinomycetes consist of fine branching filaments (about 1 ìm in diameter) during their vegetative stage, when they are similar in appearance to fungi. During reproduction, however, the filaments undergo fragmentation, sometimes forming dense colonies. For this reason actinomycetes are often classified as highly evolved and complex bacteria. Fungi produce filamentous structures (hyphae) which are about 0.5-1.0 ìm in diameter and which grow into a dense network or mycelium. They are generally less numerous than bacteria in soils ($1\text{-}4 \times 10^5$ g), although they are common in acidic soils where they can be responsible for 60-80 per cent of organic matter decomposition. All fungi are heterotrophic, living mostly in the surface layers where they play an important role in the development of soil structure. Mycorrhizal fungi live symbiotically in plant tissue, removing carbon in return for nutrients such as phosphorus. Algae are photosynthetic organisms and are therefore confined largely to the soil surface. They include Cyanophaceae (blue-green algae) and Chlorophaceae (green algae), the former playing a major role in development of the nitrogen status of soils. Protozoa are microfauna, unlike the previous groups which are microflora.

They are uni- or non-cellular organisms of 5-40 ìm in length and live in thin water films surrounding soil particles. They can be divided into amoeboid forms with silica or chitin sheaths, and rotifers with better differentiated cell structure. Protozoa are very numerous (often > 10^4 g) and are particularly important in controlling the numbers of bacteria and fungi, on which they feed. As in the case of micro-organisms, four main groups of mesofauna can be recognised—nematodes, arthropods, annelids and molluscs. Nematodes are unsegmented worms, and next to protozoa are the smallest of the soil fauna, being 0.5-1.0 mm in length. They are plentiful in soil and litter, feeding on plant remains, roots, bacteria and sometimes on protozoa.

Arthropods can be divided into a number of categories, the most significant numerically being acari (mites) and collembola (springtails). Both of these feed on plant litter, bacteria and fungi, with acari being particularly common in acidic litter where they may constitute 80 per cent of soil fauna. Other arthropod groups include myriapods, which are dominated by chilopods (centipedes) which are carnivorous, and diplopods (millipedes) which are herbivorous. Arthropods also include isopods (woodlice), beetles, insect larvae, ants and termites. Termites are prevalent in tropical soils and are particularly active soil mixers, producing structures at the surface (termitaria) up to several metres in height.

Arthropod populations can be particularly high, for example 220,000 m^2 in old grassland soils. Annelids consist largely of enchytraeid worms (potworms) and lumbriscid worms (earthworms). Enchytraeids are small (0.1-5.0 cm in length) and have a thread-like appearance. They feed on algae, fungi, bacteria and soil organic matter, and can occur in populations as high as 200,000 m^2.

Lumbriscid worms are probably more important than any other soil invertebrate in the decomposition and mixing of organic matter, and their numbers can exceed 800 m^2, although they do not tolerate highly acidic conditions, and during prolonged drought or heat they burrow deeply into the soil and become inactive. Some species, such as *Lumbricus rubellus,* live only in the surface litter layer, but most migrate between organic and mineral horizons, ingesting both constituents and ensuring thorough mixing of the soil. Large populations under grassland can consume 90 ha, and the casts which are produced are important in the development of soil structure. The main mesofaunal mollusc groups are slugs and snails (gastropods) which feed on plants, fungi and faecal remains, but these are often limited in biomass to 20-45 g/m^2.

Macrofauna include larger molluscs, beetles and larger insect larvae, together with larger vertebrates such as moles, rabbits, foxes and badgers. These often burrow deeply into the soil, feeding on smaller organisms in their path. Moles in particular are voracious feeders and consume large numbers of earthworms, insect larvae and slugs.

In terms of its composition, organic material consists predominantly of carbon, hydrogen, oxygen and nitrogen, with carbon providing the framework for organic structures; most organic residues in soils contain around 45-55 per cent carbon by weight. The elements which constitute organic material are arranged to form a variety of compounds, some of which have very complex structures. The basic building block of many organic compounds, however, is the carbon tetrahedron, where a central carbon atom with a valency of four links with other elements, notably hydrogen, oxygen and nitrogen; this structural arrangement is very similar to the basic building block of silicate minerals.

In some instances, numerous tetrahedra link together to form extremely large molecules known as *polymers*. Linkages can produce both linear and cyclic molecules, and the number of potential combinations and structural arrangements is almost endless. Consequently, organic structures and their variety are considerably more complex than those of mineral material and are less well understood, although the majority of organic compounds can, however, be isolated. The main groups of compounds are carbohydrates, proteins and amino acids, and lignin, along with smaller amounts of fats, waxes, pigments and resins. The proportions of these compounds vary according to the type of organic matter and its stage of decomposition.

Essentially, carbohydrates are hydrates of carbon with a general formula $Cx\,(H_2O)y$, although some contain other elements such as nitrogen and sulphur. Carbohydrates are the main constituent of plant material at 60-90 per cent of dry mass, and in soils 5-30 per cent of the carbon exists in this form. The simplest and most easily broken down carbohydrate compounds in soils are the sugars and starches. These comprise relatively small molecules such as glucose (a simple sugar), which is a monosaccharide. More complex, and thus more resistant to breakdown, are hemi-cellulose (a pentose sugar polymer) and cellulose (a β (1-4) glucose polymer).

These are polysaccharides which consist of monosaccharide units joined by C-O-C links. Cellulose is the most abundant carbohydrate in plants, where it may constitute more than 40 per cent of the carbon. It forms the resistant fibres of plants, the breakdown of which is catalysed by enzymes (cellulases) secreted by micro-organisms. Breakdown ultimately converts the polysaccharide compounds into monosaccharide compounds which can then be assimilated directly by the micro-organisms. Chitin is a polysaccharide of similar composition to cellulose, but it also possesses some nitrogen in the form of amino groups. It is a common constituent of insect cuticles and is also often present in fungi.

Proteins and amino acids are composed of about 50-55 per cent carbon, 20-25 per cent oxygen, 15-20 per cent nitrogen and 6.5-7.5 per cent hydrogen, although some contain small amounts of phosphorus and sulphur. They are an important source of nitrogen, with around 20-50 per cent of all organic nitrogen in soils existing as amino acids. Although they are rapidly metabolised in soils, they can persist for long time periods due to adsorption onto other constituents, such as clay, or by combination with more resistant organic components such as lignin or tannin.

Lignin is a particularly resistant component of soil organic material, and constitutes 15-35 per cent of the supporting tissues of plants. Essentially it is a complex phenolic polymer which displays a high degree of aromaticity in the development of cyclic, benzene ring structures. The aromatisation and large number of C-C bonds are the main cause of its resistance to decomposition. As a result of organic matter decomposition, fresh organic components are gradually converted, via a range of fermented products, to a material known as *humus,* which represents the end-product of this process. Humus is finely divided and amorphous with no cellular structure. Essentially it consists of insoluble, heterogeneous polymers and has an approximate composition, on an ash-free basis, of 44-53 per cent carbon, 40-47 per cent oxygen, 3.5-5.5 per cent hydrogen and 1.5-3.5 per cent nitrogen.

WATER

Soil water is derived from two principal sources—precipitation and ground water. Precipitation arrives at the soil surface in various forms; although rain, snow and hail are the main types, fog and mist may provide significant amounts of moisture, particularly in coastal and upland areas. The proportion of precipitation that reaches the ground surface depends largely on the nature and density of vegetation cover.

On surfaces devoid of vegetation, precipitation reaches the soil directly, but on vegetated surfaces a significant proportion of precipitation can be intercepted; much of this ultimately reaches the soil as canopy throughfall and stemflow, while some is returned to the atmosphere by evapouration. On reaching the surface, water can either infiltrate the soil or, if the rate of arrival of water at the surface exceeds the rate of infiltration, it will run off over the surface. The precipitation that infiltrates the soil will either be lost via evapotranspiration or drainage, or will be retained by forces which hold it in place or allow it to move only very slowly. Ground water can be derived by lateral movement from upslope, or by upward movement from the underlying rock strata. The composition of soil water is a particularly dynamic characteristic, varying dramatically even over short time periods. This behaviour arises from the intimate association between the water, small mineral and organic particles (clay and humus) and plant roots, which can involve the exchange of ions between these components. Soil water contains a number of dissolved solid and gaseous constituents, many of which exist in mobile ionic form, and a variety of suspended solid components.

Base cations (Ca^{2+}, Mg^{2+}, K^+, Na^+, NH^{4+}) may be derived from a number of sources. They are present in the atmosphere and are later dissolved in precipitation; in maritime areas, for example, precipitation often contains significant quantities of marine salts which are particularly rich in sodium and chloride ions (Na^+ and Cl^-). As well as being deposited in the soil directly, ions accumulate on the surfaces of vegetation and are subsequently washed off into the soil.

Cations can also be derived from mineral weathering and organic matter decomposition, entering the soil directly or via the soil exchange system; these processes play an important role in the buffering of soil acidity. In agricultural systems, lime and fertilizers provide yet another source of base cations, particularly Ca^{2+}, K^+ and NH^{4+}. The concentration of hydrogen ions (H^+) in soil water is a measure of its acidity, which is expressed in terms of *pH.* A major source of such acidity is carbon dioxide (CO_2) which is derived from the atmosphere, where it is dissolved in precipitation, and from the soil air where it is a product of soil organism respiration.

Carbon dioxide dissolved in water Unpolluted rain water in equilibrium with atmospheric CO^2 has a pH of about 5.6, whereas soil water in equilibrium with CO_2 in soil air is usually more acidic, with pH values often below 5.0; this is because CO_2 levels in soil air are considerably greater than in the atmosphere. Another source of acidity in soil water derives from industrial and urban emissions. In addition to these pollutants, organic acids derived from decaying organic material are an important source of soil acidity. H^+ is also released by plants in exchange for nutrient base cations, and as part of the process of nitrification where NH^{4+} is converted to NO^{3-}.

Iron and aluminium are also important constituents of soil water, particularly under acidic conditions. They are derived from mineral weathering and may exist in the soil solution either as ions (Fe^{2+} and Al^{3+}) or in the form of soluble organo-metallic complexes; significant quantities of dissolved organic carbon may also exist in this form. In some circumstances aluminium may be mobilised by mineral acids, such as sulphuric and nitric acid deposited in acid precipitation. As in the case of cations, the anions in soil water are derived from a number of sources. Nitrate and phosphate ions (NO_3 and PO_4^{3-}) are produced by mineralisation processes during organic matter decomposition, and are also derived from fertilizers. Chloride ions (Cl^-) and, to a lesser extent, sulphate ions (SO_4^{2-}) originate from atmospheric sources, including airborne marine salts and acid deposition. Bicarbonate ions (HCO_3) originate largely from the dissociation of HCO_3, and are associated with mineral weathering, especially in soils developed from carbonate-rich parent materials.

In addition to the major dissolved constituents of soil water, there are other dissolved components although these are usually relatively minor and local in their occurrence. These include organic material and silica, together with a number of pollutants such as heavy metals and radionuclides. Soil water contains not only dissolved solids but also a number of suspended constituents. These include small particles of mineral and organic material, which often result in discolouration and increased turbidity of soil water. Similarly, precipitates may accumulate in soil water, usually as a result of chemical changes as the water migrates through the soil.

For example, orange-brown precipitates of insoluble ferric iron compounds can sometimes be observed where water, in which iron compounds usually exist in the soluble ferrous form, becomes more oxygenated. The chemical characteristics of water can also influence the behaviour of fine sediments, particularly clays, in suspension. In waters with low solute concentrations, or significant quantities of monovalent cations such as Na^+, the clays tend to remain in a dispersed state and can be held in suspension for long periods of time. In waters with high solute concentrations, however, particularly where concentrations of divalent cations such as Ca^{2+} are high, the clays may undergo rapid flocculation and settling.

AIR

Air and water have a reciprocal arrangement in terms of their occupancy of soil pore space; in saturated soils, air content is low, whereas in dry soils the pore spaces are largely air-filled. Changes in water and air content are particularly dynamic because much of the water present in a saturated soil drains away rapidly, while heavy rainfall can quickly bring the soil back to saturation.

The gaseous constituents of soil air are derived largely from the atmosphere, the respiration and metabolism of soil organisms, and from the evapouration of soil moisture. Soil air is continuous with the atmosphere provided that the soil surface is not sealed due to compaction or crusting, and such continuity ensures the free movement and exchange of gases. The gases move along gradients of partial pressure, so that oxygen will tend to migrate from the atmosphere where its partial pressure is high into the soil where it is low. Conversely, carbon dioxide and water vapour will tend to migrate from the soil into the atmosphere. Movement of gases may occur by diffusion, mass flow or in dissolved form, with diffusion being by far the most significant of these processes.

The atmosphere contains approximately 78 per cent nitrogen, 21 per cent oxygen and 0.03 per cent carbon dioxide by volume. In comparison, soil air contains similar amounts of nitrogen, slightly less oxygen and more carbon dioxide. The balance between levels of oxygen and carbon dioxide in soil air depends largely on the rate of respiration of soil organisms and on the diffusivity characteristics of the soil. Respiration rates often vary seasonally in response to the availability of organic substances for consumption, and to variations in temperature and moisture conditions. Diffusion of gases occurs most effectively in dry soils with large and interconnected pores.

The presence of water tends to reduce pore continuity and consequently the soil becomes increasingly oxygen deficient or *anaerobic*. In addition to carbon dioxide, organisms release other gases into the soil, including methane (CH_4) and hydrogen (H_2), as a result of organic matter decomposition.

MINERAL PARTICLES

The principal properties of soil mineral particles in an environmental context are their size, shape, nature of surface, orientation and mineralogy. The mineral fraction of soils consists of particles that vary dramatically in size from large boulders, through cobbles and pebbles to sand, silt and clay. Most studies of soil are concerned with the <2 mm size range, which is often referred to as the *fine fraction* or *fine earth*.

It is the proportion by weight of the size categories within the fine fraction which defines the particle size distribution or *texture* of a soil. The particle size distribution of a soil is usually recorded numerically on a per centage by weight basis, or graphically using either a triangular graph, on which only three size categories can be shown, or a cumulative frequency curve which allows a fuller range of size classes to be recorded. Most soils comprise a continuous spectrum of particle sizes, and the width of this spectrum is defined by the degree of *sorting*.

Poorly sorted soils possess a wide range of particle sizes, whereas well sorted soils have a narrow range. Texture can be estimated in the field simply by rubbing the soil between thumb and forefinger. Sand grains are easily distinguished by their coarseness, while silt has a distinctive soapy feel and clay is characteristically plastic and mouldable when moist. The morphology of mineral particles relates to their shape and surface characteristics. For large particles this can be determined in the field, but for smaller material it is necessary to use a microscope. Particle shape can be expressed in two or three dimensions. Common terms adopted in two-dimensional expression include roundness, angularity and elongation, while in three-dimensional expression particles are recorded in terms of the extent of sphericity and flatness, using descriptions such as sphere, disc, rod, cube and prism.

Such descriptions can be qualitative, or by reference to a comparison chart semi-quantitative estimates can be made. A number of indices have also been devised in order to quantify shape more accurately. The nature of the surface of a particle is usually referred to as its *surface texture,* although this in no way relates to the texture of a soil as an expression of its particle size distribution. Surface texture is most commonly evaluated for the fine fraction, using a scanning electron microscope, which allows the grain surfaces to be viewed in three dimensions at very high magnification.

Orientation of mineral particles describes the disposition of the particle in three dimensions, and this can be a useful indicator of the direction of movement of soil material. The measurement of orientation is usually made with reference to particle longest axes, either in the horizontal plane as a compass bearing, or in the vertical plane as an angle of dip or plunge. In some cases, both sets of measurement are made and the results combined. Orientation can be depicted in a variety of ways including bar charts, rose diagrams and polar scattergrams.

Minerals differ markedly in their composition and consequently many physical and chemical characteristics of soils, including texture, acidity and nutrient status, can be related to their mineralogy. Mineralogical studies are also important in the examination of weathering in soils. The mineralogical characteristics of large particles can be evaluated in the field by the inspection of hand specimens, but for smaller particles laboratory methods are required. Sand- and coarse silt-sized particles are usually examined using a petrological microscope, but clay and fine silt are too small to be easily resolved in this way and their mineralogy is commonly identified using the technique of X-ray diffractometry. This allows the crystal lattice dimensions to be measured and is particularly useful in the identification of sheet silicate minerals. Samples are scanned by an X-ray beam and when dominant lattice spacing is encountered, diffraction occurs and a peak is registered on a moving chart recorder.

Each mineral has a characteristic set of peaks, although in some cases more than one mineral may have a peak in the same position. However, it is possible to distinguish between these minerals using simple chemical or heat pre-treatments which produce characteristic changes in the lattice spacing of one mineral but not in another.

AGGREGATES

Aggregation in soils is promoted by a number of physical, chemical and biotic forces. Physical forces include expansion and shrinkage associated with wetting and drying, and compaction by raindrop impact, animal trampling and agricultural machinery. Chemical forces are largely electrostatic in character and often depend on the presence of adsorbed cations in association with the negative surface charge of colloidal particles such as clay and humus.

The generation of negative surface charge on colloidal particles and the processes of cation adsorption are discussed. Multivalent cations in particular, such as Ca^{2+}, Mg^{2+} and Al^{3+}, have the ability to form an attachment with more than one colloidal particle, a process known as *cation bridging*. Similarly, attraction may occur between positive charges on the broken edges of sheet silicate particles and the negative charges on the faces of other similar particles. In soils with significant positive charge, anion bridging may occur, particularly if multivalent anions are present.

In response to the various mechanisms of interparticle attraction, particles flocculate together to form larger units known as *domains;* these may be up to 5 ìm in diameter. In addition to electrostatic interparticle forces, interaggregate attraction can occur due to the binding effects of various organic compounds, fungal hyphae and plant roots. Organic polymers, such as polysaccharide gums, together with organic mucilages and fungal hyphae, facilitate the binding of domains to form microaggregates which may be up to 250 ìm in diameter.

At the larger scale, plant roots and fungal hyphae play an important role in the binding of microaggregates to form macroaggregates or *peds* which can be easily observed in the field. Inorganic cementing agents such as carbonate and iron compounds can perform a similar binding function in soils. Because of the different scales of aggregation, it is often viewed in terms of a hierarchical model comprising building blocks which increase progressively in size. The degree of persistence of aggregates varies between the different levels in the model. Electrostatic forces predominate at the domain level and are particularly resistant to change. Similarly, at the microaggregate level, binding forces are relatively persistent, although they vary to some extent according to the organic content of the soil. In contrast, at the macroaggregate level, binding forces are often transient, being closely influenced by variations in plant cover and the development of root networks.

Aggregates, or peds, which persist during wetting/drying and freesing/ thawing cycles form the basis of soil *structure*. Soil structure is defined in terms of the size, shape and arrangement of particles, aggregates and pores. It is classified on the basis of ped morphology, class and grade. There are four main groups of ped morphology—spheroidal, blocky, prismatic and platy. Spheroidal peds are equidimensional in form and can be divided into granular and crumb types. Granular and crumb structures are most commonly found in the A horizons of soils, their development being facilitated by the presence of roots and other forms of organic material. Blocky peds are also equidimensional and may possess several curved or planar surfaces.

Frequently they are observed in B horizons, particularly in soils with significant amounts of clay, where they develop in response to the effects of regular wetting and drying cycles. Prismatic and associated columnar peds are characterised by strong vertical and weak horizontal development. The tops of prismatic peds are poorly defined, whereas those of columnar peds are often rounded in form. Like blocky structures, prismatic and columnar structures often develop as a result of wetting and drying but are found in the B horizons of less clay-rich soils. Platy structure displays strong horizontal and weak vertical development, and often develops as a result of compaction. The structures are sometimes lenticular in form, with their centres being thicker than their edges, and these are often found in soils that are subjected to regular freesing and thawing cycles. Peds which comprise more than one structural type are often observed in soils and are known as *compound peds*. For example, large prismatic peds may be made up of smaller, blocky peds. It is also not uncommon to find that structures vary between the horizons within a soil profile, such that the A, B and C horizons could possess crumb, prismatic and blocky structures respectively.

Grade, which describes the distinctiveness and durability of peds, may be expressed as structureless, or as weakly, moderately, well or strongly developed. An apedal soil is described as massive if it is coherent and as single grain if it is not. Usually, massive soils are fine-textured while single grain are coarse-textured. Weakly developed structure comprises poorly formed, indistinct and weakly coherent peds, most of which will be broken down if the soil is disturbed, while strongly developed structure comprises well formed, distinct and durable peds. The grade of structure depends on a number of soil characteristics, particularly moisture content and the quantity of aggregating colloids such as clay and humus; peds tend to be considerably more durable in dry soils and in those with high colloidal contents.

3

Forest Environment and Biodiversity

BASES OF FORESTRY

Forestry may be defined as the scientific management of forests for the continuous production of goods and services. While its practice involves many problems of an economic and administrative nature, the solution of which is essential for success, its proper development must in the first place be based on a sound knowledge of the forests and of the laws that govern their development and on an understanding of the characteristics of the individual species that make up the forests and of the interrelationship of the forests with their environment.

The forests of the world include a great many species of trees, not necessarily having any close botanical relationship, with varying adaptations to climate, soil and topography and more or less distinct reproductive characteristics, forms and rates of growth, sizes and ages at maturity, light requirements and abilities to resist damage by insects and other destructive agencies. It is with an intimate knowledge of these features of the trees and tree communities that sound forestry practice must begin.

BOTANICAL RELATIONSHIPS

Trees represent many different families within the two major divisions of the spermatophytes or seed-bearing plants — the gymnosperms and the angiosperms. Trees of the gymnosperm group are more commonly known as conifers or softwoods, while those of the angiosperm group are often called broadleaf trees or hardwoods. These names do not apply universally — the wood of some gymnosperms, for example, is considerably harder than that of certain angiosperms. For most practical purposes, however, the forests of the world are classified as softwoods, temperate hardwoods and tropical hardwoods.

The dominance of the softwoods in the northern temperate forests is of special economic significance, because man has found that these woods are particularly useful. Thus the saw timber that is used throughout the world in general building and construction is 75 per cent softwood and an even larger proportion of wood pulp comes from this class of trees. Bole or trunk. Throughout the life of the tree there are alternate periods of growth and dormancy and in general, owing to differences in the structure of the wood produced at the beginning and at the end of a growing season, the successive layers appear as a series of concentric circles when the tree is cut transversely. Where climatic conditions are such that a period of growth alternates with a period of dormancy within the year, these woody layers are known as annual rings. The yearly increase in wood volume depends primarily on diameter growth.

For the first few years after its formation the wood, then known as "sapwood," contains some living cells which store food substances and conduct water and solutions of mineral nutrients from the roots to the crown of the tree. Later the cells die, other physiological changes take place and the wood, at this stage known as "heartwood," ceases to function except in support of the crown. Quite commonly the heartwood is darker in colour, more durable and has better timber properties than the sapwood.

The physical structure and consequently the density and technical properties of the wood laid down by the cambium, vary considerably from one species to another. The rate of diameter growth may also have diverse influences on the strength and quality of the wood: while rapid growth increases the strength of certain hardwoods, it may produce soft, spongy, Jow-grade timber in some of the softwoods. The form of the bole generally varies according to the strength required to support the crown and resist wind pressures: in a closed forest the bole approaches a cylindrical form, whereas in exposed sites it is more conical owing to the need for mechanical support. There are also considerable differences between species, especially between the two major groups of hardwoods and softwoods, in the form of the crown and bole and in the vigour of development and persistence of side branches.

Usually softwoods produce a well-defined bole with relatively small branches, while the branches of many of the hardwoods are larger and more widely spreading. Since branches produce knots, for the production of clear timber it is desirable to remove the lower dead and dying branches while they are fairly small. In general, trees growing in a closed forest have fewer and smaller branches than those growing in the open. However, the rate at which lower branches are killed by shade and subsequently drop to the ground varies with the species. Important characteristics of the mature root system also vary from one species to another. Some species have deeply penetrating tap roots. Others have mainly shallow lateral roots, which are more likely to be sensitive to drought and ground fires and to render the trees more susceptible to wind throw.

Root modifications or adaptations that are related to the nutritional efficiency of certain species include:

- The presence of nodules containing nitrogen-fixing bacteria in the roots, particularly of alders and members of the Leguminosae,
- A close association between the root tips of many tree species and the mycelia of certain fungi: this association is known as "mycorrhiza."

Definition of Forestry

Although most people know what a forest is, a definition of it which suits all cases is by no means easy to give. Manwood, in his treatise of the *Lawes of the Forest,* defines a forest as "a certain territory of woody grounds, fruitful pastures, privileged for wild beasts and fowls of forest, chase and warren, to rest and abide in, in the safe protection of the king, for his princely delight and pleasure."

This primitive definition has, in modern times, when the economic aspect of forests came more into the foreground, given place to others, so that forest may, in a general way, now be described as " an area which is for the most part set aside for the production of timber and other forest produce, or which is expected to exercise certain climatic effects, or to protect the locality against injurious influences."

As far as conclusions can now be drawn, it is probable that the greater part of the dry land of the earth was, at some time, covered with forest, which consisted of a variety of trees and shrubs grouped according to climate, soil and configuration of the several localities. The conditions for an uninterrupted regeneration of the forest were favourable and the result was vigorous production by the creative powers of soil and climate. Then came man and by degrees interfered, until in most countries of the earth the area under forest has been considerably reduced. The first decided interference was probably due to the establishment of domestic animals; men burnt the forest to obtain pasture for their flocks. Subsequently similar measures on an ever-increasing scale were employed to prepare the land for agricultural purposes. More recently enormous areas of forests were destroyed by reckless cutting and subsequent firing in the extraction of timber for economic purposes.

It will readily be understood that the distribution and character of the now remaining forests must differ enormously. Large portions of the earth are still covered with dense masses of tall trees, while others contain low scrub or grass land, or are desert. As a general rule, natural forests consist of a number of different species intermixed; but in some cases certain species, called gregarious, have succeeded in obtaining the upper hand, thus forming more or less pure forests of one species only. The number of species differs very much. In many tropical forests hundreds of species may be found on a comparatively small area, in other cases the number is limited. Burma has several thousand species of trees and shrubs, Sind has only ten species of trees. Central Europe has about forty species and the greater part of northern Russia, Sweden and

Norway contains forests consisting of about half a dozen species. Elevation above the sea acts similarly to rising latitude, but the effect is much more rapidly produced. Generally speaking, it may be said that the Tropics and adjoining parts of the earth, wherever the climate is not modified by considerable elevation, contain broad-leaved species, palms, bamboos, &c. Here most of the best and hardest timbers are found, such as teak, mahogany and ebony.

The northern countries are rich in conifers. Taking a section from Central Africa to North Europe, it will be found that south and north of the equator there is a large belt of dense hardwood forest; then comes the Sahara, then the coast of the Mediterranean with forests of cork oak; then Italy with oak, olive, chestnut, gradually giving place to ash, sycamore, beech, birch and certain species of pine; in Switzerland and Germany silver fir and spruce gain ground. Silver fir disappears in central Germany and the countries around the Baltic contain forests consisting chiefly of Scotch pine, spruce and birch, to which, in Siberia, larch must be added, while the lower parts of the ground are stocked with hornbeam, willow, alder and poplar.

In North America the distribution is as follows: Tropical vegetation is found in south Florida, while in north Florida it changes into a subtropical vegetation consisting of evergreen broad-leaved species with pines on sandy soils. On going north in the Atlantic region, the forest becomes temperate, containing deciduous broad-leaved trees and pines, until Canada is reached, where larches, spruces and firs occupy the ground. Around the great lakes on sandy soils the broad-leaved forest gives way to pines. On proceeding west from the Atlantic region the forest changes into a shrub by vegetation and this into the prairies.

Farther west, towards the Pacific coast, extensive forests are found consisting, according to latitude and elevation above the sea, of pines, larches, fir, Thujas and Tsugas. In Japan a tropical vegetation is found in the south, comprising palms, figs, ebony, mangrove and others. This is followed on proceeding north by subtropical forests containing evergreen oaks, *Podocarpus,* tree-ferns and, at higher elevations, *Cryptomeria* and *Chamaecyparis.* Then follow deciduous broad-leaved forests and finally firs, spruces and larches. In India the character of the forests is governed chiefly by rainfall and elevation.

Where the former is heavy evergreen forests of Guttiferae, Dipterocarpeae, Leguminosae, Euphorbias, figs, palms, ferns, bamboos and India-rubber trees are found. Under a less copious rainfall deciduous forests appear, containing teak and sal (*Shorea robusta*) and a great variety of other valuable trees. Under a still smaller rainfall the vegetation becomes sparse, containing acacias, *Dalbergia sissoo* and Tamarix. Where the rainfall is very light or *nil,* desert appears.

In the Himalayas, subtropical to arctic conditions are found, the forests containing, according to elevation, pines, firs, deodars, oaks, chestnuts, magnolias, laurels, rhododendrons and bamboos. Australia, again, has its own particular flora of eucalypts, of which some two hundred species have been distinguished, as well as wattles. Some of the eucalypts attain an enormous height.

Utility of Forests

In the economy of man and of nature forests are of direct and indirect value, the former chiefly through the produce which they yield and the latter through the influence which they exercise upon climate, the regulation of moisture, the stability of the soil, the healthiness and beauty of a country and allied subjects.

The *indirect* utility will be dealt with first. A piece of land bare of vegetation is, throughout the year, exposed to the full effect of sun and air currents and the climatic conditions which are produced by these agencies. If, on the other hand, a piece of land is covered with a growth of plants and especially with a dense crop of forest vegetation, it enjoys the benefit of certain agencies which modify the effect of sun and wind on the soil and the adjoining layers of air. These modifying agencies are as follows:

- The crowns of the trees intercept the rays of the sun and the falling rain; they obstruct the movement of air currents and reduce radiation at night.
- The leaves, flowers and fruits, augmented by certain plants which grow in the shade of the trees, form a layer of mould, or humus, which protects the soil against rapid changes of temperature and greatly influences the movement of water in it.
- The roots of the trees penetrate into the soil in all directions and bind it together. The effects of these agencies have been observed from ancient times and widely differing views have been taken of them.

Of late years, however, more careful observations have been made at so-called parallel stations, that is to say, one station in the middle of a forest and another outside at some distance from its edge, but otherwise exposed to the same general conditions. In this way, the following results have been obtained:

- Forests reduce the temperature of the air and soil to a moderate extent and render the climate more equable.
- They increase the relative humidity of the air and reduce evaporation.
- They tend to increase the precipitation of moisture. As regards the actual rainfall, their effect in low lands is *nil* or very small; in hilly countries it is probably greater, but definite results have not yet been obtained owing to the difficulty of separating the effect of forests from that of other factors.
- They help to regulate the water supply, produce a more sustained feeding of springs, tend to reduce violent floods and render the flow of water in rivers more continuous.

- They assist in preventing denudation, erosion, landslips, avalanches, the silting up of rivers" and low lands and the formation of sand dunes.
- They reduce the velocity of air-currents, protect adjoining fields against cold or dry winds and afford shelter to cattle, game and useful birds.
- They may, under certain conditions, improve the healthiness of a country and help in its defence.
- They increase the beauty of a country and produce a healthy aesthetic influence upon the people.

The *direct* utility of forests is chiefly due to their produce, the capital which they represent and the work which they provide. The principal produce of forests consists of timber and firewood. Both are necessaries for the daily life of the people. Apart from a limited number of broad-leaved species, the conifers have become the most important timber trees in the economy of man. They are found in greatest quantities in the countries around the Baltic and in North America. In modern times iron and other materials have, to a considerable extent, replaced timber, while coal, lignite and peat compete with firewood; nevertheless wood is still indispensable and likely to remain so.

This is borne out by the statistics of the most civilized nations. Whereas the population of Great Britain and Ireland, during the period 1880-1900, increased by about 20%, the imports of timber, during the same period, increased by 45%; in other words, every head of population in 1900 used more timber than twenty years earlier. Germany produced in 1880 about as much timber as she required; in 1899 she imported 4,600,000 tons, valued at £14,000,000 and her imports are rapidly increasing, although the yield capacity of her own forests is much higher now than it was formerly. Wood is now used for many purposes which formerly were not thought of. The manufacture of the wood pulp annually imported into Britain consumes at least 2,000,000 tons of timber. A fabric closely resembling silk is now made of spruce wood. The variety of other, or minor, produce yielded by forests is very great and much of it is essential for the well-being of the people and for various industries.

The yield of fodder is of the utmost importance in countries subject to periodic droughts; in many places field crops could not be grown successfully without the leaf-mould and brushwood taken from the forests. As regards industries, attention need only be drawn to such articles as commercial fibre, tanning materials, dye-stuffs, lac, turpentine, resin, rubber, guttapercha, &c. Great Britain and Ireland alone import every year such materials to the value of £12,000,000, half of this being represented by rubber.

The *capital* employed in forests consists chiefly of the value of the soil and growing stock of timber. The latter is, ordinarily, of much greater value than the former wherever a sustained annual yield of timber is expected from a forest. In the case of a Scotch pine forest, for instance, the value of the growing stock is, under the above-mentioned condition, from three to five times that of the soil. The rate of interest yielded by capital invested in forests differs, of course, considerably according to circumstances, but on the whole it may, under proper management, be placed equal to that yielded by agricultural land; it is lower than the agricultural rate on the better classes of land, but higher on the inferior classes. Hence the latter are specially indicated for the forest industry and the former for the production of agricultural crops. Forests require *labour* in a great variety of ways, such as:

- General administration, formation, tending and harvesting;
- Transport of produce;
- Industries which depend on forests for their prime material. The labour indicated under the first head differs considerably according to circumstances, but its amount is smaller than that required if the land is used for agriculture. Hence forests provide additional labour only if they are established on surplus lands. Owing to the bulky nature of forest produce its transport forms a business of considerable magnitude, the amount of labour being perhaps equal to half that employed under the first head. The greatest amount of labour is, however, required in the working up of the raw material yielded by forests. In this respect attention may be drawn to the chair industry in and around High Wycombe in Buckinghamshire, where more than 20,000 workmen are employed in converting the beech, grown on the adjoining chalk hills, into chairs and tools of many patterns.

  Complete statistics for Great Britain are not available under this head, but it may be mentioned that in Germany the people employed in the forests amount to 2.3% of the total population; those employed on transport of forest produce. 1%; labourers employed on the various wood industries, 8.6%; or a total of 12%. An important feature of the work connected with forests and their produce is that a great part of it can be made to fit in with the requirements of agriculture; that is to say, it can be done at seasons when field crops do not require attention. Thus the rural labourers or small farmers can earn some money at times when they have nothing else to do and when they would probably sit idle if no forest work were obtainable.

Whether, or how far, the utility of forests is brought out in a particular country depends on its special conditions, such as:

- The position of a country, its communications and the control which it exercises over other countries, such as colonies;
- The quantity and quality of substitutes for forest produce available in the country;
- The value of land and labour and the returns which land yields if used for other purposes;

- The density of population;
- The amount of capital available for investment;
- The climate and configuration, especially the geographical position, whether inland or on the border of the sea, &c. No general rule can be laid down, showing whether forests are required in a country, or, if so, to what extent; that question must be answered according to the special circumstances of each case.

These data exhibit considerable differences, since the percentage' of the forest area varies from 3.5 to 50 and the area per head of population from 07 to 9.5 acres. Russia, Sweden and Norway may as yet have more forest than they require for their own population. On the other hand, Great Britain and Ireland, Germany, Denmark, Portugal, Holland and even Belgium, France and Italy have not a sufficient forest area to meet their own requirements; at the same time, they are all sea-bound countries and importation is easy, while most of them are under the influence of moist sea winds, which reduces to a subordinate position the importance of forests for climatic reasons.

Intimately connected with the area of forests in a country is the state of ownership - whether they belong to the state, corporations or to private persons. Where, apart from the financial aspect and the supply of work, forests are not required for the sake of their indirect effects and where importation from other countries is easy and assured, the government of the country need not, as a rule, trouble itself to maintain or acquire forests. Where the reverse conditions exist and especially where the cost of transport over long distances becomes prohibitive, a wise administration will take measures to assure the maintenance of a suitable proportion of the country under forest.

This can be done either by maintaining or constituting a suitable area of state forests, or by exercising a certain amount of control over corporation and even private forests. Such measures are more called for in continental countries than in those which are sea-bound, as is proved by the above statistics.

Principles of Sustainable Forest Management

Sustainable Forest Management

The past decade may come to be known by foresters as the period when they, and just about everybody else, sought to define or to prescribe sustainable forest management (SFM). Many of these efforts have painted very detailed pictures of what a well-managed forest should look like. Others have established general principles. Whatever their form, the key ingredients are sustaining multiple values through forest management, and the rights of multiple actors. The earliest initiatives to define SFM were unilateral, undertaken by industry associations or by environmental non-governmental organizations (NGOs), and consequently were mistrusted by other stakeholders. Most of the currently accepted SFM initiatives result from multistake-holder processes, and consequently reflect a range of these stakeholders' values.

They generally comprise sets of principles, criteria and indicators that have to be interpreted in detail at the local level, offering further scope for incorporating local values. They include global/intergovernmental processes (notably the UN Forest Principles 1992), regional processes (*e.g.* the Ministerial Conference on the Protection of Forests in Europe 1993) and national standards. Others are led by civil society. The International Institute for Environment and Development analysed 17 such initiatives and found that all had the following in common:

- sustaining yields of all socially valued goods and services from forests;
- ensuring positive impacts of forest use on different social groups;
- maintaining the state of the forest to enable continued production for future generations. These initiatives to define SFM all state or imply two premises.
  1. The purpose of forestry is to provide the highly varied needs of society; this requires that demands are signalled effectively in policy and markets, and necessitates means for making trade-offs.
  2. The basic principles of SFM need to be interpreted at the most local level at which specific goods and services are needed, *e.g.* the community level for recreation and firewood, the national level for major watersheds, and the global level for climate regulation. 'Sustain-ability' can be likened to liberty or justice, a goal we all understand and aspire to, but which has to be negotiated and defined locally to understand and achieve it in practice.

This all points to an emphasis on determining local values, on participation among interest groups, on using local decision-making processes, and on ensuring that forest management systems, markets and policies reflect social values.

Only recently, however, have foresters been expected to focus on social issues. There is little prior experience of foresters taking on roles of social development, although there are historical examples: the famous *taungya* system developed in Burma in the 1850s was a response to local people's needs for farmland. Yet foresters are being held increasingly accountable for social conditions (in various new regulations, voluntary codes and certification). There are many unresolved arguments.

To what extent can forestry be an instrument of social development? For which communities should foresters be held accountable for social conditions, those close to the forest or also those further away? Clearly, forest managers have at least to understand which groups are most dependent upon the forest and who could contribute most to the success or failure of forest management.

Organizational Structure of Forest

All forests are managed under the prescriptions of a *working plan*/scheme prepared on the basis of principles of sustainable forest management and recognized and innovative silvicultural practices. The authority as designated by the Ministry of Environment & Forests, Government of India, approves the *working plan* for this purpose. Generally no timber harvesting is done in any forest area without an approved *working plan*/scheme.

It is the duty of the owner/manager of the forest area to ensure the preparation of the *working plan*/scheme. While a detailed *working plan* is prepared for large areas such as forest division, working schemes are prepared for smaller areas for a specific purpose or areas like private, village, municipal, cantonment forests, etc. Even working schemes have all major elements of *working plan*; and these schemes also need the sanction of competent authority designated by the Ministry of Environment & Forests.

Organization Consists of Forests

In the Ministry of Environment & Forests, the organization consists of Director General of Forests & Special Secretary to the Government of India, Additional Director General of Forests, Inspector General of Forests, Deputy Inspector General of Forests, and Assistant Inspector General of Forests. In the field, it consists of officers of Regional Offices headed by the Regional Chief Conservator of Forests. In the States, there is no uniformity in the constitution of the organization. There are various levels of which Conservator of Forests Working Plan and Working Plan Officer (DCF) are the key functionaries. Overall situation is as under:

- Policy/Supervisor Level – PCCF/APCCF/CCF;
- Field Supervisor Level –Chief Conservator/Conservator of Forests Working Plan;
- Field Function Level – Working Plan Officer (DCF/CF) assisted by ACF, RFO, and Foresters.

The number of *working plan circles* and *working plan divisions* depends upon the workload, *i.e.*, number of *territorial divisions* for which working plans are to be prepared/revised. *Working plan* is generally revised every 10 years; and on an average, one Conservator supervises the work of four *working plan divisions*.

The *working plan* of forests other than those under the control of Forest Department like municipal, cantonment, private, village, etc., forests can be prepared by Working Plan Officer on the request of the owner by owner themselves or through outside consultants. Micro-plan of jointly managed forests is prepared by forest staff of *territorial division* as per MOU and in consultation with the communities involved. The micro-plan so prepared is integrated with the *working plan* of that *territorial division*.

Identification of Forest Area

The identification of the forest division whose normal *working plan* is to be prepared or revised is done by the Principal Chief Conservator of Forests in consultation with the senior officers of the Working Plan Organization. For other areas, this is mainly demand driven. In JFM areas, it is on the basis of a MOU between the local community and the Forests Department. The task of preparation/revision of *working plan* is undertaken by WPO for the forest divisions within his jurisdiction as an essential function of the division. In addition, the PCCF or the competent authority may assign to him the task of preparing/revising *working plan* of other forest divisions to meet the target.

Tenure of WPO and Other Supporting Staff

For the preparation of the *working plan* of a *territorial division*, normally two years time is required. It may be less or more depending upon the volume of work and technical facilities available. The officers and the staff of *working plan division* should not be transferred during the preparation of *working plan.*

Head Quarter of Working Plan Division

Head quarter of *territorial division* is generally the head quarter of *working plan division* as well. It facilitates proper coordination and smooth flow of information/records between *territorial and working plan division.*

Status and Allowances

A *working plan division* is a functional charge, and the WPO has the status and power, unless otherwise stated in any particular respect, of a Divisional Forest Officer/Conservator of Forests as the case may be. A WPO is entitled to special pay admissible under Pay Rules applicable. Special pay/allowances should also be given to other supporting staff of the WPO.

Budget and Accounts of Forests

This is regulated by the rules and regulations of the State/UT Government. It is ensured that adequate budget provision is made in time. The proposed estimate is prepared by the Working Plan Conservator a year before the *working plan* revision is due to commence and copies sent to the DFO and the Territorial Conservator concerned also. The WPO is delegated the power of drawing and disbursing officer. Office management, maintenance of records, reporting of progress of work, and expenditure are according to the procedures laid down by the State/UT Government concerned.

Co-operation of Territorial Staff

It shall be mandatory for the DFO and the Circle Conservator to extend full cooperation and assistance to the WPO for fieldwork and provide logistical support, access to official records and other information so that the WPO can prepare good quality *working plan* expeditiously.

Office and Residences

Every State and UT Government should make efforts to provide necessary office and residential accommodation to all officials of Working Plan Organization both at head quarter and field levels as per their rules and norms in force applicable to their rank and status.

Agricultural Biodiversity

In addition to its direct contributions to rural livelihoods, agricultural biodiversity may generate other rural development opportunities through eco-tourism and a variety of income generating schemes. Many humanised landscapes in Europe, South America, Australia and the Asia-Pacific regions are increasingly valued for aesthetic and historical reasons. For example, throughout the Asia-Pacific region mountainous terrain has, over the centuries, been shaped into landscapes of terraced pond fields for the cultivation principally of rice, but also of taro and other crops.

In Europe, low input, extensive farming systems such as the Dehesas in the Iberian peninsula of Spain cover some two million hectares. Dehesa systems are open savannah like woodlands used as pastures, with sclerophyllous trees, mainly *Quercus rotundifolia* Lam., and a therophytic herb layer. Dehesas are home to many endangered species of wildlife such as the Iberian lynx, the golden eagle, the little bustard and the Egyptian vulture.

These landscapes exist both as archaeological (*i.e.* preserved) sites and as living landscapes, which continue to be used and maintained by the people who created them. The conservation of these cultural landscapes is considered important by a growing number of stakeholders. For example, many low external input agroecosystems in Europe are valued by urban populations ready to pay for the experience of a holiday in rural areas. At a global level, the intrinsic value of THESE cultural landscapes, and what they can teach about enduring systems of human-nature interaction, has led to a strategy within which the identification, evaluation and conservation of specific regional landscape types are to be considered within the framework of the World Heritage Convention. The ecotourism potential of these cultural landscapes is viewed as potentially important for rural development and local employment creation, both in the developed and developing countries. However, recent evidence suggests that the potential of eco-tourism can only be realised under certain conditions. As is often the case for classical tourism, eco-tourism schemes tend not to be integrated with other sectors of the national or regional economy; and only a fraction of earnings generated actually reach or remain in the rural areas.

More importantly, the majority of the rural population is frequently bypassed economically even where some earnings remain in the tourist location, as they are used up by the related administration or appropriated by local elites and business people. At the same time, local livelihood sources and cultures are negatively affected in nearly all cases. Generating economic benefits and fostering equitable rural development is only feasible when many wide-ranging reforms,-such as restoration of land and water rights to local communities, support for new forms of tenure and rights of usufruct, strengthening of local groups and institutions, investment in technical and managerial skills and mandatory impact assessments of all ecotourism schemes-, are carried out. The necessary structural political and economic changes along these lines are difficult in many developing and developed countries.

Another potential engine for rural development is the exploration, extraction and screening of biological diversity and indigenous knowledge for commercially valuable genetic and biochemical resources (biodiversity prospecting or bioprospecting). A detailed treatment of the economics of bioprospecting is beyond the scope of this chapter. However, it may be noted here that despite frequent mention of benefit sharing agreements in commercial contracts between bioprospecting agents and sovereign states, the specific terms of benefit sharing are strictly confidential. Available evidence indicates that benefits shared with countries in which collections took place represent a small fraction of the annual Research and Development budget of the corporations involved.

Moreover, indigenous and local people receive only a minuscule proportion of the profits generated from sales of products that embody their knowledge and resources. For example, one study has estimated that less than 0.001 per cent of the market value of plant based medicines has been returned to local and indigenous peoples from whom much of the original knowledge came. And while various codes of conduct and guidelines have been developed to ensure greater equity, compensation and fair sharing of benefits between bioprospecting companies and local communities, none are internationally legally binding.

Further opportunities for rural development hinge on creating local businesses and products that sustainably use agricultural biodiversity and add value to it in the context of more localised economies. In a growing number of rural areas in Europe, North America, Australia, Japan and New Zealand the diversity of local plants and animals is being harnessed for sustainable economic development. In south east France, the regional genetic heritage programme of the Provence-Alpes-Cotes d'Azur involves a wide range of actors spread across six administrative departments,-about 31,500 square kilometres of very diverse ecosystems. Ways and means of reintegrating locally adapted, traditional animal breeds (sheep, goats, cattle and bees) and crop varieties (fruit trees, fodder plants and cereals) are being explored to generate local products, jobs, income and environmental care. Similar initiatives are reinvigorating local economies and employment in the Willapa watershed of the Pacific North West (USA). The Willapa watershed includes 275,000 hectares on the coast of Washington state and is rich in agricultural biodiversity: oysters, clams, sturgeon, crabs, salmon and dense forests. With the support of the Ford Foundation and a Chicago based community bank a range of local businesses have been set up to add value to this local agricultural biodiversity. For example, Willapa oysters are now marketed locally rather than shipped out wholesale, alder is harvested from secondary forests for high quality wood products, fish and crab are marketed with the northwest image of wholesome foods, cranberry growers produce a wide range of products retaining more of the value added from food processing within the watershed.

These forms of endogenous rural development seek to create viable and locally controlled economic activities based on locally adapted agricultural biodiversity, knowledge, skills and negotiated partnerships between civil society, government and the private sector. Initiatives to reclaim diversity for rural development often focus on regenerating local food systems and economies based on comprehensive definitions of well being and wealth, both in developed and developing countries.

These examples together with the recent "discovery" of the potential of cultural landscapes and the creativity of their inhabitants illustrate a more fundamental point. Agricultural biodiversity, together with the local knowledge, institutions and management practices associated with it, may provide a robust foundation for development in many different economic and ecological settings. Understanding the forces that have neglected or undermined the values and functions of agricultural biodiversity can help identify ways forward.

Agricultural Biodiversity or Agrobiodiversity

Agricultural biodiversity refers to the variety and variability of animals, plants, and micro-organisms on earth that are important to food and agriculture which result from the interaction between the environment, genetic resources and the management systems and practices used by people. It takes into account not only genetic, species and agro-ecosystem diversity and the different ways land and water resources are used for production, but also cultural diversity, which influences human interactions at all levels. It has spatial, temporal and scale dimensions. It comprises the diversity of genetic resources (varieties, breeds, etc.) and species used directly or indirectly for food and agriculture (including, in the FAO definition, crops, livestock, forestry and fisheries) for the production of food, fodder, fibre, fuel and pharmaceuticals, the diversity of species that support production (soil biota, pollinators, predators, etc.) and those in the wider environment that support agro-ecosystems (agricultural, pastoral, forest and aquatic), as well as the diversity of the agro-ecosystems themselves.

Agricultural ecosystems or agro-ecosystems. Agro-ecosystems are those "ecosystems that are used for agriculture" in similar ways, with similar components, similar interactions and functions. Agro-ecosystems comprise polycultures, monocultures, and mixed systems, including crop-livestock systems (rice-fish), agroforestry, agro-silvo-pastoral systems, aquaculture as well as rangelands, pastures and fallow lands. Their interactions with human activities, including socioeconomic activity and sociocultural diversity, are determinant.

Agro-ecosystems may be identified at different levels or scales, for instance, a field/crop/herd/pond, a farming system, a land-use system or a watershed. These can be aggregated to form a hierarchy of agro-ecosystems. Ecological processes can also be identified at different levels and scales. Valuable ecological processes that result from the interactions between species and between species and the environment include, inter alia, biochemical recycling, the maintenance of soil fertility and water quality and climate regulation (*e.g.* micro-climates caused by different types and density of vegetation). Moreover, the interaction between the environment, genetic resources and management practices influence the evolutionary process which may involve, for instance, introgression from wild relatives, hybridisation between cultivars, mutations, and natural and human selections. These result in genetic material (landraces or animal breeds) that is well adapted to the local abiotic and biotic environmental variation.

Agricultural Biodiversity's Contributions

Livelihood systems are diverse in rural areas and vary among different cultural groups and in different regions of the world. They commonly rely on a mix of wild foods, agricultural produce, remittances, trading and wage labour. Empirical evidence from many different locations suggests that rural households do engage in multiple activities and rely on diversified income portfolios. Contrary to received wisdom, the actual contribution of agriculture to livelihoods can be quite low in today's fast changing rural areas. In sub Saharan Africa, for example, a range of 30-50 % reliance on non-farm income sources is common but it may reach 80-90% in Southern Africa. Household decision making continually adjusts to the changing nature of the environment, local economies and governance. At higher levels, it is simply impossible to predict the relationships between agro-ecosystems and households, particularly in resource-poor areas where there is much biological and social diversity.

The tendency for rural households to engage in multiple occupations is often mentioned, but few attempts have been made to link this behaviour in a systematic way to agricultural biodiversity and its multiple functions. In reflection of sectoral interests and disciplinary specialisations, the conventional point of entry for scientific research, management and policy has been to focus on selected components of agricultural biodiversity (*e.g.* plant genetic resources). However, this approach often leads to a mismatch between standard development interventions and diverse local realities, needs and priorities. Reversing this approach requires putting people with their assets, activities, and complex livelihoods at the centre of analysis. The functions of agricultural biodiversity thus need to be situated and mapped out within a total livelihood context.

Dynamic and complex livelihoods usually rely on plant and animal diversity, both wild and in different stages of domestication.

Agricultural Biodiversity Meeting Human Needs

- There is high intra-specific genetic variation in the date palm oasis agro-ecosystems of Algeria, Chad and Egypt. The principal varieties differ from one oasis to another. In general, there are more than ten varieties of date palms in each oasis. In a well organised and maintained palm grove, the owner plants different varieties of dry and semi-dry dates that mature in different months to meet the demands of local consumption and the market. Moreover, each genetic variety confers its own unique stamp on i) the taste of the date fruit and the wine made from it ii) the texture of the edible palm centre iii) the properties of the wood from the palm trunk that is used in construction and tool making iv) the mechanical qualities of palm leaves used for ceilings and fences as well as those of the palm's fibrous leaf base used to make ropes and sacks and v) the nutritional values of date stones fed to camels.
- About 25% of the land in the small holding sector are under home gardens in Sri Lanka. The garden system is similar to the complex structure and multiple functions of the forest, though not identical to it. Endemic and naturalised species make up about 40% of Kandyan gardens (named after the city of Kandy in central Sri Lanka). Trees, shrubs, herbs, crops and animals interact and sustain themselves in association with the households. Management activities that enhance agricultural biodiversity are carried out in conjunction with domestic work or during leisure time, particularly by women. The involvement of households in managing home gardens is not limited to harvesting diverse goods. Specific management practices are frequently used to achieve better immediate use of resources, including favourable interactions. Branch pruning, coppicing, clearing excess seedlings, nurturing of naturally germinated seedlings or species emerging from garbage heaps, increasing soil nutrients, and using indigenous

knowledge to control pests and experiment with genetic variations are all widespread activities. Local definitions of culture and well being cannot be considered apart from the system because of the strong link between survival of the households and agricultural biodiversity.

- In Cameroon, at least four breeds of domestic fowl (*Gallus gallus*) are kept on a free range system in villages where 70% of the national stock live. Indigenous fowl are kept for food and income generation, for gifts and sacrifices, for breeding stock and for traditional medicine preparations. In south east Mexico, women keep as many as nine breeds of local hen, as well as local and exotic breeds of turkey, duck and broilers in their back gardens. In selecting for the best breeds, they consider eleven different characteristics,-including egg production, ease of sale, appearance, broodiness, heat and cold tolerance, growth rate and eating qualities. Using this ranking, the most preferred birds are indigenous turkeys and ducks.

A diverse portfolio of activities based on the contributions of agricultural biodiversity (*e.g.* crop cultivation, harvest of wild plant species, herding, fishing, hunting) helps sustain rural livelihoods because it improves their long term resilience in the face of adverse trends or shocks. In general, increased diversity promotes more flexibility because it allows greater possibilities for substitution between opportunities that are in decline and those that are increasing.

Many rural people, regardless of whether their agro-ecosystems are predominantly pastoral, swidden or based on continuous cropping deliberately incorporate wild resources into their livelihood strategies. Nor is livelihood diversification based on such wide use of agricultural biodiversity the exclusive preserve of rural households in developing countries. In Poland for example, wild bush and berry fruits are important for local consumption and for export, with *Vaccinium myrtillus* being the principal export species at present (over 30, 000 t/year) followed by *Rubus spp., Sorbus aucuparia, Sambucus nigra, Prunus spinosa* and *Rosa spp*..

Different types of agricultural biodiversity ("cultivated", "reared" or "wild") are used by different people at different times and in different places, and so contribute to livelihood strategies in a complex fashion. Understanding how cultivation, herding, fishing, collection, use and marketing of different types of agricultural biodiversity are differentiated by wealth, gender, age and ecological situation is essential to evaluate their overall economic value. Understanding this differentiation within communities is essential because there is great variation in wealth, ability, age and power in every rural society. For example, wild resources are particularly important for the food and livelihood security of the rural poor, women and children, especially in times of stress such as drought, changing land and water availability or ecological change. These groups generally have less access to land, labour and capital and thus need to rely more on the wild diversity available. In India, the poor obtain 15-23% of their total income from common property resources, as compared with 1-3% for wealthier households. In Zimbabwe, some poor households rely on wild fruit species as an alternative to cultivated grain for a quarter of all dry season meals. Whilst wild food species supply vital nutritional supplements to all diets based largely on carbohydrate rich staples, they are crucial sources of vitamins and minerals for children. Children are often the most frequent collectors and consumers of wild fruit.

The degree to which the resources of agricultural biodiversity are important for local people's livelihoods affects the appropriateness of policies on resource management and on incentives for conservation and sustainable use. Comparing the economics of biological diversity use with other livelihood options can help assess people's willingness to sustain biodiversity as part of a livelihood strategy. There is however no single valid economic approach for doing this. Combining economic concepts with participatory research does nevertheless allow for a more comprehensive valuation of agricultural biodiversity, recognising not only the financial value, but also the indirect and non use values. The insights thus gained into the relative and changing seasonal importance of different types of agricultural biodiversity for livelihood security can be quite startling.

For example, "wild" agricultural biodiversity may provide a significant proportion of total household incomes, particularly where farming or herding is marginal. In parts of Botswana, where unpredictable rains make farming a risky business, basket making from the wild palm *Hyphaene petersiana*, and beer brewing from the wild fruit *Grewia bicolor* provide a more secure income source, especially for women. Other local level valuation studies of agricultural biodiversity conducted in a total livelihood context show that many wild resources have significant economic value by preventing the need for cash expenditure and providing ready sources of income to cash poor households, often yielding a better income than local wage labour.

The cultural and spiritual values of some parts of agricultural biodiversity can sometimes be considered as more important than monetary values. Many rural communities designate certain biological diversity-rich areas of land or water as sacred. Sacred groves, for example, are clusters of forest vegetation that are preserved for religious reasons.

They may honour a deity, provide a sanctuary for spirits, or protect a sanctified place from exploitation; some derive their sacred character from the springs of water they protect, from the medicinal and ritual properties of their plants, or from the wild animals they support. Such sacred groves are common throughout southern and south eastern Asia, Africa, the Pacific islands and Latin America. The spiritual values of sacred places on land or water are often inextricably tied with the functions that their associated agricultural biodiversity may provide in maintaining the health of the ecosystem. For instance, in a ranking exercise conducted to show the relative importance of different values derived from savannah woodlands in Zimbabwe, villagers explained that one of the most important aspects of their woodland was the sacred areas it contained. Honouring and preserving these sacred areas according to the wishes of the ancestral spirits is essential for good rainfall.

The wide range of consumption benefits derived from the woodland were ranked lower than these spiritual ecosystem functions, as they could not exist without the rains, which in turn depend on the sacredness of the woodland. The rich tapestry of locally unique agricultural biodiversities represents, at the global level, a huge amount of diversity *between* species. Out of the 250,000 plant species that have been identified and described, some 30,000 are edible and about 7,000 have been cultivated or collected for food and the provision of other goods and services at one time or another. World-wide, several hundred animal species including mammals, fish, reptiles, molluscs and arthropods also contribute to food and livelihood security.

Diversity *within* species is also remarkable among those plant and animal species that have been domesticated for crop and livestock production by innovative rural people. The inherent variation within farmers' crop varieties (landraces) is immense for cross-pollinated species such as millet or maize. For self-pollinated crops such as rice and barley, and for vegetatively propagated crops like potatoes and bananas, individual varieties are less variable, but the number of landraces developed may be very high. Estimates of the distinct number of varieties of Asian rice (*Oryza sativa*) range from tens of thousands to more than 100,000 while some communities in the Andes grow as many as 178 locally named potato varieties.

Methods and Tools for Future Assessments

A priority challenge of future assessments is to know the state and dynamics of all tree resources both in and outside the forest. A country embarking on a planning exercise cannot confine itself solely to the trees within its forests, especially when its wood resources appear to be insufficient.

The choice of tools and methods used to describe or assess trees outside the forest depends on the scale of analysis, kind of data and degree of exactitude desired. The tools used are not generally specific or new; rather, they are combined and implemented in original ways. The inventory in Bangladesh described above is one of the numerous examples of methods developed for gathering data on TOF. The Bangladesh study gives evidence of the adaptations needed to bring conventional forest inventory procedures in line with the specific nature of this resource. In some aspects - e.g. structure, spatial distribution and extent of area cover - trees outside the forest are more difficult to assess than forest formations.

The assessment of TOF does not lend itself to the potential cost savings associated with expanded uses of remote sensing technology. Remote sensing by satellite presents more difficulties for assessing TOF resources than for assessing attributes such as forest area. However, satellite data do allow a region to be stratified on the basis of ecological criteria and land cover, providing the basis for a good working document for more specific work in the future. The most commonly used remote sensing technology for TOF resources is aerial photography, which can be used to describe spatial distribution and to distinguish TOF cover classifications, providing the appropriate scale is chosen. However, high costs prohibit widespread use of aerial photography for TOF assessments in most countries. The new 1 m resolution satellite sensors represent a possible future alternative to aerial photography.

Some TOF field inventories are modelled on forest inventory methods and keep to biological and physical criteria; others emphasize social aspects, choosing villages as the sampling units. For measurements on the ground, sampling arrangements designed for forest stands may not be the most effective arrangements for trees. Less traditional sampling plans which would theoretically be better suited to this resource should be tested on various categories of TOF, especially those covering fairly large areas. Studies of the social and economic benefits or impacts of TOF often rely on household surveys, interviews or standardized appraisals such as rapid or participatory rural appraisal. The integration of the last two approaches - biophysical inventory and socio-economic analysis - is not simple and calls for caution given the great variety of social situations that are only meaningful in the local context.

Environmental benefits or impacts of TOF might be indirectly assessed by linking measurable indicators, such as the number and type of trees, with environmental variables such as water quality or erosion. In an urban setting, tree cover might have direct impact on the ambient temperature. Measuring the environmental impact of tree management is an issue for all natural resource planning or management operations.Assessment of trees outside the forest requires geographical, ecological, biophysical, social and economic data. However, this implies that an important amount of information will have to be carefully processed. The diversity of end-uses for this information, including land use planning and analysis based on inventory, will need to be considered in data assembly and processing and in the presentation of results.

It is important to know the status of trees outside the forest at any given moment, but it is even more essential to be able to trace patterns of change over time in the same area. The two most commonly used approaches have been comparison of aerial photos taken at sufficiently long intervals and surveys among villagers/managers combined with field inventories. Some countries, such as France and the United Kingdom, have undertaken periodic inventories based on the establishment of permanent plots linked to permanent forest inventories. However, the high cost of this type of operation limits the number of countries able to adopt it. India and Bangladesh are now experimenting with options for the future.

The current trend towards decentralized authority in land use planning suggests the importance of carrying out assessments at the local level, where the geographical, historical and socio-economic context is relatively harmonious. A minimum number of common rules concerning methods and arrangements is necessary, however, if the data are to be comparable at the country level. Certainly, the technical side of assessing trees outside the forest is complex and more research is needed to better pinpoint the resource.

Forest Products to the Household Budget

Few studies directly assess the contribution of forest-earned income to the household budget; however, some present data on the income earned from different forest based activities. While links are rarely made to the ways this income is spent, it can be assumed that income which supplements the household budget is important, especially in periods when other sources are not available. For example, Asibey (1977b) relates that animal hunting and gathering provide an important source of supplemental income to some farmers. And Okafor (1979) presents data on the income earned by farmers in Anambra State from palm wine production. He shows that the income earned per day from palm wine production exceeds the Nigerian daily minimum wage (2.3 naira/day).

Kamara (1986) presents data on income earned from fuelwood sales by both rural farmers and retailers in Sierra Leone. He estimates that the returns per roan day from rice cultivation and fuelwood sales are almost equivalent, and notes that the fuelwood income is critical as it provides the first returns from land clearing and income during the agricultural slack period when few other sources of cash income are available. For urban retailers (80 per cent of whom are women), income earned from fuelwood contributes from an average of 42 per cent of total household income in Freetown (where 80 per cent of traders work full-time), to an average of 5 per cent of the total household income in Makeni.

Cashman's (1987) study from southwestern Nigeria represents another example of the importance of trees as a source of household income. She notes that in this region, women earn the majority of the household's cash. Some of the regionally important income earning activities which employ women include: palm oil processing, cola nut trade, parkia bean processing, soap making, maize processing and food sales. Oil processing is especially valued for "bulk" expenses such as school fees. In some cases, communities have come to rely and specialise in the trade and production of forest products. In Bas Mungo, Cameroon, for example, some 20,000 villagers are dependent on the sales of palm wine to nearby Douala. Similarly, in Anyama, Senegal, Vérnière (1969) relates that the cola nut trade provides the main source of income for the town's 6000 residents. Finally, in Ahwia (a Ghanaian village in the Kumasi area) all the village men depend on wood carving activities. Carvers earn a good living (about 1300 cedis/month on average).

Forest Product Exploitation and Consumption

Studies which examine and estimate the extent to and frequency with which different forest products are used also shed light on the value of these resources. Food consumption studies, and housing surveys evaluate the extent of forest product use. These case-specific studies which quantitatively assess the day-to-day use of forest resources provide analytical data which allows the utility of forest resources to be compared with the utility of other resources, and production systems.

Food consumption studies that examine the consumption of wild and forest foods or assess quantities of snack foods consumed can show the importance of forest products. Although not from the West African region, Ogle and Grivetti's (1985) study on the cultural, ecological, and nutritional aspects of wild plant use in Swaziland is an excellent example of how information from different subject disciplines can be integrated to help understand the role of wild plants in people's diets. They include a quantitative analysis of the frequency with which species are consumed. They find that over 220 wild plants are commonly (more than once a month) consumed; many agriculturalists who were interviewed claim they consume more wild plants than cultivated varieties.

Dietworst's (1987) meat consumption survey in three Nigerian village communities provides an informative analysis of the frequency of bushmeat consumption. A village study in Sine, Senegal also examines household consumption of bushmeat. Although the area is known to have poor wild animal resources, the researchers found that bushmeat was regularly consumed (on average, 12.9 g./day/per son). They added that the greatest quantities of bushmeat were consumed by children.

Faure and Vivien's (1980) study of the local uses of forest resources in the Littoral region of southern Cameroon included an analysis of the extent to which they are used in house construction. They assert that approximately 300 poles are used for constructing one house. From this, they estimate that the nine study villages consume 4,200 cubic metres of wood to this end.

While few studies actually assess the extent to which plant medicines and traditional healing practices are used Abosede Akesode (1986) provide an unusual case study which examines the use of a traditional plant cure Nigeria. Studies which examine the extent to which resources are exploited also indicate how valuable they are to local people. Moby-Etia's (1982) study of palm wine tapping in southern Cameroon estimates the amount of palm wine tapped per person per day and the amount produced per season. In Ghana, Dodah (1970) estimates that farmers in the Krobo region tap between 25 and 100 trees per season. In Nigeria, Okafor (1979) provides data on the numbers of trees tapped for palm wine by farmers in Anambra state.

Distribution of Farm

Sometimes the types of trees on farm lands and on fallow lands differ, but often valued species are maintained regardless of their location. Regionally, trees such as oil palms are often protected and selectively left in clearings and on forest fallow lands. The multitude of functions associated with these farm trees demonstrates their value to local people. There are a number of studies from the region which inventory fallow and farm trees. The density of different species on farm and fallow lands can often indicate their importance (especially when compared with density of the same species in nearby forest areas).

Okafor and Fernandez (1987) conducted an inventory of farm and fallow land trees in southeastern Nigeria. They identified 171 edible tree species. The most commonly used and highly valued species were found in compound farms rather than in outlying farm and fallow land areas. These farm tree species provided food and beverages, stakes, fodder, fuelwood, medicines, fibres and housing materials. Other tree inventories of farm and fallow species reveal similar. Herren-Gemmill compares the uses and management of fallow areas by farmers in the forest and derived savannah region of southern Nigeria. She finds that farmers in the forest zone leave a broader range of forest emergents standing when crop land is cleared.

A great number and variety of trees are also allowed to germinate during cultivation. In comparison, savannah farmers preserve only economic species. There are many regional farming system studies that focus on the functions trees serve. For example, in the Ho district of Ghana, Asamoah (1985) notes that 76 per cent of the fallow species identified provide medicines, 92 per cent are valued for their soil improving qualities, 92 per cent have domestic uses, 88 per cent are used for fuelwood, 67 per cent have commercial value and 50 per cent serve in customary rites. In another Ghanaian study, Elletey (1986) examines the reasons why farmers protect (or tolerate) farm versus fallow trees. Among the Banen of southern Cameroon, Dongmo (1985) finds that most farm trees are valued for their leaves, fruit or seeds. A few species are also left in fields for protection and demarcation.

Assessing the Value of Forest Resources by their Functions

There are a multitude of household uses for forests. Some are region specific, while others are specific to a group of people, or even a household. In examining the uses of resources by individual households, and in assessing their importance at a regional level, it is more useful to focus on the functions they serve rather than the specific species exploited. People are not concerned with trees themselves; they value functions such as cooking, warmth, food, medicine and shelter. The first part of this study broadly defined important West African tree products and their functions: as foods and materials for household consumption, as support to other productive activities, as sources of cash income, and as cultural symbols.

Information on the functions of forest resources can be found in anthropological (Bahuchet 1978) and geographic household-level studies, as well as in studies evaluating the uses of farm fallow trees. In regions where the natural forests have disappeared, useful trees are often incorporated into the farming systems (Okafor 1981, Ijalana 1983). For this reason studies of on-farm trees often reveal information on the functions that forest resources once served.

This study has identified many of the common functions forest products serve in households throughout the West African forest region:

- Food (both animals and plants) to supplement the diet and meet seasonal shortages;
- *Drinks:* Palm wine and alcohol, "water";
- Dood and cash buffers or insurance in emergency hardship periods;
- Medicines and dental chewing sticks;
- Fuel, for all household and enterprise needs;
- Marketable products exploitable for cash income (*e.g.* cola nut);
- Material for house construction: poles, bark, liana, palm roof tiles and other roof leaves, wattle slats, timber;
- Material for household, agriculture, hunting and fishing equipment;
- Yam and other crop stakes;
- Materials for crop storage containers;
- Fencing and boundary markings;
- Fodder;
- Shade;
- Inputs for processing enterprises (*e.g.* fuelwood);
- Locations for social, religious and healing ceremonies;
- Symbols of cultural and religious identity and importance.

By focusing on the functions of forest products the role of these products can be examined and information from different areas of the region can be compared. In addition, this focus allows one to examine how uses change over time and by setting. For example, in examining the function of foods gathered from forest areas, similarities can be seen across the region: forest foods provide dietary staples and supplements; fill in seasonal and emergency food shortfalls; and provide specific nutrients and culturally symbolic foods. Forest leaves, nuts and wild animals are used in sauces which accompany main meal staples. Forest fruits and insects are consumed largely as snacks, especially by children and during periods when agricultural work is most time-consuming. Mushrooms are consumed as meat substitutes, and are generally available only during the rains. Forest foods may have particular cultural value (such as the cola nut, a sign of welcome in many parts of the region). Some forest foods provide regularsupplements to the diet, others are consumed seasonally, when staple food supplies dwindle or during peak labour periods when little time is left for cooking. Still other forest foods are used only in emergencyperiods when no other foods are available. These generally differ from regularly consumed products, they are more energy-rich but require lengthy processing. What is important about the focus on function is that discussions move beyond species descriptions.

The changing uses of forest resources can also be viewed in terms of their evolving functions. For example, in some cases forest foods may no longer provide the diverse range of produce which they once did. But culturally important foods may still be widely consumed. In other instances, the growing market for some forest foods (*e.g.* Irvingia gabonensis seeds) may have changed their role in rural areas; trees may now be valued as a sources of cash income. Concomitant with these changing functions, changes in the ways in which these products are valued and managed may ensue.

In southern Benin, for example, a new technology for palm alcohol distilling was introduced which changed the way the raphia palms were used and valued. Formerly they had served a great range of household functions and had been used by anyone in the community, but with the introduction of this new technology, they were rapidly overexploited because of the nearby urban market for palm alcohol.

There can be little doubt that the functions of forests will change for households in the rural regions of southern West Africa. The relative importance of different forest products may also change: this can already be seen as some products take on more of a commercial value. But, for most of the functions listed above, contemporary evidence suggests that forest products are still important for the majority of rural households. Although substitutes exist for many non-timber forest products, few people have the resources to buy them. Perhaps for this reason, of those noted above only "material for household, hunting, and fishing equipment", "foods as a buffer during emergencies", and "drinks" appear to be of declining importance throughout the region as a whole.

The Importance of Biodiversity

Biodiversity is often used to draw attention to issues related to the environment. It can be closely related to

- The health of ecosystems.

For example, the loss of just one species can have different effects ranging from the disappearance of the species to complete collapse of the ecosystem itself. This is due to every species having a certain role within an ecosystem and being interlinked with other species.

- The health of mankind.

Experiencing nature is of great importance to humans and teaches us different values. It is good to take a walk in the forest, to smell flowers and breath fresh air. More specifically, natural food and medicine can be linked to biodiversity.

Biodiversity in the Black Sea

The Black Sea region used to be one of the most important areas for fisheries and for food and income for local people. Sturgeons (Acipenser sp.), mullets (Mugil sp.), and mackerel (Scomber sp.) as well as other species have been extensively exploited in this area. However, human activities related to agriculture, shipping and tourism now exert strong pressure on the environment, especially in the northern part of the Black Sea, and are taking their toll on biodiversity and damaging fish stocks.

Threats to Biodiversity

Large amounts of fertilizers were carried to the sea from agricultural areas along the large rivers, such as Don, Dnepr, and Dnjestr. Oil leaked into the sea from ports on the eastern shores and directly from tankers crossing the Mediterranean. Wastewater from cities and heavily visited coastal tourist towns was discharged without any purification.

Consequences for the Ecosystem

All of this leads to pollution and eutrophication meaning the enhanced blooming of planktonic algae due to increased nutrient loads. Planktonic algal blooms lower the water transparency letting only a little amount of light through the water where makrophytes usually grow. This is how the natural belt of bottom vegetation along the Black Sea coast has been destroyed. For example, the vertical distribution range of Cystoseira spp. decreased from 0-10m to 0-2,5m.

The consequences were drastic because the habitat is of major importance as a nursery for spawn and hatchlings of many marine species. This led to a drop in reproduction rates and hence fish stocks.

Biodiversity of the Coastal Zone

Coastal areas are exceptionally productive environments, rich in natural resources, biological diversity and with a high potential for commercial activity. The importance of biodiversity in the coastal zone can be demonstrated by 8 out of the 40 EU listed priority habitats of wild fauna and flora falling into the coastal habitat. Approximately a third of the EU's wetlands are located on the coast as well as more than 30% of the Special Protected Areas designated under the Directive for the conservation of wild birds. The reproduction and nursery grounds of most fish and shellfish species of economic value also comes from this area, which accounts for almost half of the jobs in the fisheries sector.

Pressure on Coastal Biodiversity

Coastal areas are increasingly vulnerable to stresses from both human activities and the forces of nature. The complexity of human activities, natural systems and ownership in the coastal zone, requires an integrated management scheme to allocate coastal resources efficiently and minimize environmental degradation. Choices have to be made between competing uses and limits of resource exploitation if escalating conflicts and resource degradation are to be avoided.

The attitudes of community and industry to the use of biological resources should change from the 'maximum yield' approach to one of 'ecologically sustainable', which recognizes the need for conservation of biological diversity and maintenance of ecological integrity. Integration of management regimes within and between different sectors to meet environmental, economic and social objectives must be realized in order to achieve sustainable development.

Integrated Approach

Integrated policies will also provide the opportunity for all the people to accept responsibility for their actions and the impact they may have on biological diversity. The development of integrated policies for managing the coastal resources is necessary:

- to coordinate activities within and between all levels of government;
- to ensure that full social and environmental consequences (and costs) of development activities are considered;
- to ensure that the public interest is properly taken into account.

Conservancy the Forest Department

During the earlier years of conservancy the forest department denied that the villagers possessed any rights of any description. The government, however, called for a report from the superintendent Mr. H.G. Ross who took a very different view of the matter. He described the most extensive prescriptive 'right' in grazing as having existed from time immemorial and he produced much evidence in support of his contention. The forest department, however, preferred to call the grazing facilities enjoyed by the people 'privileges'..

When Mr. Ross's report became the basis of notification No. 7()2 of 1880, specifying the list of villages entitled to special grazing facilities, the forest department was successful in the battle of words, so that 'rights' were not admitted, but villages included in Mr. Ross's list were permitted to exercise certain privileges. A systematic management for satisfying the basic needs of the local population thus never became an intrinsic part of the management of reserved forests. The direction in which the systematic approach did evolve was largely in the area of quantifying growing stock to guide felling to ensure steady revenue returns.

Subsequent to notification No. 702 of 1880 based on Ross's report of the villagers' rights to forest produce, notification No. 889F of 1893 very clearly spelt out the management framework for meeting local needs of grazing, fodder, fuelwood. poles and thatching grass for housing. According to this notification, the Divisional Forest Officer (DFO) was to prepare an annual list of forest areas which would be open to grazing. The list would specify which areas, in which block of the forest, would be open for grazing in that year. The grazing of cattle in the said reserve blocks was to he regulated in either of the following ways:

1. By arrangement to a fixed settlement every three years based on an enumeration of cattle they desired to graze.
2. By a yearly pass obtainable from tint. DFO.

On the basis of a list prepared in this manner the.DFO was to issue herdsmen badges specifying the number of permitted cattle, the names of villagers owning the cattle and the names of the herdsmen. Only those cattle under the charge of a herdsman and certified by the number on his badge were permitted to graze.

Village communities enjoying these grazing facilities: were also to be permitted to collect and remove headloads of fodder grass as well as fallen and dry fuel free of charge. Although operationalising this management scheme was the most important precondition for satisfying basic needs as well as protecting the reserved forests from degradation, they do not appear to have been enforced in the working plans. Conflicts over forests emerged because colonial rule ignored the demands of nature's economy and the survival economy through indifference to the conservation and basic needs role of forestry, and developed forestry only along the one dimensional criterion of commercial/ industrial requirements.

In the Kumaon region there is evidence that the needs of the empire and not of the local people led to rapid forest denudation. According to Atkinson's Gazetteer, the forests were denuded of good trees in all places. The destruction of trees of all species appears to have continued steadily and reached its climax between 1855 and 1861 when the demands of the Railway authorities induced numerous speculators to enter into contracts for sleepers, and these men were allowed, unchecked, to cut down old trees far in excess of what they could possibly export, so that for some years after the regular forest operations commenced, the department was chiefly busy cutting up and bringing to the depot the timber left behind by the contractors. While the local people were denied their traditional rights to forest resources, and while the colonial forest policy became a 'policy for deforestation', the local people were often blamed for the devastation of forests. As Pant observes:

The.tale about the denudation of forests by the hillman was repeated ad nauseum in season and out of season by those in power so much so that it came to be regarded as an article of faith.... By way of vindication of the forest policy it is claimed by its advocates that in the pre-British days the people had neither any rights in the soil nor in the forests.

The violation of people's ancient rights to forest resources through the colonial forest policy led to popular opposition to the forest policy. Their resentment was first manifested in 1906 in the state of Tehri Garhwal. On 27 December 1906, the forest surrounding the Chandrabadin temple about 14 miles from Tehri town was earmarked for reservation. The next day 200 villagers gathered to protest against state interference in their forests over which they claimed full and extensive rights.

In 1907, a mass meeting was held in Almora to protest against the forest policy which authorised the government to declare all forests and 'wastelands' ('benap'or unmeasured land) as reserved forests. As people's agitation increased because they were unable to get a response, they set fire to government forests and resin depots in 1916. The Kumaon Association was also established in that year to look into the forest problems of Kumaon, with G.B. Pant as its general secretary. Increasing people's protests forced the government to set up a 'Forest Grievances Corranittee' to enquire into forest protests in Kumaon and Garhwal. Though the committee reclassified forests to pacify the villagers, yet people's rights were not protected. As Pant concluded in The Forest Problem in Kumaon, The policy of the Forest Department can be summed up in two words, namely, encroachment and exploitation.

The Government has gone on pushing forward, extending its own sphere and scope and simultaneously narrowing down the orbit of the rights of the people.... The memory of the 'San assi' boundaries (1880 predemarcation) is green and fresh in the mind of every villager and he cherishes it with a feeling bordering on reverence; he is simply unable to see his way to accepting the claim of the Government to the benap lands comprised within his village boundaries and regards every advance in that line as nothing short of encroachment and intrusion. Let the san assi boundaries be vested with their real character instead of being looked upon as merely nominal, and, to remove misgivings, let the areas enclosed within these boundaries be declared as the property of the villagers and all the benap lands included within these areas be restored to the village community, subject to such conditions to impartibility, etc., as may be desirable in the public interest.

It is a matter of common knowledge that a large number of memorials were sent by the villagers at their own instance, about the year 1906, asking the Government to restore the areas within the san assi boundaries to them:

the unsophisticated villager spontaneously reiterates the same demand today. This is the minimum demand of the people and there seems to be no other rational and final solution. The simple fact should not be forgotten that man is more precious in this earth than everything else, the forests not excepted, and, also, that coercion is no substitute for reason, and, however stringent and rigid the laws may be, the forests cannot be preserved in the midst of seething discontent against the unanimous wishes and sentiments of the people.... The collective intelligence of a people cannot be treated with contempt, and even if it be erratic, it can come round only by being allowed an opportunity of realising its mistake. If the village areas are restored to the villagers, the causes of conflict and antagonism between the forest policy and the villagers will take the place of the present distrust, and the villager will begin to protect the forests even if such protection involves some sacrifice or physical discomfort.

The contradictions between people's basic needs and the state's revenue requirements, however, remained unresolved, and in due course these contradictions intensified. In 1930 the people of Garhwal launched the non cooperation movement to draw attention to the issue of forest resources. Forest satyagrahas to resist the new oppressive forest laws were most intense in the Rawain region The King of Tehri was in Europe at that time. In his absence, Dewan Chakradhar Jayal resorted to armed intervention to crush a peaceful satyagraha at Tilari. A large number of unarmed satyagrahis were killed and wounded, while others lost their lives in a desperate attempt to cross the rapids of the Yamuna river. Years later, the martyrs of the Tilari massacre provided inspiration for the Chipko movement when people pledged themselves to protect their forests.

From ancient times forests have been the foundation of our cultural and material life. We reaffirm our birthright to crow susterence and livelihoods from forests while protecting them. Frorn time to fume, our forest rights have been violated through brute force leading to a disintegration of our cultural and economic life. Sometimes the mirage of petty reforms and privileges hove been put before us. But only a few vested interests hove gained from changes in forest management.

Governments will come and will go. But it is our firm belief that our happiness and prosperity are based on a harmonious relation between our forests and ourselves. This relationship must be allowed to continue forever. Today, we remember the martyrs of Tilari and offer homage to them. Their peaceful and non-violent movement and sacrifices give us a timeless inspiration to protect our forests and forest rights.

Forest Composition and Environments

Interactions between tree species and between them and disturbance regimes generate distinctive assemblages of species, forest dynamics and landscape patchworks. These are best illustrated by examples at the ends of the range of disturbance frequency.

Highly Disturbed Environments

Where disturbances are frequent, large scale or severe, forests tend to be dominated by intolerant and fast-colonising species. Three examples illustrate the range of circumstances.

Floodplain Forests

Floodplain forests are influenced by channel movement, which destroys mature stands but reworks the deposits into new shoals. In northern temperate regions, these are colonised mainly by *Salix, Alnus* and *Populus* species, which grow into even-aged, often monospecific stands. The sequence of channel movements is manifested as a pattern of elongated even-aged stands, whose age increases with distance from the channel. Tolerant species colonise beneath these pioneer stands and, given sufficient time, develop into mixed old-growth. In northern temperate deciduous regions the principal long-term dominants are *Ulmus, Fraxinus, Quercus, Carya, Sassafras* and, in the Pacific North-west, *Picea sitchensis* and *Pseudotsuga menziesii.* At any one time, the pioneer stands tend to predominate near the present channel, and the mixed old-growth tends to survive in elongated patches at some distance from the river.

Ice-dominated Forests

The classic type is the wave-regenerated *Abies* forests of the eastern USA and Japan. Exposed mature forest degenerates when foliage is stripped by ice and wind from trees that have already lost the vitality of youth. Death of exposed trees exposes others, leading to a 'wave' of mortality, which moves steadily through the forest in the direction of the prevailing wind. Regeneration starts within the degenerating stands and grows vigourously in the lee of slightly older stands. The forest as a whole takes the form of a series of parallel waves, which move through the forest at 1-3 m annually on a return time of 60-70 years. This perpetual recycling ensures that *Abies balsamea* remains dominant and that the longer-lived *Picea rubens* is perpetually excluded.

Fire-dominated Forests

The most widespread form of highly disturbed forest is dominated by fire. Most boreal forests are naturally fire-dominated, but so too are Mediterranean forests and the forests that fringe extensive grassland and desert regions. Each region has a suite of species that undergo a characteristic succession after fire, for example in Scandinavia *Betula-Pinus* develops into *Pinus* dominance, which is then succeeded by *Picea.* Exceptionally, some patches remain unburned, which allows the pioneers to be completely displaced, although generally fire returns in good time to ensure that pioneer species remain a permanent feature of the forest. In fact, there is much variation in return time, associated with variation in topography, ground vegetation and the configuration of water bodies. Nevertheless, fires were frequent enough to maintain most boreal forests as young or maturing stands, not old-growth.

Similar fire-dominated regimes control other forests, such as the *Eucalyptus* forests of Australia and the *Pinus*-dominated forests of the coastal plain of the south-eastern USA. However, in western North America many tree species not only withstand fires once they have achieved a moderate size but also grow to great size and age, thus generating the monumental forests of *Sequoiadendron giganteum, Sequoia sempervirens* and *Pseudot-suga menziesii* in which old individuals may bear the scars of several fires. On the margins, the boundary between mesic- and fire-dominated forests advances and retreats according to the history of fires.

Relatively Undisturbed Environments

At the other end of the range are mesic forests growing in relatively undisturbed environments. These are not disturbance-free; instead, catastrophic disturbances are rare enough to allow most of the forest to develop into old-growth, where disturbances are small-scale events. Over much of the north temperate zone, a distinction can be drawn between *Fagus* forests and mixed deciduous forests. *Fagus*-dominated forests are found mainly in northern latitudes and submontane elevations, where *Fagus* spp. often share dominance with conifers, for example *Tsuga canadensis* in the eastern USA, *Abies alba* and *Picea abies* in central Europe.

European beech, *Fagus sylvatica,* is not long-lived, but casts dense shade, is capable of regenerating in the small transient gaps generated in beech forests and is almost fireproof, so that it is able to both dominate the site and perpetuate this dominance.

However, it is prone to disaster, in the sense that mature stands are vulnerable to drought and high winds. A few intolerant species, such as *Salix caprea,* maintain a foothold in large gaps, while the shade-tolerant *Acer pseudoplatanus,* like *Acer saccha-rum* in the eastern USA, fills gaps, competes in advance regeneration and can grow into mature stands. Diversity is maintained partly by the tendency of species not to regenerate under themselves. Thus, *Fagus grandifolia* and *Acer saccharum* in the USA and *Fagus syl-vatica* and *Abies alba* in Europe have been reported to alternate, thereby maintaining a small-scale mosaic of different dominance, though in the former case coexistence has been ascribed to different responses to light intensity. Occasionally, such stands are destroyed by storms, whereupon pioneer species dominate the regrowth, although beech thrives in the underwood and eventually restores its position.

Mixed deciduous forests are those which lack the dominating influence of beech. At their greatest development, for example in the southern Appalachians, they comprise a mixture of several dozen species, each with the capacity to occupy the canopy or subcanopy. A wide range of genera are represented, notably *Acer, Quercus, Fraxinus, Carya, Aescu-lus, Betula, Castanea, Tilia, Magnolia, Carpinus* and *Halesia,* forming rich mixtures in which no single species becomes absolutely dominant. Such forests are rarely devastated by any single disturbance and for most of the time are renewed by gap-phase regeneration on a small scale. Pioneer species regenerate in the larger gaps and may dominate the canopy after periods of enhanced gap creation. An example is *Lirioden-dron tulipifera,* which not only forms the tallest trees in the Appalachian forests but can also live for 500 years, enabling it to perpetuate itself through long periods lacking disturbance.

Within these complex mixtures, each species has a distinctive pattern of growth, longevity and regeneration. Furthermore, there is a tendency for individuals of one species to be replaced by another species. The forests comprise a small-scale mosaic of groups of different canopy and underwood species, each with its particular successional trend. The trends in one patch are countered by opposite trends in other patches, thereby retaining the mixture. Recruitment of particular species tends to be irregular, depending on particular combinations of mast-years and disturbances. Composition remains fairly constant overall but at a small scale changes perpetually, except where individual trees replace themselves vegetatively, *e.g. Tilia* spp.

Single-species groups are common in mixed forests. These may develop in response to small differences in site conditions and/or the chance coincidence of an episode of gap creation with heavy seed production by a particular species. Simulations have recently shown that neighbourhood effects may also play a part. Where there is a high probability that canopy trees will be replaced by individuals of the same species, 10 generations is enough for this feedback to generate small-scale single-species patches. The scale of the patches increases substantially where minor environmental differences result in 5 per cent alterations in recruitment probabilities.

Interactions and Intermediate Conditions

The forests described above are stereotypes that represent the ends of the range of variation. In practice, many forest types are intermediate in some respect, although their dynamics and composition can be related to the stereotypes. Some examples illustrate the range of possibilities.

- Forests in which partial stand destruction is commonplace can develop into complex mosaics of tolerant and intolerant species. An example is the *Pseudotsuga menziesii* forests of the Pacific North-west, where saplings and underwood are frequently burned but the overstorey is unaffected. Moreover, light fires tend to be patchy, burning some places but leaving other patches untouched. This allows the overstorey of *Pseudotsuga menziesii* to be infiltrated by *Abies amabilis,* which develops as patches in a mosaic with groups of younger *Pseudotsuga menziesii* in larger burned patches.

- Complex interactions can develop between disturbance regimes. For example, in the southeastern USA *Pinus* forests were maintained naturally by frequent fires. However, where stands happen to remain unburned, *Quercus* and other broadleaved species colonise, form an underwood and eventually dominate the canopy. Since the broadleaved stands are relatively fireproof they tend to remain undisturbed, thereby generating a patchwork of broadleaved and *Pinus*-dominated stands whose pattern is determined by disturbance history. This capacity for two forest types to develop on one site type is inherent in any region where disturbance becomes less likely as succession proceeds.
- The characteristics of *Fagus sylvatica* in central Europe determines much of the forest pattern. Its capacity to coppice enables it to form distinctive low scrub woodlands with *Sorbus aucuparia* near the treeline, where wind and snow maintain a chronically disturbed environment. In this case, the *Fagus sylvatica* is effectively undisturbed. Elsewhere, in steep slopes in gorges, the stands are chronically destabilised by the fall of large trees growing on insecure rootholds, and this breaks the dominance of beech and enables a mixture of tree species to survive, including *Tilia, Ulmus* and *Fraxinus.*

Disturbances have long-term effects on forest composition. For example, drought rarely destroys stands entirely but confers an advantage on species that survive droughts. Ultimately this is expressed as a distinctive assemblage on drought-prone sites, such as outcrops and convex slopes with thin soils. In this instance, disturbance tends to generate adapted forests, which thus reduces the incidence of disturbance. Alternatively, species that depend on disturbance may both invite it and be ready to survive. The best examples are the fire-dependent *Pinus banksiana* and *Pinus contorta* forests, which are inherently inflammable, particularly those that exhibit serotiny, where long-lived cones release their seed only after fire and which falls into a seed bed free of litter and competing vegetation. A more extreme example is seen in the *Pinus mugo* of avalanche tracks, where repeated disturbance has provoked the evolution of a species that can withstand the disturbance. The multitude of interactions between trees and disturbance regimes is expressed as different assemblages on different site types and as a range of successional states on each site type. Disturbance should be seen as an integral component of the forest type, not an external destructive force. Collectively, disturbances enhance regional diversity by creating distinctive forest types, maintaining a range of successional states, and by enhancing the amount of edge in the landscape.

Influence of People

The relationships described so far have dealt with natural conditions. In practice, most forest types are overwhelmingly influenced by people:

- Secondary successions to forest on land previously cleared are a response to severe disturbance, *i.e.* the complete loss of tree cover and the alternative use of land for cultivation or pasture. Pioneer trees are almost always intolerants. The pattern of secondary succession depends on seed sources, so that colonisation may be extremely patchy. Colonisation often takes place in waves, for example when initial colonisation by wind-dispersed trees is followed by bird-dispersed trees once perches have been created.
- Coppice systems enable shrubs and intolerant trees to form a higher proportion of a stand than they would naturally maintain. Despite the constant disturbance of felling, individual trees can sprout again indefinitely and thus live well beyond their natural span. In terms of individuals and small-scale patterns, coppices are probably more stable than the natural forests from which they were derived.
- In woodland pastures, regeneration is inhibited by grasing and browsing, although established individuals have little competition and can expand to great sizes. Moreover, the constant lopping of branches for fuel and fodder, followed by regrowth from the pollard, enables individual trees to grow to great ages. Here, too, the effect of management disturbance is to generate unnaturally high stability.

Finally, there is the problem of *Quercus petraea* and *Quercus robur* in Europe. These are long-lived intolerant species that have been abundant in the pollen rain over very long periods, which implies that the forests were considerably disturbed, albeit at long intervals. They have been maintained at unnaturally high levels by traditional management, which favoured them as timber trees in coppices and wood-pastures. Under natural conditions it is possible that oaks were maintained by the high level of browsing and grasing imposed by populations of large herbivores, such as deer, horses and cattle. In fact, it has been proposed that natural forest was more like savannah before prehistoric hunters reduced and domesticated the larger herbivores. However, this seems far from proven, although the possibility remains that herbivore populations maintained a controlling presence on the structure and composition of some types of temperate forest, particularly perhaps those on wet ground and heavy fertile soils (where ground vegetation would be vigourous) and steep dry slopes (where tree growth might be poor).

4

Marine Biodiversity

INTRODUCTION

India has a vast extent of coast line of about 8129 km and an Exclusive Economic Zone (EEZ) of 2.02 million km^2, which are home to a diversity of coastal and marine ecosystems, comprising nationally and globally significant biodiversity rich areas. India's wealth of biodiversity is found in highly diverse marine and coastal habitats. Until today marine diversity is less known than terrestrial biodiversity due to the logistic difficulties of explorations, underwater surveys and collections. Oceans support a large human population in India directly and indirectly.

The Indian Ocean accounts for 29% of the global oceans, 13% of the marine organic carbon synthesis, 10% of the capture fisheries, 90% of the culture fisheries, 30% of the local reefs, and 10% of mangroves. It is pertinent to note that India has 246 estuaries draining a hinterland greater than 200 km^3 besides coastal lagoons, mangroves and backwaters. Over 200 species of diatoms, 90 dinoflagellates, 844 algae, 14 seagrasses and 39 mangrove species have been reported from Indian waters. The marine algae include Rhodophyta, Chlorophyta, Phaeophyta and Xanthophyta.

Tamil Nadu has the maximum number of species of seaweeds followed by Gujarat and Maharashtra. A number of products such as agar, alginates, carrageenan, liquid fertilizers and bio-active compounds are produced from seaweeds. Foraminiferans (> 500 species), sponges (> 480 species), coelenterates (> 840 species), polychaetes (> 250 species) and echinoderms (> 765 species) have been studied by several researchers. Crustaceans form one of the biggest groups including both commercially important and lesser known organisms. Commercially important groups such as molluscs (> 3370 species) and fishes (>2546 specie-s) are also well documented by the researchers of India. 5 species of sea turtles are also reported from India. Twenty five species of marine mammals belonging to the orders Cetacea and Sirenia are reported from Indian waters. The coastline of Bay of Bengal and Arabian Sea continues to be a rich fishing ground in the south Asian region and

India is one of the world's largest marine product nations. Indian marine ecosystems are all known for their high biological productivity, which provide a wide range of habitat for many aquatic flora and fauna. It also provides important food resources and critically major services to human beings. Therefore, sustainability of these fragile ecosystems should be our primary concern.

THREATS AND CONTROLS:

It is often argued that the changes in biodiversity will be mainly restricted to land and consequently attention to biodiversity changes in the ocean is limited. However, humans do impact the already to a considerable degree, especially in the coastal areas but increasingly in the open ocean as well. Over the last thirty years, broad controls have been proposed or developed related to the five major threats. Controls can be categorized with threats. Many nations have commendable statutes and policies; however implementation failures are widespread.

Amongst these five major threats to marine biodiversity, fishing has, until the present time, been the most damaging on a global scale. The destructive impacts of fishing stem chiefly from overharvesting, habitat destruction, and bycatch. Over the coming century the threats posed by increasing atmospheric greenhouse gases pose huge dangers to the marine environment. At smaller scales, other threats (particularly pollution and habitat damage) are dominant at different localities. Coral reef, mangrove, estuarine, seagrass, mud-flat, and sponge-field habitats have been (and are being) extensively damaged. River passage, essential for anadromous and diadromous species, has been impaired or destroyed around the globe.

MARINE BIODIVERSITY CONSERVATION APPROACHES

Aquatic conservation strategies support sustainable development by protecting biological resources in ways that will preserve habitats and ecosystems. Methods like Restoration of habitat, Restricted fishing in particular areas, Declaration of biosphere reserves, Control on the establishment of industries, chemical plants and thermal power plants, Publish the list of threatened or endangered species, Regulation of waste water discharge, Increasing public awareness will helps to conserve the marine ecosystem to a large extent.

Several initiatives were taken by the Government of India focusing on the conservation of wetland, mangroves and coral reef and the management through implementation of law and continuous monitoring. The Wild Life Protection Act of India provides legal protection to many marine animals. A National Committee on mangroves, wetlands and coral reefs was constituted in 1993 with a mandate to advise the government on relevant policies and programmes.

The mangrove and coral reef areas were declared as ecologically sensitive areas under the Environment Protection Act, 1986. The Coastal Regulation zone (CRZ) notification came in 1991 prohibiting developmental activities and disposal of wastes in the fragile coastal ecosystems.

Some of the national parks and sanctuaries in India were declared exclusively as marine protected areas in the 1980s and 1990s. There are a total of 31 major marine protected areas in India. They cover coastal wetlands, especially mangroves, coral reefs and lagoons and have been notified under Wildlife Protection Act, 1972. The three notified Biosphere Reserves in India are the Sundarban Biosphere Reserve, Great Nicobar Biosphere Reserve and the Gulf of Mannar Biosphere Reserve. IUCN has defined a Marine Protected Area as 'any area of intertidal or subtidal terrain, together with its overlaying water and associated flora, fauna, historical and cultural features, which has been reserved by law or other effective means to protect part or all of the enclosed environment'. The country took one more step ahead with the promulgation of the Biological Diversity Act of India 2002 and the Biological Diversity Rules 2004.

A National Biodiversity Authority with state –level Boards and district-level Management Committees was established under this Act. The main functions of the Authority is to advise the government on matters related to the protection and conservation of biodiversity, sustainable use and equitable sharing of its components, Intellectual Property Rights, etc.

WETLANDS BIODIVERSITY

Wetlands have been called "biological super systems" because they produce great volumes of food that support a remarkable level of biodiversity. The combination of shallow water, high levels of nutrients, and high primary productivity (the amount of biomass produced) is ideal for the development of organisms that form the essential base of our planet's food web.

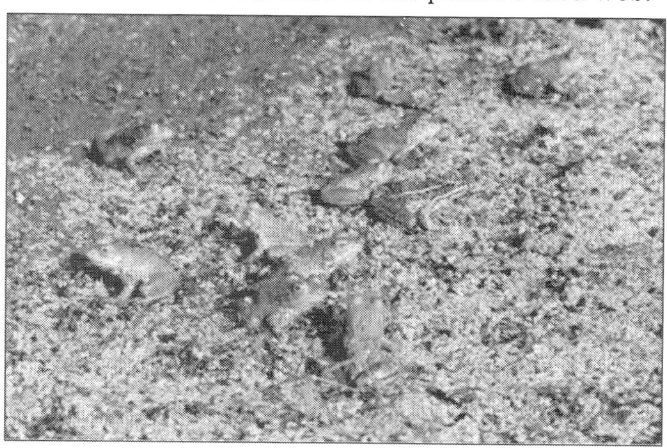

The food web supports myriad species of birds, fish, amphibians, shellfish, and insects. In Illinois, countless species depend on wetlands, including:

- 105 bird species depend upon, or are strongly associated with, wetlands in Illinois; an additional 169 bird species use wetlands in Illinois opportunistically for nesting, foraging, and resting.
- 46 of the 59 mammal species in Illinois use wetlands to some extent.
- 37 of the 41 amphibian species in Illinois depend upon wetlands at least part of the year.
- 47 of the 60 reptile species found in Illinois use wetlands to some extent.

Wetlands — marshy areas of land where the soil is saturated with water — are crucial incubators of species diversity, as important as tropical rain forests and coral reefs. They exist on all continents, save Antarctica, and include salty coastal flats, such as estuaries, and inland systems. Scientists classify wetlands into bogs, swamps, marshes, and other types, depending on geography, soil, and plant life.

Wetlands help filter pollutants and soil runoff from upstream sources, which helps keep rivers, bays, and oceans downstream clean. In this way, healthy wetlands help mitigate the negative effects that human and farm waste, and some byproducts of industrial pollution, have on our Nation's water. Wetlands help control inland flooding and forestall wave erosion along shorelines; they diminish drought damage.

Fig. Mangroves in Indonesia.

Coastal wetlands protect spawning and feeding grounds for valuable fish and shellfish. Mink, otter, and other mammal species thrive in North American wetlands, as do myriad plants and insects, amphibians, and reptiles. They are vital nesting grounds for birds. According to the U.S. Environmental Protection

Agency (EPA), wetlands are essential to the survival of about a third of the endangered animal and plant species in the United States, and about half of these species make use of wetlands at some point in their life.

Unfortunately, especially in this century, the utility of wetlands has been poorly understood. They have been regarded as useless, or worse — seen solely as breeding grounds for mosquitoes, other insects, or as sources of odours. Vast numbers of wetlands have been drained or destroyed to accommodate agriculture, dams, and human habitation.

Fig. Mangroves in Indonesia.

It's estimated that over half of the wetlands in the continental United States have been lost since the 18th century; and wetlands elsewhere have fared no better. With the destruction of wetlands has come destruction of biodiversity, both in the wetland areas themselves and downstream. For instance, nitrogen fertilizer runoff from farms has overwhelmed the capacity of some wetlands to filter pollutants, creating "dead zones" in areas such as the Gulf of Mexico, where algae blooms fuelled by this and other nutrients have run riot and displaced a once thrivingly diverse ocean ecology. Wetlands in a sense are a biodiversity laboratory.

For one, the diversity of conditions in wetlands set the environmental parameters that allow for, even encourage, the evolution of novel survival strategies. According to the EPA, for example, many bog species have "special adaptations to low nutrient levels, waterlogged conditions, acidic waters, and extreme temperatures.

" Vernal pools, ponds in winter and mud flats in summer, often include rare species that weather the drought as seeds, eggs, and cysts, and then grow into mature form when the ground is watery again. Mangrove swamps are full of shrubs and trees that have adapted to salty water.

Fig. Alaka'i Swamp Forest in Kauai, Hawaii.

Wetlands are important to natural cycles involving water, nitrogen, and sulphur. Their plants and rich soil may provide one buffer against global climate change, by storing carbon instead of releasing it into the atmosphere as carbon dioxide. In dollar terms, the services wetlands provide are invaluable. According to the EPA, the Congaree Bottomland Hardwood Swamp in South Carolina performs water purification functions equivalent to a five million dollar wastewater treatment plant. Wetlands act like giant sponges, storing, then slowly releasing ground water, melted snow, and floodwater. Because urban buildings and pavements release water runoff quickly, wetlands downstream from urban areas perform valuable flood control services.

Fig. Honduras.

In some cases, wetlands have been destroyed to create artificial flood control. Hardwood wetlands along the Mississippi river once stored 60 days' worth of floodwater. Now, due to filling or draining, they store 12 days' worth. Attitudes are changing. In the 1970's, the U.S. Army Corps of Engineers determined that draining 8,500 acres of wetlands near Boston would result in $17 million of flood damage per year - as a result, those wetlands were never drained. Even the finest wetlands, however, will ultimately be degraded or destroyed if too much pollution, silt, and non-native species are sent their way from upstream.

Fig. Kenya

Many familiar animals — ducks, falcons, bears, deer — make use of wetlands. Some species of migratory fowl are completely dependent on wetlands. Most commercial fish breed and nurture their young in coastal marshes and estuaries. Such familiar species as striped bass, shrimp, oysters, clams, and crabs can't survive without wetlands. Consequently, wetlands are essential sources of food for burgeoning human populations. Other wetland harvests include blueberries, cranberries, wild rice, and timber, not to mention plants that are sources of medicine.

On the Southeastern shore of the United States, almost all commercial fishing depends on healthy estuary wetlands. In 1991, commercial fish and shellfish taken from the state of Louisiana's coastal marshes contributed $244 million to that state's economy. Fur-bearing animals — muskrat, beaver, and mink — add millions more. Wetlands are popular with hunters, fishermen, and tourists: nearby towns enjoy economic benefits as a result. In the United States, the Wetlands provision (1987) of the Clean Water Act established a "no net loss" policy for managing wetlands. In theory, this means that filled-in wetland areas should be offset with restored wetland acreage. In practice, this portion of the law has been slow to be implemented due to the differing interests of communities, environmentalists, and property owners. In addition, the government has offered tax deductions to people who donate or sell wetlands for conservation purposes, and has taken other measures to try to preserve them. Watershed conservation programs that include federal, state, local, and indigenous tribal governments have proved to be useful in managing streams, rivers, and wetlands.

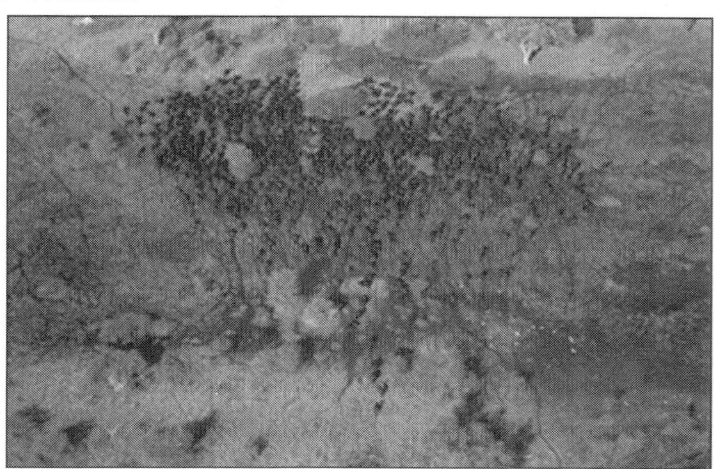

Fig. Aerial View of Large Herd of African Buffalo in the Okavango Delta.

It's difficult to conserve and restore wetlands; the short-term interests of landowners and farmers frequently clash with the long-term benefits of a sustainable natural ecosystem. In the United States, there is particular concern about wetland areas of the Mississippi River, the Missouri River, and the Everglades swamp in Florida. Large parts of these riverine systems have been re-engineered, dammed, and channelled. Previous to this century, 150 species of fish are thought to have cavorted in the lower Mississippi River; now there are 90. Four-fifths of the 8.8 million hectares of wetland forest near the river were cleared. The Lower Mississippi River Conservation Committee is hoping to restore 50 percent of filled-in secondary channels, restore 32,000 hectares of drained wetlands, and reforest 52,000 hectares of wooded wetlands, as a start.

In 1986, the U.S. Congress created the Missouri River Fish and Wildlife Mitigation Project. Over the years, the U.S. Army Corps of Engineers has used funds for this project to buy land from residents and use it to restore natural floodplain habitat. Governments at all levels are becoming involved in wetlands restoration. The city of Cleveland, Ohio, is planning to spend millions to restore wetlands in the region to compensate for an airport expansion that will destroy a watery area. According to the Associated Press, in Bath, Michigan, "a run-down, garbage-strewn park has been restored as a wetland, home to native Michigan wildlife." In the New Orleans area, efforts are under way to restore wetlands, using aerial photographs from the 1930s and '40s to figure out where the wetlands were before they were destroyed.

Fig. Conversion of Wetlands for Oil Palm Plantations in Costa Rica.

As part of a major effort to restore the Florida Everglades, Florida government officials and scientists are planning to construct five artificial wetlands on 17,200 hectares of land. In July, 2000, according to the Washington Post, "the Senate Environment and Public Works Committee reported out—a bill authorizing the first 10 of 68 planned projects meant not quite to restore the Everglades—that will never happen, but to revive and allow them to flourish once again." Plans are under way to acquire and restore about 10,000 hectares of wetlands in the Kankakee River Basin of northwest Indiana. The Grand Kankakee Marsh used to extend over 200,000 hectares. About 4 billion (thousand million) dollars has been spent cleaning up Boston Harbour, and more is being done. Scientists are trying to come up with new answers, figuring out, for instance, how farmers could use computers to determine how to farm with the least amount of wetland-disrupting fertilizer.

Fig. The Pantanal in Brazil.

Efforts in many parts of the world also play a vital role in preserving natural wetlands. The Australasian Wader Studies Group in conjunction with Wetlands International has completed five years of surveying and shorebird counting activities in China during the migration season, a step towards developing information about the importance of inland wetlands in China to bird species. Tasek Bera, the largest natural freshwater body in Malaysia, has been designated a Wetland of International Importance under the Ramsar Convention, raising the profile of an important cradle of many species.

Often, the best advocates for wetlands are groups of concerned citizens and scientists willing to do exploration and to inventory declining species, and educators able to train local groups of people to better manage wetlands. Governments are becoming more sensitive to the need to protect wetland areas. A variety of approaches to wetland conservation, undertaken in many parts of the world, are effective in reversing or forestalling damage and in preserving biodiversity.

Fig. Caiman in the Pantanal.

In the end, however, natural systems are so complex that they are difficult to restore, once highly polluted and degraded. In her book "The Work of Nature," ecologist Yvonne Baskin writes, "Certainly, restoration efforts to return previously damaged lands to ecological service should be encouraged. There are millions [of] hectares of rangelands, forests, and marsh on every continent that might be returned to health and productivity if complex environments could be rebuilt as skilfully as they are being dismantled." She points out, though, that "it would be far less costly to preserve robust natural systems rather than count on being able to piece them back together after they've been torn apart."

DIVERSITY AND PRODUCTIVITY OF WETLAND PLANTS

A variety of topographic gradients exist in wetlands and these influence the nature of the colonising vegetation. Gradients exist between terrestrial uplands and flooded basins, lakes or river beds. In coastal situations they occur in relation to tidal fluctuations which produce great habitat variability on the shoreline, as they do across lagoons and the zones of nearshore coral reefs. Wetland vegetation may respond to the topography and hydrology with a distinct zonation pattern formed by the dominant plant species, particularly in tidal situations, or produce a complex mosaic of plant communities around minor local variations in height.

Fig. Zonation patter in a coastal wetlands, Carriacou: from land in foreground through salina, black mangrove, red mangrove and tidal channel to the open sea. (Photo: Peter Bacon)

Further variability is introduced to inland wetlands by seasonal fluctuations in the rainfall or inundation pattern. The area covered by a wetland may expand and contract with the seasons and thus produce a border of plant communities adapted to alternate flooded and dry conditions. The 'varzea' wetlands of the Amazon floodplain, for example, extend for hundreds of kilometres and show a distinct change in the degree of adaptation of the plant species as one goes from the permanent river channel to the upland terra firma. In temperate wetlands, spring flood and summer drawdown introduce a similar variability in terms of the nature and availability of plant habitats. As a result, wetlands support diverse plant communities, particularly the inland wetlands associated with major drainage systems in both tropical and temperate regions. The US National List of plant species that occur in wetlands compiled by Reed included 6,728 species.

This diversity is reflected at individual sites, such as the 243 km^2 Cache River-Cypress Creek wetland, a Ramsar site in Illinois, USA, where 138 woody plants, 251 non-woody vascular (flowering) plants and 11 ferns were present. However, under more extreme conditions, such as the arctic Tundra, high mountain peat bogs and hypersaline saltmarshes in the dry tropics, the diversity is lower, even though a range of highly specialized plants will be present. Many wetland plants, or hydrophytes, grow in dense and prolific stands. For example, in Papyrus Cyperus papyrus swamps in Lake Naivasha, Kenya, Jones reported a harvestable standing crop of 30 tonnes per hectare compared with only 10 tonnes per hectare of grass from the finest European pastures. After harvest, this amount of Papyrus biomass was replaced in about nine months. Table shows that many other types of wetlands are highly productive. The ready availability of water, which transports nutrients and removes waste products, and the frequent association between plant roots and microscopic organisms able to use nitrogen, allow wetland plants to grow rapidly and produce large quantities of organic matter. In tropical wetland plants, such as mangroves, this primary production can go on all year and reach levels comparable to the most intensively mechanised agricultural production, for example sugar cane crops. Plants play a critical role in the structure and productivity of coral reefs in nearshore wetland environments. In many areas, the reefs can be described as 'cor-algal' reefs because of the close association between the corals (animals) and species of algae (plants). Other algae living in the coral tissues aid in the production of organic matter and are largely responsible, thus, for the high productivity of the reefs.

Fig. An Aerial View of Seasonal Drawdown Zone on a Lake Margin.

Table. Productivity of Selected Wetland Ecosystems.

Wetland type (above ground	**Location hectare per year**	**Annual production , tonnes per only).**
Estuarine mangrove	Sri Lanka	12
Tidal saltmarsh	Louisiana, USA	14
Riparian forest	Louisiana, USA	14
Freshwater (reed) marsh	Denmark	14
Freshwater (Papyrus) marsh	Kenya	30
Freshwater (reed) marsh	Wisconsin, USA	34
Tropical seagrass bed	Caribbean	70

DIVERSITY OF ANIMALS IN WETLANDS

The species diversity and high production levels of wetland plants support even more diverse animal communities. The vegetation distribution patterns and water level fluctuations make a range of continuously changing wetland habitats available at different times of the year to aquatic, terrestrial and arboreal animals. Wetlands support a wide variety of grazing and browsing animals, including several large mammals such as African Buffalo Syncerus caffer and Hippopotamus Hippopotamus amphibius in Africa, Capybara Hydrochaeris hydrochaeris and manatee Trichecus spp. in the Neotropics, Asian Water Buffalo Bubalus bubalis in Asia and Moose Alces alces in North America and Eurasia.

Many species of rodent, such as the beaver Castor spp., Muskrat Ondatra zibethicus and Nutria Myocastor coypus in North America and Europe also depend on wetlands. A number of invertebrates, particularly snails and crustaceans, and some fish, such as Grass Carp Ctenopharygodon idella, graze on water plants and convert these to animal biomass, in some cases impoverishing wetland vegetation. Herbivorous diets are often generalized, but some South American fish feed exclusively on fruits from swamp forest trees and, thus, aid in seed dispersal. On coral reefs, a variety of green, red and brown seaweeds provides food for a great diversity of invertebrates and fish. Some of these, such as damselfish (Pomacentridae), behave like gardeners by protecting and trimming the plants that they hide among and feed upon. The range of plant species in the different wetlands, and their flowers, fruits and seeds, ensures a rich diversity of associated animals.

However, much of the vegetable material produced by wetland plants does not enter food chains directly. In mangrove swamps, for example, only about 10% of leaf production is grazed by snails, crabs and insects, with the remaining 90% falling into the water where it decomposes. Decomposition is brought about initially by microbes, largely marine bacteria and fungi, which break up the leaves and other plant parts. Microbes not only reduce the vegetable matter to smaller and smaller particles of detritus, but they increase the protein content by their presence and make the particles increasingly attractive to a wide variety of aquatic invertebrates. Similar processes occur in other types of wetlands, particularly many inland and tidal types in which decaying plant materials tend to accumulate before they can be consumed. The litterfall of dead leaves, flowers, fruits and twigs may be up to 17 tonnes per hectare per year in riverine and estuarine wetlands. The result of such litterfall is the production of a complex detritus-based food web which supports a great diversity of invertebrates, fish and amphibians, with fishes, frogs and toads being characteristically associated with wetlands. Larger predatory reptiles, birds and some mammals feed on the abundant food resources supported by decomposing plant parts. Characteristic wetland predators include crocodilians, freshwater turtles, the Anaconda Eunectes murinus, otters, dolphins and waterbirds.

Wetlands of the Ebro Delta in Spain support 48 resident species of fish, 29 species of amphibians and reptiles, 27 mammals and 46 resident or migratory birds and this important Ramsar site forms one of the European case studies.

Many different kinds of birds with a wide range of feeding and breeding habits are found in wetlands. Among the 104 species recorded in the Black River Morass, Jamaica, were 11 seabirds, 36 waterfowl, 7 birds of prey, a kingfisher and 49 forest birds; while 251 species have been found in the Cache River Basin, Illinois, USA. According to Weigers some 40 species of birds commonly breed in the somewhat restricted wetland forests in Western Europe. In the case study of the St. Lucia estuarine system of South Africa (a Ramsar site), some 350 species of birds are reported, including 90 species of waterfowl, such as ducks, geese, two species of flamingo and 15 species of herons and egrets.

Fig. An aquatic algal-invertebrate community on a mangrove root, Jamaica

Many wetlands have such abundant food resources (both living plants and their decomposition products) that they can be utilized by species other than the permanent residents. Entry by 'visitor species' serves to increase further the diversity of animals that may be seen in wetlands from time to time. The life cycles of many species of marine shrimps include a period spent feeding in coastal estuaries or marshes.

Several marine fish spawn in mangrove swamps or use these habitats as a nursery for their young because of the ready availability of small food materials and the security provided by mangrove roots. In addition, mangrove swamps are used for nursery and feeding by a range of coral reef-inhabiting species, while the reefs provide sheltered conditions along the coast which encourage mangrove establishment; this suggests that these associated wetland types are mutually supportive. Migration into wetlands to benefit from food or favourable habitat conditions does not occur only in aquatic species, such as shrimp and fish. Many freshwater environments show seasonal fluctuations in water level which influence grazing and other feeding behaviour. Seasonal drawdown in water level permits the movement of animals, including livestock and their herders, into wetland basins, where they utilize the abundant, lush plant resources. In the Nariva Swamp, Red Brocket Deer Mazama americana, Collared Peccary Tayassu tajacu and Agouti Dasyprocta aguti and smaller rodents migrate from the swamp margins and interior islands during the dry season and occupy habitats populated by aquatic species at other times. In effect, any area of the swamp basin will support two different faunas at different times of year, thus increasing the diversity of animals which can be supported by one set of resources. Movement of herders into wetlands as flood waters retreat and fresh grazing areas become available.

Many wetlands provide habitat for other important faunal components, serving as resting and feeding stations along migratory flyways for ducks, waders and shorebirds which benefit from the diversity of food organisms. The seasonal influx of passage migrants serves to increase the biodiversity of many wetland sites. In their study of coastal wetland habitats in Surinam, South America, Swennen and Spaans found more than 75% of the foraging waterfowl were migrants of northern origin, with only a minority being local resident species. For the eight families studied in an area of just 736 ha of these rich and varied coastal wetlands they found 15,678 waterfowl belonging to 40 species dependent on the wetlands during the tropical part of their life cycle. This example shows that the migratory component of the bird life of wetlands is important, not only in terms of species diversity but in numbers of individuals. Similarly, the 24,000 ha Cache River Basin in North America provided wintering habitat annually for nearly 200,000 Canada Geese Branta canadensis, 35,000 Snow Geese Anser caerulescens and 26,000 ducks which would breed further north. The value of wetlands as habitat for migratory birds is documented many times in the regional case studies.

Fig. Wetlands support a variety of waterbirds - pelicans and sandpipers at Banc d'Arguin National Park, Mauritania.

THE ECONOMIC VALUE OF WETLAND BIODIVERSITY

Wetland plants are a major source of materials on which large numbers of people depend, particularly in the subsistence economies of tropical countries. In addition to the variety of goods produced, the quantities exploited are impressive. Mangrove trees annually produce 7,400 m^3 of charcoal and 400 tonnes of bark for tanning in Panama and 120,000 m^3 of firewood in Honduras, while 80% of households in Nicaragua use mangrove wood for cooking.

Throughout the world, wetlands produce a range of animals of commercial importance, particularly as food, skins and for sport and the ecotourism business. Thus, inland wetlands in Africa produce over 1.5 million tonnes of fish annually, with a further 1.0 million tonnes from coastal marine areas. At least one million fishermen and perhaps five million workers in processing, transportation and market activities depend on these fisheries. Twenty percent of commercial fish in Australia are caught in mangrove swamps; 45% were strictly dependent on mangrove resources, while 35% of mangrove dwelling species were food for commercial marine species.

Table. Economic uses of Tropical Wetland Plants (not in Order of Importance).

Construction Materials (Housing & Industry)	Medicines (from Fruit, Sap, Bark, Leaves)
Scaffolding, House beams & rafters, Flooring & paneling, Thatch & matting, Chipboard, Furniture, Fencing, Bridges, Posts, Tool handles, Water pipes, Packing boxes, Boats, Dock Pilings, Railroad ties, Mine pit pros	Diuretics, Purgatives, Astringents, Febrifuges, Vitamins (mainly B group) Treatments for: Arthritis, Leprosy, Catarrh, Rheumatism, Skin rashes, Hemorrhage, Hemorrhoids, Snake bite, Tuberculosis
Fuel	**Textile & Leather Craft**
Firewood, Alcohol, Charcoal, Wood (curing fish, smoking rubber & firing bricks), Peat	Synthetic fiber (rayon), Dyes for cloth, Tannin for leather preparation (tanning)
Fishing Materials	**Agricultural, Horticultural & Aqua Cultural Products**
Poles for fish traps, Branches - fish attracting devices, Floats, Fish poisons, Dye for nets, Tannin - net & line preservation	Fodder, Fish feeds, Green manure, Peat/compost/fiber, Landscape plants, Plantings for coastal protection, Ornamental pond plants, Insect repellent
Food & Beverages	**Miscellaneous**
Sugar, Vinegar, Honey, Alcohol, Cooking oil, Tea substitutes, Fermented drinks, Masticatories, Condiments from bark, Vegetables from fruit, propagates & leaves	Contraceptives, Aphrodisiacs, Cigar substitutes, Drilling lubricant, Matchsticks, Paper (various kinds), Hairdressing oil, Waxes, Incense, Glues

World trade in crocodilians from tropical and sub-tropical wetlands peaked in the 1960s at over 10 million skins per year, declining to a present volume of 1.5 million. In Venezuela alone the harvest of caiman skins and meat was valued at US$9.0 million in 1989. The major part of bird hunting in all parts of the world is based on wetland habitats, with significant numbers taken in some areas. Of the millions of fish, waterfowl and mammals hunted in North America, all the fish and more than 50% of other groups come from wetlands.

In the Caribbean, parks and protected areas containing wetlands have functioned as major tourist attractions for many years, particularly for their bird life. They include the Caroni Swamp, Trinidad, the Flamingo Sanctuary, Bonaire, the Lagoons of Humacao, Puerto Rico, and the Virgin Islands National Park, which includes extensive shorebird and coral reef habitat. Annual recreational values of Caroni Swamp and the Virgin Islands National Park have been valued at US$1.0 and US$23.4 million respectively.

The major part of the foreign exchange earnings of the Turks & Caicos Islands comes from tourism based on coral reef diving. The large sport hunting industry and the rapidly expanding ecotourism sector have a multiplier effect on the economy through expenditure on transport, food, camping gear, hunting and fishing gear, license fees, photographic supplies, visitor facilities and related goods and services. Wetland faunas are, thus, of major economic importance globally.

Fig. Crocodilians Play an Important Ecological Role in Wetlands: their Meat and Skins are of Commercial Value as Well.

Table. Estimated Numbers of Wetland Birds Killed by Hunters in Selected Parts of Asia.

Country/Region	Bird Group	Numbers per annum
China (Lake Shengjn)	Coots	3,900
	Ducks	14,300
	Geese	200
Indonesia	Ducks	2,400
	Herons	38,000
	Others	3,600
Japan (1981)	Ducks	694,646
USSR (E. Siberia)	Ducks	3,420,000
West Java (Cirebon)	Crakes and rails	170,000

Not all wetlands produce all types of resources, of course, but most will produce a wide variety, particularly larger sites like the Pantanal of Brazil, the Florida Everglades, the Kafue Flats in Zambia and the 6,000 km^2 Sundarbans in Bangladesh and India which are the subject of a case study. The economic value of many wetlands is decreased by the presence of noxious animals, particularly mosquitoes, sandflies and midges, some of which act as vectors of disease. There can be a tremendous diversity and abundance of these insects: some 84 mosquito, 13 sandfly and 21 horsefly species are reported from the freshwater Nariva Swamp, Trinidad, while 183,000 individual mosquitoes of a single species were caught in a light trap one night in a Cayman Islands mangrove swamp. However, their nuisance potential must not be allowed to serve as an excuse to destroy their important role as links in aquatic, and to a lesser extent terrestrial food webs, particularly their larvae which are eaten by commercially important fish.

Links Between Wetlands and Other Habitats

In addition to direct economic values, through the provision of a range of goods and services, wetlands are of great indirect value through linkages with associated aquatic ecosystems. As indicated above, many species use wetlands for nursery purposes. In addition, the transfer of organic matter and biota by downstream flow or tidal export influences nutrient status and food webs outside the wetland itself.

In Australia, Banana Prawns Penaeus merguiensis require mangrove-lined estuaries if they are to complete their life cycles; in Colombia, the Cienaga Grande lagoon is thought to be responsible for rearing 70% of the fish harvested on the Caribbean coast; the organic matter and nursery environment of the Laguna de Terminos, Mexico, support a coastal fishery producing annually 15,000 tonnes of shrimp, 13,000 tonnes of shellfish and 122,000 tonnes of fish.

Mention has been made above of migratory waterfowl utilization of wetlands as staging posts, an example of a wetland in one country supporting the biodiversity and commercial harvest of resources in another, often in a different biome in a distant country. In Jamaica, the close association between mangroves and coral reefs, in terms of exchanges of nutrients and biota between the two wetland types, suggests that the presence of mangroves greatly influences the health and productivity of the reefs which are the mainstay of the artisanal fishing industry.

The Consequences of Loss of WetLand Biodiversity

It is obvious from the large number of resource organisms mentioned earlier, that loss of wetland species has economic implications. The livelihood and culture of large numbers of people, in almost every country of the world, will be endangered if wetland resources become further depleted. A major portion of fisheries production, most hunting, much forest production and a significant part of ecotourism will be lost worldwide, as well as elements of heritage and environmental quality. It is important to stress, however, that it is not sufficient just to protect the populations of plants and animals that are directly exploited: their health and survival, or sustainability, depend on maintaining the whole complex of biodiversity that characterizes wetland ecosystems. Commercially exploitable wetland plant and animal species will be available only if the biological processes which produce them are maintained. These include primary production, nutrient cycling, pollination, flowering, fruiting, decomposition, food web interactions, grazing, predation, immigration and emigration, to name a few. Hundreds of inter-related organisms take part in this gamut of processes and it is this diversity of wetland species which keeps these ecosystems in ecological equilibrium and makes them so productive. Loss of any link in the web of biodiversity will reduce the goods, functions and attributes of a wetland site.

Decline in a wetland will impact on associated systems: loss of nursery habitat could reduce coastal fishery yields or loss of a wetland on a flyway could disrupt waterfowl migrations, threatening the capacity of individual birds to reproduce and eventually the survival of populations or species.

MARINE BIOLOGY

Marine biology is the scientific study of organisms in the ocean or other marine or brackish bodies of water. Given that in biology many phyla, families and genera have some species that live in the sea and others that live on land, marine biology classifies species based on the environment rather than on taxonomy. Marine biology differs from marine ecology as marine ecology is focused on how organisms interact with each other and the environment, and biology is the study of the organisms themselves.

Marine life is a vast resource, providing food, medicine, and raw materials, in addition to helping to support recreation and tourism all over the world. At a fundamental level, marine life helps determine the very nature of our planet. Marine organisms contribute significantly to the oxygen cycle, and are involved in the regulation of the Earth's climate. Shorelines are in part shaped and protected by marine life, and some marine organisms even help create new land. Marine biology covers a great deal, from the microscopic, including most zooplankton and phytoplankton to the huge cetaceans (whales) which reach up to a reported 48 meters (125 feet) in length.

The habitats studied by marine biology include everything from the tiny layers of surface water in which organisms and abiotic items may be trapped in surface tension between the ocean and atmosphere, to the depths of the oceanic trenches, sometimes 10,000 meters or more beneath the surface of the ocean. It studies habitats such as coral reefs, kelp forests, tidepools, muddy, sandy and rocky bottoms, and the open ocean (pelagic) zone, where solid objects are rare and the surface of the water is the only visible boundary. A large proportion of all life on Earth exists in the oceans. Exactly how large the proportion is unknown, since many ocean species are still to be discovered. While the oceans comprise about 71% of the Earth's surface, due to their depth they encompass about 300 times the habitable volume of the terrestrial habitats on Earth.

Many species are economically important to humans, including food fish. It is also becoming understood that the well–being of marine organisms and other organisms are linked in very fundamental ways. The human body of knowledge regarding the relationship between life in the sea and important cycles is rapidly growing, with new discoveries being made nearly every day. These cycles include those of matter (such as the carbon cycle) and of air (such as Earth's respiration, and movement of energy through ecosystems including the ocean). Large areas beneath the ocean surface still remain effectively unexplored.

SUBFIELDS

The marine ecosystem is large, and thus there are many sub-fields of marine biology. Most involve studying specializations of particular animal groups, such as phycology, invertebrate zoology and ichthyology. Other subfields study the physical effects of continual immersion in sea water and the ocean in general, adaptation to a salty environment, and the effects of changing various oceanic properties on marine life. A subfield of marine biology studies the relationships between oceans and ocean life, and global warming and environmental issues (such as carbon dioxide displacement). Recent marine biotechnology has focused largely on marine biomolecules, especially proteins, that may have uses in medicine or engineering. Marine environments are the home to many exotic biological materials that may inspire biomimetic materials.

Related Fields

Marine biology is a branch of oceanography and is closely linked to biology. It also encompasses many ideas from ecology. Fisheries science and marine conservation can be considered partial offshoots of marine biology (as well as environmental studies).

LIFEFORMS

Microscopic Life

Microscopic life undersea is incredibly diverse and still poorly understood. For example, the role of viruses in marine ecosystems is barely being explored even in the beginning of the 21st century. The role of phytoplankton is better understood due to their critical position as the most numerous primary producers on Earth. Phytoplankton are categorized into cyanobacteria (also called blue–green algae/bacteria), various types of algae (red, green, brown, and yellow–green), diatoms, dinoflagellates, euglenoids, coccolithophorids, cryptomonads, chrysophytes, chlorophytes, prasinophytes, and silicoflagellates. Zooplankton tend to be somewhat larger, and not all are microscopic. Many Protozoa are zooplankton, including dinoflagellates, zooflagellates, foraminiferans, and radiolarians. Some of these (such as dinoflagellates) are also phytoplankton; the distinction between plants and animals often breaks down in very small organisms. Other zooplankton include cnidarians, ctenophores, chaetognaths, molluscs, arthropods, urochordates, and annelids such as polychaetes. Many larger animals begin their life as zooplankton before they become large enough to take their familiar forms. Two examples are fish larvae and sea stars (also called starfish).

Plants and Algae

Plant life is widespread and very diverse under the ocean. Microscopic photosynthetic algae contribute a larger proportion of the worlds photosynthetic output than all the terrestrial forests combined. Most of the niche occupied by sub plants on land is actually occupied by macroscopic algae in the ocean, such as Sargassum and kelp, which are commonly known as seaweeds that creates kelp forests. The non algae plants that survive in the sea are often found in shallow waters, such as the seagrasses (examples of which are eelgrass, Zostera, and turtle grass, Thalassia). These plants have adapted to the high salinity of the ocean environment. The intertidal zone is also a good place to find plant life in the sea, where mangroves or cordgrass or beach grass might grow. Microscopic algae and plants provide important habitats for life, sometimes acting as hiding and foraging places for larval forms of larger fish and invertebrates.

Marine Invertebrates

As on land, invertebrates make up a huge portion of all life in the sea. Invertebrate sea life includes Cnidaria such as jellyfish and sea anemones; Ctenophora; sea worms including the phyla Platyhelminthes, Nemertea, Annelida, Sipuncula, Echiura, Chaetognatha, and Phoronida; Mollusca including shellfish, squid, octopus; Arthropoda including Chelicerata and Crustacea; Porifera; Bryozoa; Echinodermata including starfish; and Urochordata including sea squirts or tunicates.

Fish

Fish have evolved very different biological functions from other large organisms. Fish anatomy includes a two–chambered heart, operculum, swim bladder, scales, fins, lips, eyes and secretory cells that produce mucous. Fish breathe by extracting oxygen from water through their gills. Fins propel and stabilize the fish in the water. Well known fish include: sardines, anchovy, ling cod, clownfish (also known as anemonefish), and bottom fish which include halibut or ling cod. Predators include sharks and barracuda.

Reptiles

Reptiles which inhabit or frequent the sea include sea turtles, sea snakes, terrapins, the marine iguana, and the saltwater crocodile. Most extant marine reptiles, except for some sea snakes, are oviparous and need to return to land to lay their eggs. Thus most species, excepting sea turtles, spend most of their lives on or near land rather than in the ocean. Despite their marine adaptations, most sea snakes prefer shallow waters nearby land, around islands, especially waters that are somewhat sheltered, as well as near estuaries. Some extinct marine reptiles, such as ichthyosaurs, evolved to be viviparous and had no requirement to return to land.

Seabirds

Seabirds are species of birds adapted to living in the marine environment, examples including albatross, penguins, gannets, and auks. Although they spend most of their lives in the ocean, species such as gulls can often be found thousands of miles inland.

Marine Mammals

There are five main types of marine mammals:

1. Cetaceans include toothed whales (Suborder Odontoceti), such as the Sperm Whale, dolphins, and porpoises such as the Dall's porpoise. Cetaceans also include baleen whales (Suborder Mysticeti), such as the Gray Whale, Humpback Whale, and Blue Whale.
2. Sirenians include manatees, the Dugong, and the extinct Steller's Sea Cow.
3. Seals (Family Phocidae), sea lions (Family Otariidae—which also include the fur seals), and the Walrus (Family Odobenidae) are all considered pinnipeds.
4. The Sea Otter is a member of the Family Mustelidae, which includes weasels and badgers.
5. The Polar Bear (Family Ursidae) is sometimes considered a marine mammal because of its dependence on the sea.

MARINE HABITATS

Marine habitats can be divided into coastal and open ocean habitats. Coastal habitats are found in the area that extends from the shoreline to the edge of the continental shelf. Most marine life is found in coastal habitats, even though the shelf area occupies only seven per cent of the total ocean area. Open ocean habitats are found in the deep ocean beyond the edge of the continental shelf

Alternatively, marine habitats can be divided into pelagic and demersal habitats. Pelagic habitats are found near the surface or in the open water column, away from the bottom of the ocean. Demersal habitats are near or on the bottom of the ocean. An organism living in a pelagic habitat is said to be a pelagic organism, as in pelagic fish. Similarly, an organism living in a demersal habitat is said to be a demersal organism, as in demersal fish. Pelagic habitats are intrinsically shifting and ephemeral, depending on what ocean currents are doing. Marine habitats can be modified by their inhabitants. Some marine organisms, like corals, kelp and seagrasses, are ecosystem engineers which reshape the marine environment to the point where they create further habitat for other organisms.

Intertidal and Shore

Intertidal zones, those areas close to shore, are constantly being exposed and covered by the ocean's tides. A huge array of life lives within this zone. Shore habitats span from the upper intertidal zones to the area where land vegetation takes prominence. It can be underwater anywhere from daily to very infrequently. Many species here are scavengers, living off of sea life that is washed up on the shore. Many land animals also make much use of the shore and intertidal habitats. A subgroup of organisms in this habitat bores and grinds exposed rock through the process of bioerosion.

REEFS

Reefs comprise some of the densest and most diverse habitats in the world. The best–known types of reefs are tropical coral reefs which exist in most tropical waters; however, reefs can also exist in cold water. Reefs are built up by corals and other calcium–depositing animals, usually on top of a rocky outcrop on the ocean floor. Reefs can also grow on other surfaces, which has made it possible to create artificial reefs. Coral reefs also support a huge community of life, including the corals themselves, their symbiotic zooxanthellae, tropical fish and many other organisms.

Much attention in marine biology is focused on coral reefs and the El Niño weather phenomenon. In 1998, coral reefs experienced the most severe mass bleaching events on record, when vast expanses of reefs across the world died because sea surface temperatures rose well above normal. Some reefs are recovering, but scientists say that between 50% and 70% of the world's coral reefs are now endangered and predict that global warming could exacerbate this trend.

Open Ocean

The open ocean is relatively unproductive because of a lack of nutrients, yet because it is so vast, in total it produces the most primary productivity. Much of the aphotic zone's energy is supplied by the open ocean in the form of detritus. The open ocean consists mostly of jellyfish and its predators such as the mola mola.

Deep Sea and Trenches

The deepest recorded oceanic trenches measure to date is the Mariana Trench, near the Philippines, in the Pacific Ocean at 10,924 m (35,838 ft). At such depths, water pressure is extreme and there is no sunlight, but some life still exists. A white flatfish, a shrimp and a jellyfish were seen by the American crew of the bathyscaphe Trieste when it dove to the bottom in 1960.

Other notable oceanic trenches include Monterey Canyon, in the eastern Pacific, the Tonga Trench in the southwest at 10,882 m (35,702 ft), the Philippine Trench, the Puerto Rico Trench at 8,605 m (28,232 ft), the Romanche

Trench at 7,760 m (24,450 ft), Fram Basin in the Arctic Ocean at 4,665 m (15,305 ft), the Java Trench at 7450 m (24,442 ft), and the South Sandwich Trench at 7,235 m (23,737 ft).

In general, the deep sea is considered to start at the aphotic zone, the point where sunlight loses its power of transference through the water. Many life forms that live at these depths have the ability to create their own light a unique evolution known as bio-luminescence. Marine life also flourishes around seamounts that rise from the depths, where fish and other sea life congregate to spawn and feed.

Hydrothermal vents along the mid–ocean ridge spreading centers act as oases, as do their opposites, cold seeps. Such places support unique biomes and many new microbes and other lifeforms have been discovered at these locations.

DISTRIBUTION FACTORS

An active research topic in marine biology is to discover and map the life cycles of various species and where they spend their time. Marine biologists study how the ocean currents, tides and many other oceanic factors affect ocean lifeforms, including their growth, distribution and well–being.This has only recently become technically feasible with advances in GPS and newer underwater visual devices.

Most ocean life breeds in specific places, nests or not in others, spends time as juveniles in still others, and in maturity in yet others. Scientists know little about where many species spend different parts of their life cycles. For example, it is still largely unknown where sea turtles and some sharks travel. Tracking devices do not work for some life forms, and the ocean is not friendly to technology.

This is important to scientists and fishermen because they are discovering that by restricting commercial fishing in one small area they can have a large impact in maintaining a healthy fish population in a much larger area far away.

BIOLOGICAL CHARACTERISTICS OF MARINE ECOSYSTEMS

Brackishwater estuaries are the meeting point of the fauna from three different ecosystems. Brackishwater species, *e.g. Acetes indicus, Raconda russeliana, etc.* live, grow and spawn in the same environment, while the marine and freshwater fauna use the brackishwater low-saline, nutrient-rich area as nursery grounds and visit only for a short time.

LIFE CYCLE PATTERN OF PENAEID SHRIMPS

Penaeid shrimp live and spawn in offshore waters beyond the 40-metre depth zone. The larvae enter the estuarine habitat with the tidal currents. There they find an environment suitable for feeding and living at that age. As they grow bigger, their physical and biological demands change gradually and, so, they move back towards the sea. When they reach the parent stock they have grown to adulthood and have matured to participate in the spawning process, thus completing the life cycle.

During the different phases of their life cycle, the penaeid shrimp population encounter a variety of fishing gear in different areas and depths of the sea starting with shrimp fry collecting gear at the PL stage, followed by the estuarine set bagnet (ESBN) and beach seine in the open brackish waters and estuaries, when they are at the juvenile stage. The survivors are then captured by the marine set bagnet (MSBN) at post-juvenile and pre-adult stages, by the trammel nets at adult size and, finally, the residual population is caught by the shrimp trawlers in the marine environment. Thus, members of a single stock are exposed to a multigear fisheries exploitation system.

ARTISANAL FISHERIES RESTRICTING OFFSHORE RECRUITMENT

The artisanal fishing gear operated in the open brackishwater environment - the ESBN, the pushnet and the beach seine - catches only the post-larvae and juveniles of the marine fauna in huge numbers, thereby restricting their recruitment in the open sea at an adult stage. The degree of such restriction by size and number would be evident from Figure which shows *P. monodon* (Tiger shrimp) harvested by different interactive fishing gear. These gear, in this fashion cause imbalance in the food chain and in the overall ecology of marine habitat.

INDUSTRIAL FISHERY REDUCING RECRUITMENT AND KILLING SPAWNERS

The trawler fleet, although not permitted by rules and ordinance to fish at depths shallower than 40m, normally fish upto 30 m and even upto 20m depth. As their gear is non-selective, they too harvest sizes of fish and shrimp which fall under the post-juvenile and pre-adult categories, thereby restricting adult recruitment of a part of the population.

Penaeid shrimp spawn throughout the year, with two peak seasons (January-Feb and July-August) when more than 90 per cent of the population in the spawning ground show the signs of full ripeness. To ensure a larval population for the next generation, the brooders need to be saved from fishing mortality during there periods. But the trawler fleet randomly harvests the brood shrimp during the peak spawning season. Thus, the larval population is getting reduced year after year.

EXPLOITATION PATTERN AS AGAINST MSY OF VARIOUS STOCKS

Analysis of the catch effort as well as the biological data for the growth and mortality during the last decade (under DOF) provides information on the pattern of exploitation of various fish and shrimp stocks by different fishing gear. Some gear were found to be totally destructive, while the others were either partially harmful or proved to be biologically sustainable.

EXPLOITATION BY INDUSTRIAL (TRAWL) FISHERY

The effort in the trawl fishery during the last decade and a half has been around 5000-6000 standard fishing days, to produce 3,500-6,000 t. of shrimp. The MSY of penaeid shrimp is 7,000 t. and the optimum effort for producing this amount is 7,000-8000 standard days. In some years it was around the level of MSY. Some effort was lost due to a major cyclone in April 1991. Thereafter, till date shrimp production has been much below the MSY level. At present there are 53 trawlers (41 shrimp and 12 finfish) which gave a total annual effort of about 5,000 standard days. This is at least 1000 days less than the optimum, mainly because ten of the trawlers are too old and their present average effort is 62 days/year/trawler (250 days standard; Rahman and Khan, 1995). If these old trawlers are made effective, the total effort would reach the optimum level. Whitefish landed by the trawler fleet is in the range of 8,000-12,000 t, which is only 20 per cent of the actual catch; while 80 per cent, equivalent to 35-45,000 t, is discarded dead in the sea. Even if the discarded amount is considered as production, the MSY is not being achieved. The MSY is 85,000 mt. Although the trawl fishery is below the optimum effort, tiger shrimp, the targeted species, has been over-exploited as the gear is non-selective. On the other hand, the brown shrimp, *M. monoceros,* shows an upward trend.

EXPLOITATION BY SEMI-INDUSTRIAL FISHERIES

Introduction of engine and nylon twine prior to and after liberation have made it possible to exploit even more the unexploited resources. One such development is the small mesh drift gillnetting for *Hilsa.* Production was raised by at least 1,00,0001 over-night. Till date (1995), no report on overfishing of *Hilsa is* available. Then there are the bottom long-lining and marine set bagnet fishing operations with motorized boats. These fisheries do not provide sufficient indication of overfishing, but capture of *Hilsa* spawn and *jatka* is a concern for management. Over-fishing, however, is noticed in the brooder exploitation of Indian salmon and long Jewfish (*Lakha and Lambu*) with large mesh driftnet (LMD).

EXPLOITATION BY ARTISANAL FISHERIES

Detailed catch assessment and biological information on the pattern of exploitation by this fishery is available (with DOF). It is evident that this fishery is the most destructive fishery of the natural resource. The natural mortality, fishing mortality and exploitation pattern of the 19 most significantly occurring species of this fishery, covering three different ecosystems. It can be seen that the species of brackishwater origin, *i.e. Acetes indicus* (the sergestid shrimp), *Raconda russeliana* and *Setipinna taty* are not over-harvested, rather they are underfished to some extent, while almost all species of marine and freshwater origin which visit the brackishwater area for nursery and breeding purpose are seriously overfished (growth overfishing). It can be seen in Figure that all the shrimp are caught by this gear before the adult stage and are thus not permitted to join in the spawning process.

More than 2035 million *Bagda* post-larvae are collected annually, which is only a little over one per cent of the total catch of the fishery. The rest of the catch equivalent to about 200 billion PL of the shrimp/fish and zoo plankton is thrown on the sand to die. This is serious growth overfishing.

The trammel net is at present operated only on a limited scale along the Teknaf coast. Since this gear is operated with country boats, the fishing is on artisanal scale, but the technique is modern. This is a selective gear and biological studies show that the exploitation pattern is below the optimum level, though the size at first capture is definitely above the optimum. The present management is largely concentrated on the industrial trawl fishery. The brackishwater and marine fisheries are not managed very much.

MULTIGEAR AND MULTISPECIES FISHERIES MANAGEMENT

Bangladesh has the problem of not only tropical multispecies fisheries but also multigear fisheries. Any single fishing operation catches a number of species at different sizes and ages on the one hand. On the other hand, members of the population of one species belonging to a single stock become available to different interactive fishing gear, in a sequential order. So each of them need to be accounted, as it is not possible to carry out an assessment on the basis of a single fishery alone. For instance, a large increase in artisanal fishing pressure significantly lowers the number of shrimp reaching the industrial trawling grounds and results in lower overall catches in the industrial fleet. This is unrelated to change in industrial fleet size. A situation like this preceded a temporary collapse of the shrimp stocks in the gulf.

The situation in the Bangladesh coastal shrimp fishery is similar and the penaeid shrimp stock is under pressure from all sides. It can be seen that the ESBN, pushnets and beach seine harvest the members of the same population at sizes much lower than the size at first maturity and, as a result, about 99 per cent of the population do not get a chance to participate in the spawning process. It was estimated that out of the total penaeid shrimp harvested, ESBN comprises 55.87 per cent, the trawler fleet 29.70 per cent, MSBN 14.30 per cent beach seine 0.09 per cent and the pushnet only 0.04 per cent by weight. The same production if converted into number shows a reversed situation *i.e.* the trawler fleet takes only 1 per cent, ESBN 3.4 per cent and the pushnetters alone take 94.6 per cent. Thus it is evident that the enhanced production in the trawl fishery largely depends not on the trawl fishery alone but on the management of the three artisanal fisheries.

As a regulatory approach, a 30 mm codend mesh size regulation was introduced in the ordinance (not enforced) for ESBN, assuming that the juveniles would escape. But with the availability of valid scientific information now, it is evident that this gear targeted the juveniles and that increased mesh size virtually results in no catch. So there is no alternative to complete withdrawal of this gear from the estuarine environment. Overnight withdrawal of this gear would not be possible, as it would force 50,000 fisherfolk, who live below the poverty line, into starvation. So, alternate employment, to ensure their livelihood, needs be identified inside or outside the fisheries sector. New fisheries development and expansion of the presently sustainable fisheries with the direct participation of ESBN fisherfolk would be necessary.

Gradual withdrawal of effort in the fishery needs to be pursued. As a first step, closure of fishing during the peak recruitment periods, *i.e.* July to September and February to April in the Cox's Bazar area would substantially reduce the juvenile mortality, since the Cox's Bazar district alone accounts for 46 per cent of the total destruction of juvenile penaeid shrimp. Large mesh driftnets (LMD) have been operated in the Cox's Bazar region for the last two decades, targeting the Indian Salmon (*Lakhua*) and Long Jewfish (Lambu). These species have been fished so much that they are nearly extinct now. So, use of this gear needs to be stopped for a few years and the situation monitored. Beach seine in estuaries should be banned.

FISHING METHOD AND REDUCTION OF MORTALITY IN SHRIMP PL FISHERY

Since the valuable coastal shrimp culture industry is dependent on wild seed, in the absence of hatcheries, and since this industry earns more foreign exchange than the capture fishery at present, it may not be possible at this stage to stop this fishery. So, with a view to catalyze reduced collection from Nature, some management measures. The by-catch in the Tiger shrimp PL fishery is about 200 billion plankton. These are destroyed, with no benefitters, and result in reduced standing stock in the sea year after year. An alternate method of collection must be tried, targetting only tiger shrimp PL. For example, a water hyacinth type of method may be experimented within which particular shrimp larvae will stick to a flora introduced. This will save much of the bycatch.

Upto 60 per cent of the PL collected from Nature die during sorting, transporting and stocking, forcing harvests of PL from Nature. If this mortality could be reduced substantially, 50 per cent of the PL could be left behind in the sea to give enhanced production. Strong extension and motivation campaigns are necessary for this. PL's are overstocked in ponds for no substantial gain. So fishing for PL in the months other than the peak stocking season should be restricted. At present, now 98 per cent of the shrimp culture industry is dependent on natural seed. Only two private hatcheries are in production. Many more hatcheries are needed to support the rapidly expanding culture industry. Transition to semi-intensive farming should not be permitted without the direct support of hatcheries. Otherwise both the goose and the golden eggs will undoubtedly disappear.

EXPANSION AND EXTENSION OF THE TRAMMEL NET FISHERY

The trammel net fishery has proved to be the biologically most sustainable fishery. At present this fishery is in operation at Teknaf coast only. Feasibility and technical demonstration of this gear in the western part of the coast may help expand this fishery in the other areas to produce more fish and would enhance economic as well as socio-economic benefits. This gear is presently operated with traditional country boats and hence cannot go far. There is need to extend this fishery vertically upto 40 m depth of water and there is room for such expansion. But it would need study on the biological and economic feasibility and hopefully would add to the sustainable increase in production and would create an alternate source of income for the ESBN fisherfolk. Technological improvement may also be required.

Bottom long lining for croaker off the Cox's Bazar coast has created a good employment opportunity for the fisherfolk in that area. This is an export oriented venture and very feasible. This fishery needs to be expanded in the other areas of the coast as a substitute employment for the ESBN fishermen, as well as to produce more fish.

The marine *behundi* fishery is a seasonal activity and localized in Sonadia (Cox's Bazar), Dubla (Khulna) and Sonarchar (Patuakhali). This fishery catches a good percentage of pre-adult shrimps, but its target species is ribbonfish and Bombay Duck which are harvested mainly in the adult phase. Some of the larger ESBN may be encouraged to operate in deeper waters from other island bases (*e.g.* St. Martin Island) with mechanized boats in order to reduce effort in the ESBN fishery. This fishery has recently turned export-oriented and new-comers are joining in every year. It would be necessary to take immediate steps to stop new entry and utilize the space for partial rehabilitation of the ESBN fishery.

MANAGEMENT OF THE INDUSTRIAL MARINE FISHERY

Optimum fishing effort for MSY is more than 7000 standard days, but the effort at present is at least 1000 days less than that. If the ten inefficient trawlers are made efficient, effort will be at optimum level or else new effort would need be added. One shrimp trawler is equivalent to 3.5 fish trawlers in terms of shrimp catch efficiency. So 1000 standard days is equivalent to four shrimp trawlers or 14 fish trawlers. If all the trawler effort is converted to finfish trawlers, it is estimated that the total production of both finfish and shrimp would be at MSY level. The benefit would be mainly in the assurance of landing all whitefish caught, because fish trawlers hardly make any discards. As a result, an additional amount of at least 50,000 t. of fish would be landed. The present rule for season closure must be implemented to ensure breeding in the peak season and the 40 m depth zone restriction should be enforced strictly to facilitate post-juveniles being recruited as adults. Competition has increased over the years for investment in the demersal and artisanal fisheries sectors, both of which are already overburdened, while the pelagic resources, *e.g.* the Tuna and Tunalike fish resources have remained untapped. *Hilsa is* the only pelagic species categorized are present in the fish catch statistics of Bangladesh. Eight species of Tuna and Skipjack and a number of potential species of Mackerels, Shark, Ray, Sardines, Anchovies, Shad and several species of cephalopods. Soles and flat-fish, lobster etc. are available in Bangladesh waters. Development of these resources will open a new era for the Bangladesh fisheries sector, to boosting production sustainably. But a detailed survey and assessment of stock and MSY would be necessary, along with feasibility and technological demonstration, for transfer to the private sector.

Being encouraged by the positive indication of the yield as well as the belief revenue per recruit in the trammel and trawl fisheries that in the ESBN fisheries, a 'long-term prediction' analysis was done with a view to find the comparative economic gain from various fisheries if only one is allowed to operate and all the other fisheries are suppressed, or if only the ESBN is suppressed and the others allowed to operate as they are.

The analysis indicated that if the trawl fishery is kept and all other interactive fisheries (but not pushnet) are suppressed, there would be substantial gain in weight and about 300 per cent gain in value of the catch, while there would be 250 per cent gain in value of the catch. There would be 250 per cent gain in value of the catch if only ESBN fishery is not in operation.

On the other hand, trammelnet showed an extremely high gain (about ten times) in revenue when all fisheries (including trawlers, but not the pushnet fishery) are suppressed, but a smaller gain in yield and a large gain in revenue (300%) if only the ESBN is suppressed. So, from different magnitude analyses, it is evident that withdrawal of the ESBN fishery would not only maintain a healthier stock but would give a substantial difference in economic return. The pushnet (larval) fishery was kept out of this analysis because data necessary for such analysis were not available. But since about 95 per cent of the exploited population is taken by the larval fishery alone, suppression of this fishery would definitely give a manifold higher economic return. This perception would definitely lead fishery managers/planners to avail of the opportunity to achieve manifold higher economic returns from the same stock by only changing the traditional fishing attitudes.

In 1983, the Government of the Peoples Republic of Bangladesh enacted the Marine Fisheries Rules, 1983, in accordance with the provisions of the Marine Fisheries Ordinance, 1983. Under the provisions of the Ordinance, the Marine Fisheries Wing of the DOF is authorized to deal with matters relating to marine fisheries exploitation, including licensing and monitoring of operations of fishing vessels. The Marine Fisheries Rules amended in 1993 provide for licensing and monitoring of artisanal mechanized fishing boats as well. Under the Ordinance, the officers of the Marine Wing of the DOF have been empowered to check, seize or take appropriate action required for surveillance and enforcement of the rules of the Ordinance. These activities are performed from the Marine Fisheries checkpost established under the DOF Marine Wing at Patenga bay front, Chittagong.

The Ministry of Industry is currently authorized to accord permission in consultation with the MOFL, for the acquisition of fishing trawlers. The mechanized fishing vessels are registered with the Mercantile Marine Department (MMD). To patrol the EEZ, the DOF has procured two modem gunboats and placed them under the operational control of the Bangladesh Navy. Besides the MOFL, other ministries directly involved are the Ministry of Land, Ministry of Industries, LGRD and MOEF.

MANAGEMENT OF THE MARINE FISHERIES RESOURCES

Fisheries administration and management primarily remain under the control of the MOFL, headed by a cabinet minister. The Department of Fisheries (DOF) is the key organization responsible for development and management of fisheries. The Bangladesh Fisheries Development Corporation (BFDC) was established in 1964 with a view to promoting the fishing industry and developing landing, preservation and processing facilities, particularly in the marine sector. A part of the survey and exploratory work was once included in the mandate of the BFDC, but is now carried out by the Marine Fishery Survey Project, DOF. The Fisheries Research Institute (FRI) was established in 1984, as an autonomous body under the administrative control of the MOFL.

Research stations and ancillary facilities of the DOF were, subsequently, transferred to the FRI by an administrative order of the Government. The mandate of the FRI is to plan and undertake adaptive research programmes to develop suitable technology for fish farmers and fishery managers. But survey work producing information on resource monitoring and management remains with the DOF for practical reasons. This work is at present done by a permanent set up of the DOF, called the Marine Fishery Resources Management and Monitoring unit. The unit is based in Chittagong and has two marine and brackishwater research vessels and other equipments. Several

NGOs and fishermen's cooperatives are involved in marine fisheries development activities in the country. The Bangladesh Jatiyo Matshyajibi Samabay Samity (BJMSS), for example, had direct involvement in marine fisheries development, but is now ineffective. Among the NGOs, Codec, Caritas, and Proshika-MUK are directly involved in the development of the coastal fisherfolk community.

INFRASTRUCTURE AND SERVICE FACILITIES

Infrastructural and service facilities are inadequate. In the absence of proper landing centres artisanal fishermen land their catches at scattered places which do not have processing, marketing or fast transportation facilities. Only the industrial trawler fleet (public and private) lands at defined places. Some of the mechanized boats catching *Hilsa* land at a few landing centres of the BFDC. The other private landing places do not have adequate ice, freshwater, berthing bunkering facilities.

The BFDC operates four landing centres, in Cox's Bazar, Khulna, Barisal and Patuakhali. It has modem landing, preservation, ice, water, berthing and bunkering facilities at Chittagong when it uses it for its own fleet. It also extends services to private operators. Such landing centres need be developed in every coastal district and in other important landing areas.

INSTITUTIONAL STRENGTH FOR FISHERIES DEVELOPMENT, MANAGEMENT AND RESEARCH

The DOF Marine Wing has two establishments. One is for survey and monitoring and to provide management information. But its few scientific staff are hardly enough to carry out the task of operating two research vessels as well as do land-based work for routine collection and processing of industrial, semi-industrial and artisanal, statistical and biological data. The other establishment is the law enforcement and legislative unit. This too does not have adequate manpower or training to undertake its task along the entire coastline, particularly in respect of the marine artisanal fishing operations. The BFDC's activities have been reduced over the last decade and it now maintains only the previously established facilities, which provide hardly any opportunities for further development.

The FRI has two stations for marine and brackishwater research, one in Cox's Bazar and the other in Paikgacha, Khulna. The scientific staff is limited, and inadequate training and exposure limit proper programmes for marine and brackishwater research.

RULES OF THE MARINE FISHERIES ORDINANCE

The Marine Fisheries Ordinance and subsequent rules were prepared more than a decade ago. Since then, more knowledge about the exploitation dynamics of marine fishery resources has been gathered.

In the light of this knowledge and the changes in fishing pressure, as discussed in the text, certain amendments in the rules are called for:

- The provision exists (although not enforced) for a 30 mm codend mesh size limit for ESBN. At present, 8-12 mm mesh is used. Results of recent experimentation and investigation reveal that, the 30 mm codend would bring in no catch, as they target the juvenile. Further it will be non-remunerative for the fisherfolk and will also involve further cash investment for change with no benefit to, the fishers. So mesh size increase or improvement of gear design would not be helpful for management. This is why the complete withdrawal of this gear from the estuarine habitat is necessary. This needs be included in the Marine Fisheries Rules.
- As an immediate measure for reduction of effort, ESBN operations in the Cox's Bazar district should be stopped during the July-September, and February-April period.
- Operation of LMD in shallow waters should be banned until further orders.
- The marine set bagnet should use 45mm codend mesh size.
- Net making/mending factories should produce nets not below 45mm mesh and they should be encouraged to make nets suitable for trammelnetting.
- Limitations should be imposed on the boatbuilding yards not to build boats (mechanized) for fisheries without the permission of the Government.
- Aging shrimp trawlers should not be replaced with shrimp trawlers, but with fish trawlers, in the light of the discussions in the text.
- Rule 7, Clause h: (landing of 30 per cent of the total catch) is rather a vague term and also a very small figure. It needs to be replaced as "50 per cent of the fish catch, equivalent to five times of the shrimp catch", because a shrimp trawler's finfish by-catch is 8-10 times its shrimp catch.
- Rule 6 prescribes the licence fee for fishing vessels. Fees for resource monitoring and management (including operation of research vessels) should also be included, as this is now a routine and continuing programme under the revenue budget. Some recompense should now come from the beneficiaries.

RESOURCE MANAGEMENT AND SURVEILLANCE WORK

Other management and legislative measures which are not under the direct control of the DOF/MOFL need interministerial and interdepartmental decisions and arrangements. Some of these measures which deserve special attention.

- There are waterbodies under the control and ownership of the ministries other than fisheries. Some of these waterbodies/areas are directly 'managed' by administrative units such as the Ministry of

Land, Department of Forests, and MOEF. These are revenue oriented managements which collect produce or lease the water areas. But since fisheries resources are a living renewable resource, biological management, based on research findings and scientific information, would be necessary, irrespective of which agency/organization owns the land, water or the fish. The Department of Fisheries must be entrusted with the responsibility for such management because it also has the capability to handle fisheries-oriented duties.

The various water development activities have altered much of the ecological habitat concerning fisheries, so any development or management activity (non-biological) in such waterbodies, for purposes other than fisheries, should be done only in close consultation with the MOFL/DOF to ensure a healthy environment for the growth of the fish population. A high power steering committee for Integrated Coastal Zone Management, with a respected position for MOFL/DOF, needs to be established.

- To minimize the degradation of the coastal environment by industrial, agrochemical, oil and other pollution, integrated research - to qualify and quantify the toxic effect on fish - needs be undertaken immediately. Such research should identify the pesticides which are not water soluble or do not have a toxic effect on aquatic life and limit the other brands of fertilizer and pesticides. Waste treatment facilities need be established in the coastal districts and rules need be reformulated to take care of aquatic life. Legislation must be improved. Co-ordination is necessary between the MOFL and MOEF.
- Mangrove afforestration programmes should be undertaken jointly by the DOF and DOF (Forest) to save the environment from further degradation.
- In the Marine Fisheries Ordinance subsequent rules have been made for licensing of artisanal fishing boats by the DOF. The Mercantile Marine Department (MMD) deals with the registration. But it is evident that all boats are not registered and there is no enforcement to make them comply The MMD also looks after the safety of life at sea and issues vessel health certificates after checking the safety equipment. Fishermen, it is reported, are not interested in observing formalities of registration and licensing with two similar departments. The DOF has the capability to check craft health and safety equipment. The two functions need to be put under the DOF for easy monitoring and enforcement.
- There are gear and area conflicts between artisanal and industrial fishermen. There are reports of sea piracy and resource use conflicts. The Coast Guard is required to solve these problems. Resources survey, research, monitoring, surveillance and management in the

area of marine fishries is a very big task, involving activities not only within the sector but also with other sectors. So the capabilities of the organizations under the MOFL, particularly in respect of marine and brackishwater fisheries, need to be strengthened greatly.

DEPARTMENT OF FISHERIES (DOF)

The DOF performs two main functions in respect of marine fisheries viz:

1. It does resource assessment, monitoring of the fishing dynamics and stock assessment for sustainable gear development and expansion. The findings of this routine work enables the DOF to offer continuous advice to fisheries planners on the options and strategies for management available to them.
2. The DOF implements and enforces the rules under the Marine Fisheries Ordinance 1983. To do this, the DOF has two units, but these are managed by a limited scientific and enforcement staff, which is inadequate for the present task, leave alone any expansion in the future.

The Marine Fisheries Resources, Monitoring and Management Centre has recently decided to established a permanent set-up. The proposal document identifies some urgent programmes for facilities and manpower that will enable the entire coastline, as well as the EEZ to be covered. Implementation of the proposal will help strengthen the resource assessment and management capacity of this centre. At present the DD(M) is given the task of licensing mechanized boats, but its manpower is inadequate for the job. To date, only 1000 boats have been licensed, licences for the rest are delayed due to shortage of manpower. A separate directorate of marine fisheries would be needed to enhance the expansion and comprehensive planning of the marine fisheries sector. The FAO proposal for marine wing strengthening may be reviewed and considered. The proposal identifies the manpower and facilities required to strengthen the marine wing of the DOF.

MARINE INVERTEBRATES

Invertebrates are animals that do not have a bony internal skeleton, although many do have hard outer coverings that provide structure and protection. More than 90% of all animals are invertebrates and they are classified into at least 33 major groups, or phyla. Nearly every phylum of invertebrates has members that live in the oceans. Six phyla of invertebrates that are commonly found in the oceans are: Porifera (sponges); Cnidaria (corals, jellyfish, and sea anemones); Annelida (segmented worms); Molluska (snails, clams, mussels, scallops, squid, and octopuses); Arthropoda (crabs, shrimp, barnacles, copepods, and euphausids); and Echinodermata (sea stars, sea urchins, and sea cucumbers). Each phyla is divided into smaller groups called classes, which is then split into families and then species.

PHYLUM PORIFERA

The sponges are the least complex multicellular animals. They generally live attached to a surface and have three basic shapes, encrusting, vase–like, and branching. Sponges live in intertidal (between the tides) zones as well as in the deep ocean. They can be a few inches (centimeters) to 10 feet (3 meters) in diameter. There are nearly 10,000 species of sponges and all but two families are only found in ocean environments. The general body of a sponge is a central cavity surrounded by a fleshy body riddled with holes. The cells lining these holes pump water into the central cavity. As the water moves through their bodies, the sponge absorbs nutrients, and filters out particles from the water as their food. The name Porifera means "hole bearer," reflecting the many holes in the animal's bodies.

PHYLUM CNIDARIA

The phylum Cnidaria includes jellyfish, sea anemones, and corals. The word Cnidaria comes from the root word knide, which means "nettle." It refers to the special stinging cells that the animals in this group have for protection and predation (hunting an animal for food). These cells contain coiled threads that are fired at predators and prey. The threads may contain substances that paralyze or sticky substances that entangle their target.

Cnidarians have two body plans: polyp and medusa. Corals and sea anemones are typical polyps. They are jar-shaped animals with a mouth at the opening of the jar. Tentacles rim the mouth and are used to pull food into the stomach, which is located on the inside of the jar. Sea anemones have a basal disc (tooth-like structure), which is located where the bottom of the jar is, and it is used to burrow (dig a tunnel or hole) into the sand or rocks. Corals have a skeleton made of the mineral calcium carbonate, which cements the individual coral polyps together into a colony (group). Jellyfish have a medusa shape, and this can be visualized as a polyp turned upside-down so that it looks like a bell. Jellyfish swim by contracting their bodies and forcing water out of the bell. They have long tentacles (appendages) surrounding their mouths that are used to capture prey.

PHYLUM ANNELIDA

The phylum Annelida includes worms that are segmented, meaning that the body is made up of sections. Each body part may have a specialized purpose, such as reproduction, locomotion, or sensing the environment. These segments are apparent on the outside of the worm's body and make it look ringed. The word Annelida comes from the word annelus, meaning "ringed." The most familiar member of Annelida is the earthworm, which is not a marine species. The class Polychaeta (meaning "manyfooted") is the largest class of Annelida, with about 10,000 species, most of which are marine (of the ocean). Almost all of them have paired appendages on their segments that can be used for swimming, burrowing, or walking. Polychaetes often have very well developed heads with a variety of sensory organs that detect prey by touch, vision, and smell. They range in appearance from very colourful to very plain-looking. Some swim through the water, others crawl on the sand or rocks, and others live cemented to the seafloor, building and living in tube-like structures.

PHYLUM MOLLUSKA

The mollusks are an extremely large phylum with over one hundred thousand species, most of which are marine. Most mollusks have a head, a foot, and a body that is covered by a shell–like covering called a mantle. The three most common classes of mollusks are the snails; the clams, oysters, scallops, and mussels; and the squid and octopuses. The gastropods are the largest class of mollusks with over eighty thousand species. They include snails, slugs, abalone, and limpets.

Most gastropods crawl along the seafloor among rocks, grazing on algae (tiny rootless plants that grow in sunlit waters). Some, however, hunt for their food among plankton, which are organisms that drift through the ocean. Other gastropods filter water for food particles. The majority of gastropods live in coiled shells, which provides protection from predators and protection from the force of waves. The shells are also used to protect animals that live in the areas between low and high tides from becoming too dry. The bivalves, meaning "two doors," are mollusks that have two shells like clams, scallops, mussels, and oysters. These animals generally live in sediments (sand, gravel, and silt) on the bottom of the ocean and gather food by filtering particles out of the water.

Many clams are able to burrow into sand or clay. Scallops can swim by forcing water through their shells. Oysters tend to cement themselves to hard surfaces. Mussels produce tough strings called byssal threads that attach their shells to surfaces in wave-swept areas. The most complex of the Mollusca phylum are the squid, octopuses, nautiluses, and cuttle fish. Each of these animals has a head that is surrounded by tentacles. They swim through the water using propulsion from their tentacles, much like swimmers kick their legs underwater, and creep along the bottom of the ocean using their tentacles as legs. Cuttle fish and squid have a shell like other mollusks, but it has been internalized. In octopi, the shell is completely absent. The members of this class have excellent eyesight and are considered intelligent.

PHYLUM ARTHROPODA

Arthropods are the most numerous invertebrate phylum with over one million species identified. Some scientists expect that there may be as many as ten million arthropods on Earth. All arthropods have a strong external skeleton that protects them from predation and supports their body structures. They have a type of muscle called a striated muscle that allows them to move quickly. They also have legs, antennae, and other appendages that are jointed. The phylum Arthropoda has three major divisions or subphyla. The subphylum Crustacea includes about thirty thousand different species, most of which live in the ocean. Crustaceans include many different types of marine animals that are divided into several classes. Brine shrimp, which are important fish food and live in very salty water, belong to the class Branchiopoda.

The class Maxillopoda includes barnacles and copepods. Barnacles are specialized crustaceans that spend the adult part of their life cemented head-down on hard surfaces like rocks, piers, the bottoms of ships, and even the undersides of whales. Their legs have developed into featherlike appendages that they use to generate water currents to bring food particles into their mouth. Copepods are small shrimp-like animals that are extremely important to the planktonic food web, the network of plankton that form the base of the food chain in the oceans. They are the most numerous animals in the ocean, sometimes reaching densities of more than a million per cubic yard (meter).

The class Malacostraca includes shrimp, lobsters, crabs, and euphausids. There is an enormous amount of diversity among the members of this class, which includes about twenty–five thousand different species. Some malacostracans spend their lives swimming among plankton, others walk along the ocean floor scavenging for food, while others live in burrows and attack prey that come nearby. Many members of this group live and feed off of fish or even other crustaceans. This class is very important to the economy, both as food for humans and as pets in the aquarium industry.

PHYLUM ECHINODERMATA

All of the six thousand members of the phylum Echinodermata are marine. The root word echino means "spiny" and the root word derma means "skin." The name Echinodermata refers to the bony structures called ossicles found in the skin of these animals. Echinoderms do not have well developed sensory organs or brains. They all have a water vascular system that is used to circulate nutrients and gasses through their bodies.

All echinoderms share the same general appearance, which is based on five similar sections that radiate out from a central point. There are five classes of echinoderms: feather stars, sea stars, brittle stars (sea stars), sea urchins, and sea cucumbers. Brittle stars are fairly uncommon.

There are about six hundred species of brittle stars, living mostly in shallow waters. Sea stars (commonly known as starfish) use their water vascular system to operate suction cups located on the bottoms of their legs. These suction cups are called tube feet and they are used both for predation and for gas exchange. Sea stars eat by gripping both shells of a clam or mussel with its tube and pulling the prey open. Then it inverts its stomach inside the shells and digests the victim. Ophiuroids usually look like a small disc surrounded by five long worm-like arms.

They are called brittle stars because when they are attacked, they will simply detach the arm that has been the target. Later, the ophiuroid will regenerate its arm. Sea urchins are usually pin cushion–shaped and covered with sharp spines. These spines are used for locomotion as well as for defence. Between the spines, sea urchins have special appendages called pedicellariae that look like tiny claws. These are used for capturing prey and for cleaning. The mouth of the sea urchin is on its underside and is composed of five tooth-like plates.

Sea cucumbers live on the sea floor and look like cucumbers with five soft ridges. Many live on coral reefs and are extremely colourful. Sea cucumbers feed by extruding feathery appendages that can capture prey that swim too close. When attacked, many sea cucumbers will suck in water and then use the water pressure to eject their internal organs, including their digestive and respiratory systems.

The predator becomes confused among all the tissues in the water and may even become entrapped in some of the sticky material. After such an attack, a sea cucumber will regenerate its internal organs over a period of several weeks.

MARINE MAMMALS

Mammals are vertebrates (animals with a backbone) that share characteristics of nursing their young with milk, breathing air, having hair at some point in their lives, and being warm–blooded.

Marine mammals are the species of mammals that depend on the oceans for all or most of their lives. There are about 115 different species of marine mammals. Marine mammals vary from the small sea otter to the giant blue whale. Some of them live in groups, like dolphins, while others are solitary, like polar bears.

All marine mammals share four characteristics:

1. They have a streamlined body shape that makes them excellent swimmers.
2. They maintain heat in their bodies with layers of fat called blubber.
3. They have respiratory (breathing) systems that allow them to stay underwater for long periods of time.
4. They have excretory (waste) systems that allow them to survive without drinking freshwater. Instead they obtain the water they need from the food they eat. Marine mammals belong to three groups called orders. An order is a classification of a group of organisms, which eventually splits into species. Marine mammals are in the orders Cetecea, Carnivora, and Sirenia.

ORDER CETACEA

There are about seventy–seven species of cetaceans, which include whales, dolphins, and porpoises. All of these animals live their entire lives in the water. Cetaceans probably evolved from the hoofed mammals that were similar to horses and sheep. Their front legs became fins that are used primarily for steering, and their hind legs became extremely small flippers or flukes (tail fins) so as to streamline the animals for swimming. Cetaceans swim by moving their strong fluke up and down. They are grouped into two categories or suborders: Odontoceti, the toothed whales, and Mysteceti, the baleen whales.

Suborder Odontoceti

The toothed whales account for about 90% of all cetaceans, including dolphins and porpoises as well as the orca (killer whale) and the sperm whale. Toothed whales hunt for prey, which they capture with their teeth. They also have one external hole called a blowhole for breathing. The toothed whales have a large brain for their size and are considered to be some of the most intelligent animals. They hunt using sophisticated echolocation, which is the method of detecting objects by listening to the reflected sounds that it calls out.

Suborder Mysteceti

The baleen whales have no teeth. Instead, they have bristly plates called baleen, which hangs like a curtain from their upper jaws. Baleen is made from a protein similar to that which makes up human fingernails. When it is eating, the baleen whale sucks in huge amounts of water and then forces the water out through its baleen, which acts like a sieve. These whales diet primarily on tiny animal plankton, animals that drift through oceans. Microscopic plankton are concentrated behind the baleen and then swallowed. Even though baleen whales feed on some of the smallest animals in the world, baleen whales are some of the largest animals in the world. The blue whale, the fin whale, and the gray whale all weigh more than 2 tons (9 metric tons). Baleen whales are also distinguished from toothed whales because they have two external blowholes instead of one.

ORDER CARNIVORA

The Carnivora include animals that prey on each other (meat-eaters) like dogs, cats, bears, and weasels. There are two suborders of carnivores that have marine representatives: the pinnipeds and the fissipeds.

Suborder Pinnipedia

Pinnipeds include seals, sea lions, fur seals and walruses. They all have flippers that can be used to move around on land as well as on water. Although they swim much more efficiently than they walk on land, they do give birth to their young on land. Seals account for nearly 90% of all pinnipeds. There are nineteen species that live in all of the oceans and even in a few lakes. Most seal species live near Antarctica and in the Arctic Circle. Seals do not have an external ear, although they can hear very well.

They propel themselves with their rear flippers and use their front flippers for steering. Sea lions and fur seals do have small external ears. Their hind legs are more flexible than those of seals, so they can move around better on land. They propel themselves with their front flippers when swimming. Walruses are the largest of the pinnipeds. They do not have an external ear, but they can use their rear flippers for moving on land. The canine teeth (pointed, in the front) of walruses are enlarged into tusks. Walruses swim along the bottom of the ocean using their tusks like runners as they look for clams to eat.

Suborder Fissipedia

The suborder Fissipedia includes cats, dogs, raccoons, and bears, as well as two marine mammals: sea otters and polar bears. Sea otters are about 4 feet (1.2 meters) long, the smallest of the marine mammals. Their favourite food is sea urchins, which they eat by lying on their back and smashing them on a stone that is balanced on their chests. Polar bears wander long distances across sheets of floating ice in the Arctic hunting for seals and whales. They can swim between patches of ice using their powerful forepaws like oars.

ORDER SIRENIA

The sirenians, also called sea cows, evolved from the hoofed land mammals, like the cetaceans. The Sirenia include the manatees, which are large plant-eating marine mammals, and the dugongs (commonly known as sea cows). They are the only plant-eating marine mammals, eating sea grasses and algae (tiny rootless plants that grow in sunlit waters) in warm waters. They grow to be quite large, up to 15 feet (4.5 meters) and weigh 1,500 pounds (680 kilograms).

ENDANGERED MARINE MAMMALS

The Endangered Species Act, an animal that could become extinct in all or part of its range is endangered. An animal's range is the entire area where it lives. Of the 115 species of marine mammals, 22 were considered endangered as of 2004, including the blue whale, the gray whale, the finback whale, the Hawaiian monk seal, the stellar sea lion, the marine sea otter, and all four species of dugongs and manatees.

Much of the threat is due to human activity. People hunt these animals for their pelts, blubber, and meat, and destroy their habitat from overfishing and mining.

MARINE AQUACULTURISTS

As with the development of our species in going from being hunter-gathers of foods to their controlled growth and reproduction, marine aquaculturists have begun to make inroads in successfully breeding and rearing of marine fishes. Indeed, a few species (the Clownfishes, subfamily Amphiprionae and Neon/ Cleaner Gobies, genus Gobiosoma) are largely produced in captivity nowadays, doing more than reducing correspondingly their wild-collection. Captive-produced marines, as with freshwater have proven hardier, more adaptable to aquarium conditions. They much more readily take prepared foods and are often specific pathogen free.

Though the vast majority (likely more than 99%) of marine fishes used for ornament are still wild-collected, there are great improvements in capture of healthy broodstock, knowledge of modes of their reproduction, induced and natural spawning and breeding, culture of useful foods, and overall husbandry techniques which foretell the possibility of captive production of any and all species. Fishes are induced to spawn by altering environmental parameters, particularly temperature and photoperiod, and hormonal manipulation. These variables are generally constant for a given species, but vary widely amongst cultured fishes.

SPAWNING, BREEDING

Some fifty-four families of fishes include livebearing species that utilize internal fertilization with either no further nourishment from females (ovoviviparity) or placenta-like structures between their young and their bodies (viviparity). Forty of these fish families are cartilaginous fishes (sharks and rays) and the remaining marine ones of little notice (currently to marine aquarists). Additionally there are fishes that engage in internal fertilization, placing their sticky embryos at a later time. None of these species are either presently utilized by aquarists. Hence, we are almost exclusively concerned with spawning species, those that broadcast their sex cells into the environment when considering the captive production of ornamental marine fishes.

History

Early records at trials at culture of ornamental marine fishes likely don't exist. Many "casual aquarists" of the sixties experimented with home and some commercial production of Clownfishes. Their accounts are anecdotal at best. Friese (1971) reminisces over hobbyist accounts of partial successes of breeding, rearing marine fishes (seahorses, pipefishes, some gobiids, a few pomacentrids) in the 1960's, and cites "recent" work to that time on multigenerational breeding done at the Wilhelma Aquarium in Stuttgart, Germany by Dr. W. Neugebauer of *Hippocampus, Dunckerocampus* and *Amphiprion* spp. Hoff (1985) relates the early beginnings of commercial ornamental marine fish culture in the United States with his involvement with Instant Ocean Hatcheries, Aqualife Research (moved to Walker's Caye in the Bahamas from Florida, founded by Martin Moe in 1972, who is still engaged in active research) and Sea World (San Diego, run by Chris Turk) who have all ceased production.

These early ventures likely failed commercially as Hoff states for Instant Ocean Hatcheries, due to too much overhead cost. They all were able to sell all (mainly Clownfishes) they could produce. Current commercial facilities in operation include Tropic Marine Centre in the U.K., ORA (Oceans Reefs Aquariums) in Florida in the U.S.A, and C-Quest in Puerto Rico, Mangrove Tropicals (clowns and Dottybacks) and Ocean Rider (seahorses), both of Hawaii and South Australian Seahorse Marine Services (*Hippocampus abdominalis*). Experimental/university labs that are engaged in research involving marine pet-fish species include The Aquarium Complex/Marine & Aquaculture Research Facilities Unit (MARFU) at James Cook University in Queensland, Australia, Guam Aquaculture Development & Training Center, Hofstra University Marine Laboratory in Hempstead, New York, various Sea Grant locations in the United States (principally in Hawaii),

Production Considerations

Several critical elements must come together in aquaculturing captive ornamental fishes. Selection of suitable species, securing of broodstock, conditioning, the actual physiology of gamete production and release through environmental and/or hormonal manipulation, provision of grow-out facilities and supplying of appropriate foods.

Species Selection

Criteria for selecting potential species for captive propagation include compatibility, colour and markings, size, inherent hardiness to captive conditions/shipping and handling, and interesting behaviour.

Compatibility is key as there is a huge range of desirable behaviour amongst reef fishes. Some are absolute terrors that even with successive generations of captive breeding are proving troublesome. The best example here is *Pseudochromis steenei*, the Lyreback Dottyback. This fish is likely responsible for many many aquarists outright quitting the hobby due to its predaceous nature.

Broodstock

Many breeding attempts are made impractical if not impossible by the selection of inopportune broodstock. Sufficient numbers of potential spawners or breeders must need be secured to afford the luxury of their controlled upbringing and possible sacrifice to determine viability.

Conditioning

Keeping broodstock in optimized, stable environments is absolutely key to captive breeding programmes. As such culture facilities do their utmost to properly house their breeders. Most are kept in recirculated, centralized filtered tanks, with automated controls for water quality, including dosers for values like pH, probes for temperature, and regular testing for metabolites like nitrate, phosphate.

Diet

Bogenschutz and Clemens (1967) were amongst the first to scientifically explore links between nutrition and gonadal development. They demonstrated gonadal regression induced by restricted diets, accompanied by an inversion of the basophil/acidophil ratio of the mesoadenohypophysis and a reduction of gonadotropin content. These conditions were reversed with adequate food. These authors stressed the interplay between photoperiod and diet and pointed out that optimum benefit from photoperiod manipulation can be overridden by poor diets.

Often special foods with vitamins, HUFAs (Highly Unsaturated Fatty Acids), and minerals added are specifically made, tailored to the individual species. Some excellent businesses like England's Tropic Marine Centre use this opportunity to tie in sales of their proprietary frozen food line (Gamma Foods). Use of dried foods and relatively hollow nutrient content foods like adult brine shrimp is discouraged for conditioning breeders.

DISEASE PREVENTION

As part of a working plan of maintaining broodstock attention must be paid to preventing the introduction of disease-causing organisms. To mention them again, TMC (Tropic Marine Centre) should be cited for the exclusion of such pathogens. Through careful quarantine, use of antibody treatments, their entire facility is specific pathogen free.

Gametogenesis & Reproductive Behaviour

Marine fishes produce and release sex cells based on maturity of the individuals, their nutrition and overall health, triggered by cues from the environment (temperature, light/dark duration, tides, presence of conspecifics, mates...) that in turn influence their hormonal/endocrine systems. Along with endocrine control there is a steady, intimate, more sudden interplay of the fishes' nervous system, feeding in and coordinating activity through their eyes, hearing, lateralis system, senses of smell, and memory. Conditioning and triggering of actual spawning involves combining knowledge of modes of reproduction, social factors such as sex ratios, environmental manipulation and possibly direct/exogenous hormonal administration.

CURRENT STATUS OF MARINE FISHERIES

In general, considered on a single-species basis, many marine fisheries are fully exploited or overexploited, while relatively few seem to have the potential for increased exploitation. In general, this is true both for the United States and globally, especially in estuarine, nearshore, and continental-shelf fisheries, which produce approximately 75 percent of the world's fish catches.

The primary source of global information about the condition of fisheries is the Food and Agricultural Organization (FAO) of the United Nations. In the United States the task of carrying out assessments has been primarily the responsibility of the National Marine Fisheries Service (NMFS) of the National Oceanic and Atmospheric Administration, U.S.

Department of Commerce. Regional fishery management councils (established under the authority of the Magnuson-Stevens Fishery Conservation and Management Act of 1976) also are involved in the assessment of fish stocks in federal waters. Interstate fishery commissions in the Atlantic, Gulf, and Pacific regions work with states and the NMFS to conduct assessments of migratory fish stocks in state waters. Although assessments and statistics from FAO and NMFS provide only an imperfect characterization of the status of global and U.S. fisheries, the assessments—corroborated by many kinds of evidence—appear to provide a reasonably accurate description of the overall picture.

GLOBAL OVERVIEW

Fishing is an important source of food, recreation, community development, wealth, and cultural values in many countries. Although thousands of freshwater and marine fish and shellfish species are used globally, a relatively small number of these species provide the major fraction of the global marine catch. The 10 marine species that provided the greatest catch in 1993 accounted for 35 percent of the commercial marine catch and the top 20 species accounted for 46 percent of the global marine catch.

Recent estimates indicate that the global first-sale revenues from fishery products are approximately $U.S. 95 billion annually and that fishery products account for about 20 percent of the animal protein consumed by humans. Fisheries provide direct and indirect employment to about 200 million people worldwide. Fisheries are especially important in developing countries, which increased their proportion of global catch from about 40 to 65 percent from 1973 to 1993. The net value of fishery products exported from developing countries totaled $16 billion in 1994, greater than the exports of coffee, bananas, rubber, tea, rice, and many other commodities that developing countries have traditionally relied on for foreign exchange.

Global marine fish production increased at an average rate of about 3.6 percent per year from 1950 to 1995, from about 18 million to about 91 million metric tons, including mariculture production. In the same period, the world's population increased from 2.5 billion to 5.7 billion people, an average annual increase of 1.8 percent. In 1995 total fish production (both freshwater and marine, both through culture and through fishing) was approximately 112 million t, of which marine landings accounted for approximately 84 million t. In 1996 the total production reached approximately 116 million t; the increase was due mainly to an increase in freshwater aquaculture production, mainly in China. The supply of fish and fish products for human consumption (including freshwater fish and aquaculture products) reached roughly 14 kg per person annually in 1995. By 1995, mariculture accounted for 6.7 million t, 7.4 percent of the total global marine fish yield; freshwater aquaculture provided 14.6 million t. Approximately 31.5 million t (28 percent) of world fish production was used for animal feed—including feed for mariculture—and other products that do not contribute directly to the human food supply in 1995.

In addition to fish that are caught and processed, a substantial number of fish and other organisms are caught and discarded—usually dead—at sea. Discards are a result of bycatch, which results because fishing gear and methods are not selective enough to catch only the target species, and of high grading, the discarding of smaller or less desirable fish in favour of larger or more desirable fish that are caught later. Some bycatch is retained, but the remainder is discarded when the species, size, quality, or condition of the fish reduce their value, or when fishery management regulations prohibit their retention. Alverson *et al.* (1994) estimated that commercial marine fisheries around the world discarded an estimated 27 million t of nontarget animals in the early 1990s, an additional biomass about one-third as large as total landings.

In addition to fish mortality caused by landings and discards, fishing can cause additional *unaccounted* mortality. Potential causes of unaccounted mortality include illegal or misreported landings; escapement or avoidance mortality that occurs when fish are injured by fishing gear but are not captured; and ghost fishing mortality, caused by lost gear (*e.g.*, traps and gillnets) that continues to catch fish. The magnitude of unaccounted mortality is unknown but may be high for some fisheries. For example, the Scottish Fishermen's Federation estimated illegal or misreported landings, probably of groundfish and crustaceans, to be 100 to 200 percent of the reported catch.

Myers *et al.* (1997) concluded that discards of young undersized fish were an important reason that the fishing mortality of northern cod off Canada's maritime provinces was consistently underestimated in the 1980s, leading to the overfishing that caused the collapse of the fishery. Although these examples are not necessarily representative of all fisheries, they show that the total mortality resulting from fishing can easily be underestimated. The FAO (1994a) and Garcia and Newton (1997) have concluded that the relatively stable catches of the early 1990s indicated that capture fisheries are near, or have reached, their sustainable limit based on existing fishing techniques and market systems.

The increase in catch between 1950 and 1994, occurred because of steadily increasing demand for fishery products, resulting in increased fishing capacity and effort. Fisheries were developed or expanded on formerly less-exploited or unexploited species and populations. While global assessments indicate that there are still opportunities to expand some fisheries, most are fully exploited or beyond, based on single-species considerations. Some have been so depleted that they are producing much less than their long-term potentials. Global fishing capacity is much greater than needed for sustainable marine fisheries, again based on single-species considerations.

More than 25 years ago, Gulland (1972) estimated that the potential sustainable yield from traditional fishery resource species (excluding Antarctic krill and oceanic mesopelagic fishes) was about 100 million t, with a practical limit (due to imperfect management and multispecies interactions) of about 80 million t, similar to catches that have been achieved in recent years and to recent FAO assessments.

Maximum potential global marine fisheries yield has also been estimated by Schaefer (200 million t, 1965), Ryther (100 million t, 1969), Idyll 400–700 million t, 1978), Houde and Rutherford (more than 300 million t from all marine ecosystems, 1993), and others. Based on those estimates, one might conclude that we have not yet reached maximum global fisheries yield. However, most of the highest estimates—admittedly upper limits in some cases—were made using unrealistic assumptions about food-web structure, effects of bycatch, feedback effects of fishing on other fish populations and marine ecosystems, and the technical and economic feasibility of new fisheries.

Limits to Global Production

Three kinds of information suggest that marine fish catch is near, at, or above its maximum sustainable level: estimates of theoretical limits imposed by available primary production, information on the degree of utilization of fish populations, and information on the catch per ton of fishing vessel.

Food-Web Limitations

The capacity of the ocean to produce fish is limited in part by the amount of marine phytoplankton produced annually. Fishery landings tend to be higher from ecosystems with higher levels of primary production, especially marine areas characterized by fronts, convergence, and upwelling areas. Satellite and in situ measurements of phytoplankton concentrations and in situ measurements of nutrients, water temperature, irradiance, and primary production allow estimates of the primary production of the global ocean, as well as regional estimates.

An upper limit to the ocean's potential fisheries yield has been estimated many times by applying knowledge of the amount and location of global primary production, trophic level of the catch, and the transfer efficiency of biomass among trophic levels.

Pauly and Christensen (1995) used global catch data—which they divided according to trophic level—to estimate the flow of carbon up through the trophic levels of global marine ecosystems. They estimated that the transfer efficiency between trophic levels was about 10 percent, and concluded that about one-quarter to one-third of total primary production in coastal and continental shelf waters is needed to support recorded landings plus discards.

Houde and Rutherford (1993) used relationships between catches and primary production and between fish production and primary production to estimate a partitioned global fisheries production for estuaries, coastal zones, and upwelling areas. They estimated a total global fisheries production in those ecosystems of 543 million t, from which 111 million t might be removed as yield. (Note that Houde and Rutherford's estimate of the total potential yield of more than 300 million t was based on an estimated production of more than 1,300 million t in all marine ecosystems. They considered open-ocean production to be technologically difficult to use.) These estimates suggest that landings are near or beyond their sustainable limit, particularly if fish production lost as discards and unaccounted mortality are considered. Another line of evidence suggesting that global marine catch might not increase, even by fishing at progressively lower trophic levels, is provided by analyzing changes in the mean trophic level of marine fishery landings.

The average trophic level of fish catches from the 1950s for the northeast Atlantic and from the 1970s for the northwest Atlantic. This reflects a decrease in the proportion of long-lived carnivores in the catch relative to shorter-lived smaller pelages and invertebrates. Fishing down the food web, while overfishing higher trophic forms, does not necessarily lead to increased total catches. As fishing takes animals lower in the food web, an increasing portion of the total catch may consist of animals for which there are no current markets or that are so diffuse that the cost of their capture does not warrant the expense (*e.g.*, some large zooplankton species). In addition, the loss of predators (*i.e.*, animals higher in the food web) can lead to an increase in competitors of the target species. The average trophic level of landed species can drop rapidly as catches of top predators or decline as observed in most other FAO areas analyzed in this fashion.

MARINE MIGRATIONS AND FOOD CHAIN

THE CALIFORNIA COAST

California's famous coast, the golden shore, the stuff of songs, stretches roughly 1,100 miles from the dry Mediterranean climes of the south to the lush rain forests of the north. In sunny San Diego, weather forecasts are notoriously boring; according to legend, they are sometimes recorded days or more in advance. In the north, redwood and Douglas fir forests drink in the driving rain and tickle water out of the fog, seeing little sun for months out of the year.

In between, plant and animal communities hover in their preferred ranges in overlapping steps, slowly shifting in the same way habitats shift as you climb a mountain, or as you look higher along a tidal edge. Because of this broad spectrum of varied ecological zones, California boasts some of the greatest diversity of all the fifty United States. Its sheer length also makes it a perfect place to patiently, slowly, and meticulously observe the way nature responds to a warming world.

An hour past dawn, the central California sun burns through the morning haze. The waters of Monterey Bay are rising after a 6:21 a.m. low tide. Rafe Sagarin, curly hair tousled by the unruly wind, gingerly crosses the exposed granite tidepools like a teenager negotiating a cluttered bedroom floor. This intertidal tangle is a library to Sagarin, who knows precisely where reefs of tube snails set their mucous nets, striped sunburst anemones open their tentacles to the tides, and barnaclelike limpets farm algae for their supper.

Sagarin can also tell you that this tidal community is not what you would have seen in the 1930s. In fact, it looks a lot more like southern California. Sagarin knows because he has counted every critter along a line between two brass bolts in the rocks, and compared them to a similar count done 60 years earlier. That change is so big, so obvious, and so important that an article he and colleague Sarah Gilman wrote about it was published in the prestigious scientific journal *Science* while they were still undergraduates.

That research, which used their counts combined with temperature and other measurements going back threequarters of a century, found that the most likely cause for the massive changes in the kinds of species there was warming temperature. It is exactly what one would expect to happen as the world's climate changes. In 1993, Sagarin and Gilman were juniors at Stanford University, among two-dozen students who took a term away from the main campus to study marine biology at the university's Hopkins Marine Station on the Monterey Peninsula. Upon arrival, the undergraduates were treated to presentations about areas of research at the station, and projects they would have the opportunity to participate in.

It was not what Sagarin expected. "I came down here and I had a naïve view of marine science," he says. At twenty, he had thought the discipline consisted of learning about the animals that inhabit the different niches along the salt-water edges. But at Hopkins he found that most people were molecular biologists or neurobiologists, considering the marine species on a tiny, chemical scale. Sagarin says he did not understand what most of the faculty were talking about when they discussed the minutiae of their work. One of the faculty members that spring was Chuck Baxter, a marine ecologist who had worked from the Hopkins station since the 1970s. Deep-voiced and white-bearded, Baxter looks like a silver-haired William Shakespeare with a dash of Ernest Hemingway thrown into the mix. The professor's seniority and drive made him a central figure in such projects as helping found the now-famous Monterey Bay Aquarium and developing a major film documentary to bring the complex lives of snails and clams to a public that sees them mainly on dinner plates.

Pulling no punches, Baxter offered the students the same challenge he had offered their predecessors for 5 years. He wanted to know "why there were so many damned *serpulorbus*"-a prevalent tube snail from southern California-in the Monterey area, Sagarin recalls. The tube snail covered everything down south, but until a few years earlier had been nearly impossible to find from Hopkins north.

Baxter knew the tube snails from the 1950s, when he was an undergraduate at UCLA and later as a graduate student living in Santa Monica. *Serpulorbus* covered every surface, from rocks to docks that were splashed by Pacific waves. Like most such animals, larval *serpulorbus* float through the water, eventually landing on and cementing themselves to rocks or other surfaces that are intermittently inundated by the tides. Once settled into place, they begin growing long, tubular shells, looking rather like the extended blur you see when something runs in slow motion.

As the *serpulorbus* population grows, the overlapping tubes completely cover almost any surface that will stand still long enough, like calcified kudzu. The reefs they form help other species move into areas where they would otherwise have no toe hold. Sea urchins, for example, cannot latch onto the tough granite of the Hopkins intertidal area. But they can wedge themselves into *serpulorbus* reef and become established that way.

Researchers in the 1960s who came to Hopkins looking for tube snails to study were disappointed, and had to head south for their work. When Baxter arrived in 1974, he saw a few *serpulorbus* in the rocky intertidal area outside the station's offices. "I guess you could find one every 10 feet," he recalls. When his research took him up to Half Moon Bay and Pescadero, about 40 miles to the north, he did not see any *serpulorbus* at all. That started to change within a few years of Baxter's arrival. "Sometime in the early 1980s, the numbers began to pick up," Baxter says. "By the mid to late 1980s they were very abundant." The tube snails completely covered the rocks behind the Monterey Marina's breakwater, just as they had in southern California 40 years before.

The new abundance of *serpulorbus* was not the only change Baxter had noticed in his decades at the station. Earlier researchers had documented the seaweed cover on the rocks; by the late 1980s much of that was completely gone or had shifted substantially, taking over what had been habitat for barnacles and other critters. Baxter also saw that one predatory snail in the whelk family, *ocinebra*, seemed to have increased; it was another species he knew from his southern California days.

While casual observations make interesting stories, they are not the same as rigorous science. They are, Baxter says, "anecdotes that you can't do much with." To turn his eyeballed observations into respectable research would take a concerted effort to actually measure any changes, compare them with other kinds of data about the physical environment, and then try to determine the factors that might have pushed the change forward.

Baxter had a couple of starting points. He had found a small handful of older studies documenting all the plants and animals along specific sections of the rocky tide pools just downhill from the Hopkins station's lawn. The station had been collecting data including air and water temperature for decades; some statistics even went back as far as 1917. And the intertidal area between Hopkins and the open waters of Monterey Bay was California's first marine reserve, protected from fishing and other extraction since the 1930s, so the species found there would be essentially unaffected by human tastes.

"For about five years it had become very obvious to me that some very dramatic changes had taken place and were continuing to take place in the intertidal off Hopkins," Baxter says. That was the source of his challenge to the students who came for the undergraduate research course each year. What had changed? How much had it changed? What factors were associated with those changes? And, did the changes in snails and the seaweed simply represent a shift in a few individual species, or was something broader and more significant happening?

For years, the students were not interested. Like the other researchers at Hopkins, they favoured quick, experimental science-the molecular or neurobiology-over tediously counting critters in a tidepool. Most looked at Baxter's suggestion as research unlikely to yield spectacular findings.

It was not until Baxter's last semester as a full-time member of the Hopkins faculty that he got a bite: Rafe Sagarin and Sarah Gilman. They started with a study done in the 1930s by a Stanford Ph.D. candidate named Willis Hewatt, who had pounded heavy-duty brass bolts into the slow-eroding granite and meticulously counted every anemone, whelk, snail, sea star, barnacle, limpet, and other critter along the 108 yards in between.

The area where Hewatt did his work is a little triangle of pools made up of huge granite rock and ending at a bouldered point, streaked with white from the seabirds it hosts. The outer rocks provide some shelter from the more turbulent-and cold-waters of Monterey Bay, about 100 yards from shore. It is a perfect spot for looking at the creatures that thrive in the intertidal zone, the band along the rocks that is submerged at high tide, exposed at low tide, and wet by spray and waves in between. In the same way zones of life change as you climb a mountain-or as you drive north or south along a coastline-marine life varies by how often it is exposed to air and scalding sun, or how deeply it is covered by water.

The first problem Gilman and Sagarin encountered was finding which way Hewatt's line ran. The location of the first bolt was obvious, but finding the second proved difficult. When they finally did find it, the line ran in a different direction than they had initially thought. Armed with the location of both bolts, they meticulously constructed a map of squares in between, precisely repeating what Hewatt had done. In all, they counted 125,590 animals, representing 136 different species.

Once the counting began, Gilman and Sagarin quickly saw dramatic changes in the makeup of the intertidal animal and plant community. What they found was a substantial species shift in the intervening 60 years that was best explained as a northward migration of species away from warming water-exactly the kind of response predicted by models of human-caused climate change. When Hewatt first looked at the tidepools, there were no tube snails or sunburst anemones at all, although both were prevalent further south. When Gilman and Sagarin did the same, they found hundreds of the southern anemones, and as many as 229 tube snails crowded into 1 square yard. In all, ten of eleven species previously identified as southerners increased significantly. Six of eight northern species decreased significantly. Those changes showed up regardless of what the animals ate, how they reproduced, or where they sat in the taxonomic hierarchy. Meanwhile, daily temperature records showed waters there had warmed about 1.8°F since Hewatt squatted in the surf.

They looked at all the masses of data in front of them, and considered different reasons for the changes. The best they found was the increase in temperature. The combination of counting so many species and having such a rich set of temperature data going so far back means that their findings withstood substantial statistical scrutiny. In other words, it was almost impossible that what they found could be simply chance. "If you had a table with a diagonal line and some marbles," Sagarin says, "and you just dropped them randomly, there's no way you'd ever get the pattern we saw." Gilman did some separate work looking at one seaweed species growing on the granite. Using a study from the 1960s by Peter Glynn, now a researcher at the University of Florida, Gilman did a similar rock-by-rock comparison of turfy red algae called *Endocladia muricata*.

"The areas where Glynn worked just became deserts, in a sense, because there was no algae there," she explains. "But if you looked a foot lower, most of what used to be up in his research area shifted a little bit lower. We think that has something to do with temperature changes, because as it warms up, everything at low tide becomes a little warmer and dryer. So it sort of makes sense that things would tend to shift a little bit lower so they're exposed to air for shorter periods of time."

The snails, sea stars, and anemones Gilman and Sagarin counted were all small. But the shifts they recorded-and the implications of those changes-were anything but. These were "massive community reorganizations," Sagarin says. And the implication of such a finding is that "it seems that species respond to climate change in the present day. It's not just going to happen at some point in the future."

"It was one of the first studies to suggest that you could go out and find [climate-related] changes already," Gilman adds. "But it was completely unexpected to me. I thought I was just counting snails for a summer project." The results have held up over time. Sagarin, who received his Ph.D. from the University of California at Santa Barbara in 2001, has resampled the Hewatt line regularly. And while he has not found new changes with the same kind of drama as those he and Gilman found in 1993, he also has not found anything to indicate that the species are shifting back.

"Changes in the short term-every few years-are much smaller than the dramatic, aggregate changes that we saw when we looked 60 years later," he says. "So it just suggests that, even though each of these organisms is changing in its own time frame, real ecosystem evolution takes a long time."

Baxter points out that what his students found in the intertidal area by Hopkins are the same kinds of community changes one observes climbing up into the Sierra Nevada mountains. And if climate is affecting the globe, and therefore the Sierras, the same kinds of changes are happening there on a different scale. Scientists believe it takes something on the order of 300 years to see the tree line shift; with *Endocladia*, the timeline is probably about 10 years. That, he says, makes studying intertidal areas ideal for observing the ways a community of plants and animals responds to a warming world.

"This shows that, at a level we wouldn't have expected, the animals at Hopkins have reacted and responded to a small change in climate by totally restructuring their community," he says. "Organisms are more susceptible to small changes than we ever would have predicted. I think that it is clear that we are in the middle of a great experiment."

NOBODY NOTICED

The findings at Hopkins were remarkable in themselves, but also significant because the changes had been under way for decades and no one noticed. The first sunburst anemones appeared in 1947. Seaweeds that once covered the rocks-not just Gilman's *Endocladia*-completely disappeared. "The astounding thing is that we didn't know it was going on," Baxter says with a growl. "This is a marine biological station!" If not for Baxter's tenure and tenacity, the changes might still be going unnoticed. Those slow shifts are easy to miss or to write off as a temporary change. Without a basis for comparison or long personal experience, population shifts in small ecosystems are nearly invisible.

The best way to avoid missing those changes is to track small details over long periods. But in the world of scientific research, there is a simple fact: Monitoring-the kind of meticulous counting that Sagarin and Gilman did, repeated over time-simply is not glamorous. It is not quick, it generally does not attract money, and it does not interest top researchers who can more easily find funding and fame with other kinds of studies. Even dramatic findings such as those at Hopkins and elsewhere do not often inspire funding agencies such as the National Science Foundation. But when trying to understand how a warming climate evolves-and the effects of such warming on all natural processes-monitoring is the best and possibly the only way to document reactions to a phenomenon that every year becomes more obvious and less theoretical. Researchers examining what long-term data are available have uncovered alarming and portentous evidence of a changing world.

One area where monitoring has been funded is in the southern portion of the California Current, a 600-mile-wide swath of south- ward-flowing water running along the western U.S. coastline, roughly from Oregon down through California. Part of a great subtropical gyre of currents in the northern Pacific Ocean, the California Current is the eastern end of a swirling seawater highway that circles from Japan to Oregon, south past California just into Mexico, across to the Philippines, and back up to Japan to begin again. Along the way the water changes. As it crosses the Pacific toward North America, more water comes in through rain than leaves through evaporation, so the overall current becomes less salty. As it comes down the U.S. coastline, it meets cold, salty water heading north on another current, mixing in great meanders and eddies that can be up to 300 miles wide. The current turns west again south of California to start the process again.

In 1949, a combination of state and federal organizations began monitoring physical, chemical, biological, and meteorological facets of the California Current under the auspices of the California Cooperative Oceanic and Fisheries Investigations programme, known as CalCOFI. It was designed in part to track many factors affecting commercially important fish species such as mackerel and sardines. The data gathered under CalCOFI include air temperatures, wind speeds, nutrient levels, salinity, water temperature on the surface and deep below the surface, and the abundance of larval fish and zooplankton-the smallest marine animals. The early monitoring cruises brought researchers as far north as the Oregon border, but the surveys were scaled down to meet budget demands in 1970. But those first 20 years of data were enough to show that what happens in the south and what happens in the north tend to be the same. The data since 1970 cover the area between San Diego and Santa Barbara. It is the largest, longest-term data set of its kind on the West Coast.

What happens if you watch those data change over the years? Your findings might echo those of John McGowan, an oceanography professor at the Scripps

Institution of Oceanography in San Diego: As water temperatures have risen, the base of the marine food chain off the coast of California has crashed. And one by one, the fish and birds farther up that food chain are crashing, too. Life in the ocean begins with tiny plants known as phytoplankton. Like all plants, phytoplankton need light to drive photosynthesis and nutrients to feed the process. Although it is somewhat counterintuitive, the richest and most nutritive ocean waters are the coldest and heaviest. Strong winds do the work of stirring the system and pulling the nutrient-rich waters up toward the light.

The first problems showed up in conjunction with El Ninos, shortterm changes in ocean temperatures that tend to increase the warm water along the western U.S. coastline, reducing the food that boosts the phytoplankton. But researchers like McGowan noticed a difference between early El Ninos and the later ones. Numbers of zooplankton-the tiniest animals in the food chain, which depend on the phytoplankton-dropped during the El Nino of 1957 to 1959 and then quickly rebounded. But after subsequent El Ninos during the 1983 to 1984 and 1997 to 1998 seasons, the zooplankton did not come back.

In 1995, going back through the accumulated years of data, McGowan reported a staggering finding in *Science*: Zooplankton numbers in the California Current had dropped by 70 percent. The CalCOFI data show a sharp increase in California Current water temperatures in 1977-at the same time the zooplankton numbers crashed. "It's the largest change ever measured in plankton productivity in the ocean," McGowan says. "This enormous change in the zooplankton in the California Current could not be detected from year to year. It took several decades before we discovered this big drop, by at least 70 percent or even up to 80 percent."

If you pull out the bottom stone in a pyramid, you expect the structure to come tumbling down. With that huge loss at the base of the food chain, reverberations throughout the system that depended on it were inevitable. Since McGowan's study came out, declines of species throughout the area have been attributed to the loss of zooplankton and the warming water.

The crash showed up in fish, although it is often tough to tell if such declines come from too many nets or too little fish food. But even when researchers look at species for which human markets have no appetite, they find precipitous declines. The larvae of *Leuroglossus stilbius*-a fish of so little market value that it does not even have a name in English-historically are the third most abundant in the California Current.

Counts of its larvae dropped 50 percent after 1977. Another similarly ignored species with no common name, *Stenobranchus leucopsarus*, saw its larvae drop 42 percent after the sharp temperature rise. Its larvae are typically the sixth most abundant in those waters.

In 1967, aerial surveys found 70 square miles of kelp forests along the long California coastline. In 1989, that number dropped 42 percent. By 1999, the most recent year for which data are available, the total plummeted to just 17.8 square miles, down 75 percent from the 1967 survey.

BREAKING THE TOP OF THE CHAIN

But the most dramatic decline came to the sooty shearwater, a predatory seabird at the top of the marine food chain. "In the 1960s and 1970s they were present in the tens of millions," McGowan says, "the largest population of pelagic [marine] seabirds in the entire California Current. They dominated it. Millions and millions of them." The birds feed on juvenile fish and larger zooplankton. Researchers began looking at the birds regularly in 1987. By the 1990s, the population of sooty shearwaters-like the guillemots in Alaska-had crashed, with numbers down 90 percent.

"The decline of the sooty shearwater is very dramatic," McGowan says. "And that clearly is not due to man harvesting it, or at least not directly." As of 2003, water temperatures in the California Current are back down to their long-term average, but zooplankton numbers have not changed. In fact, McGowan says, samples taken in February 2003 showed the lowest abundance ever recorded.

Eight years after first reporting the zooplankton die-off, McGowan thinks he knows why it occurred: The warm surface layer of the ocean got so deep that the nutrient-rich waters below could not get close enough to the light to help the phytoplankton grow. McGowan compares that oceanic phenomenon to the familiar warm layer on a lake or swimming pool. When you go in, he explains, "Sometimes your feet are cold and your upper body's warm. Warm water floats on top of cold because it's less dense. And because there's this sharp layer between the warm upper water and the cold lower water, it makes it difficult to stir."

In the California Current, McGowan and his colleagues have found that the line between warm surface water and cooler, nutrientrich water became substantially deeper in 1977, when everything heated up. The deeper that line goes, the harder winds need to blow to mix up the nutrients. And the winds have not changed. That means the basic driver of ocean life is on an extremely low setting. "Less productivity, less plankton and birds and squid and fish," he says.

Although McGowan's work is limited to southern California, the earliest CalCOFI data showed that what happened on the south end of the California Current was very similar to what happened further north. That means the die-offs he has documented could be repeated all the way up the West Coast. "The whole system pumps up and down and up and down," he says. "It's one system, in spite of all of the eddies, in spite of all the meanders, in spite of all the species that are involved. We know that the temperature change is still synchronous, all up and down the coast. That has been extensively tested."

And that is perhaps the most important implication of McGowan's research: The ecosystem crash documented in southern California could happen along with warming in any part of any ocean anywhere. Some British studies are turning up evidence of deepening surface layers of warm, nutrient-poor water, McGowan says. U.S. agencies and the Intergovernmental Panel on Climate Change have documented higher ocean surface temperatures around the world.

"This 1977 regime shift that we thought was something that happened in the North Pacific, there's evidence globally that temperatures took a big jump up, " McGowan says. "It wasn't just California, the West Coast, the North Pacific or the Gulf of Alaska. It was everywhere." "It's a very, very serious problem," he adds. "And the seriousness comes from the fact that we really don't know what the consequences will be." The crashes McGowan has seen "could be a generalized response of the whole ocean-all the world's oceans-to heating. It might look something like what we see in the California Current."

As with Gilman and Sagarin's research, McGowan cannot definitively prove that increased carbon in the atmosphere caused heating that was the direct and sole cause of the changes he has seen. But his research continues to suggest temperature increases as the most likely driver for his observations.

RISING WATERS

Climate change appears to be making big-picture changes to transcontinental currents and the rocky interface of shore and ocean, but rising sea levels also combine with runaway development to devastate local environments. In the last remaining salt marshes ringing San Francisco Bay, a bird and a secretive mouse are steeling to fight their own battles with climate change. The problem there is not that the water is getting too warm, but that it is getting too high.

As glaciers melt and oceans rise, so too will the waters of the famous bay. That would not necessarily be a problem, since the marshland plants and animals in theory could simply move upslope, in much the same way Sarah Gilman's seaweed moved down toward cooler water. But two endangered species-the California clapper rail, a secretive bird that does not much like to fly, and the salt marsh harvest mouse, which lives without drinking fresh water-are stuck between rising water and asphalt, the latter covered by multimillion-dollar development. If or when the water rises, they will not be able to afford the new rents. Marge Kolar, who manages the Don Edwards San Francisco Bay National Wildlife Refuge, says that 150 years ago the water was ringed by 50,000 acres of muddy tidal flats and 190,000 acres of lush tidal marsh. Today only three-fifths of the tidal flats remain, and a mere onefifth of the marshland.

The clapper rail and harvest mouse live nowhere else on earth, and need the marshes to live, hide from predators, and feed. The clapper rail was one of the first birds put on the federal list of endangered species, and only about six hundred individual birds are still alive to perpetuate the species. No one knows how many of the mice there are. Their population is assumed to have declined as much as their habitat has: 79 percent.

From the hill above the refuge's visitor centre, the marshes look like geometric farm plots outlined by sandy dirt roads. Here, however, the fields are ponds used by Cargill Salt to evaporate water and harvest salt, and the roads are earthen levees 5 to 6 feet tall that keep the waters of the bay out.

Before those walls went up, cordgrass grew out of the mud ringing the bay at low tide. As it eased away from the water's edge, the cordgrass gave way to salty, bitter pickleweed, whose round, segmented stems grow like anemone arms, historically feeding the native Ohlone Indians and still the major source of water for the harvest mouse. Further above the tide line, the pickleweed in turn gives way to the shrubby, yellow-flowered gumplant that the clapper rail runs to when it is hiding from predators. It is possible to reverse time and turn salt ponds back into marshland, Kolar says. It just takes breaching the levees and letting the bay waters go to work. As an example, she points just east of the refuge's visitor centre. The refuge bought the land from Cargill and, in 1985, breached the levees, and let tide water flow back in. Although it did not have the complexity of the more mature marsh nearby, by 2000 harvest mice were in it, munching the new pickleweed and making themselves at home.

But the busy commuter street just behind the recovering marsh, Thornton Road, is a sharp dividing line between the ecological reality of the harvest mouse and the economic reality of the area's famed computer industry. Just behind it rise the shiny offices of Sun Microsystems, which paid Cargill $477,000 per acre for the old salt pond at the peak of the Silicon Valley boom. Kolar is quick to say that the refuge did not pay anything like that amount of money when it bought the land for restoration. In the spring of 2003, Kolar's refuge and the state of California joined forces to buy another 16,500 acres of salt pond from Cargill at bargain, nondevelopment prices in a bad economy-slightly above $6,000 per acre. The purchase nearly doubles the total protected area around the bay, bringing it up to 38,500 acres of total refuge land. Kolar says the U.S. Fish and Wildlife Service and two state agencies are beginning to plan for restoration, and they will take rising water into account.

Nevertheless, the purchase became controversial because details about pollution and the true value of the land were kept from the public. Newspaper reports say the $100 million deal was based on outdated economic assumptions-especially prickly given the collapse of the Silicon Valley economy-costing taxpayers unnecessary millions. And while it may be physically easy to turn salt ponds back into marsh, factors other than development and rising sea levels are complicating restorations. Especially in the south part of the bay, near San Jose, the land ringing the bay is slowly subsiding. Between the dropping land

and the rising water, the native endangered species are getting further squeezed. For them, the simple solution is to break the salt-pond levees. But doing so now carries another political cost: Those levees currently protect the city of San Jose from encroaching bay water.

The new land purchase will do the clapper rail and the harvest mouse little good if it is inundated and there is nothing but asphalt when the cordgrass, pickleweed, gumplant, and assorted species that depend on them try to move upward away from the rising water. But exactly what is happening on the edges of the bay is still murky. At the same time as the water is rising and the land subsiding, development and erosion are constantly spilling more sediment into the bay. Those sediments fill in the very areas where the mouse and the rail live, with the possible effect of creating a new ring of habitat.

The big question is which will come first: rising water pushing these endangered creatures onto asphalt or rising sediments building them new homes. "Will [sediment deposition] keep up with sea-level rise?" Kolar asks. "That's a question people haven't answered yet." In the world of science, nothing is ever proven. Researchers say their findings "show" a certain thing, and may discredit some previously held notions. McGowan, Sagarin, Gilman, and Baxter cannot say that the crashing food chain or the migration of intertidal species they documented is "caused" by a warming world, although they all seem to believe that is the case. But all the pieces together begin to form a picture. If it is not a picture of global warming, it is certainly a picture of what global warming could bring.

5

Aquatic Biodiversity: Threats and Conservation

Biodiversity or Biological Diversity a sum of all the different species of animals, plants, fungi, and microbial organisms living on Earth and the variety of habitats in which they live. Each species is adapted to its unique niche in the environment, from the peaks of mountains to the depths of deep-sea hydrothermal vents, and from polar ice caps to tropical rain forests. According to the definition of the Convention on Biological Diversity, biodiversity is the variability among living organisms from all sources, including terrestrial, marine and other aquatic ecosystems and the ecological complexes of which they are part; this includes diversity within species, between species and of ecosystems.

Aquatic biodiversity can be defined as the variety of life and the ecosystems that make up the freshwater, tidal, and marine regions of the world and their interactions. Aquatic biodiversity encompasses freshwater ecosystems, including lakes, ponds, reservoirs, rivers, streams, groundwater, and wetlands. It also consists of marine ecosystems, including oceans, estuaries, salt marshes, seagrass beds, coral reefs, kelp beds, and mangrove forests. Aquatic biodiversity includes all unique species, their habitats and interaction between them. It consists of phytoplankton, zooplankton, aquatic plants, insects, fish, birds, mammals, and others.

IMPORTANCE OF AQUATIC BIODIVERSITY

Aquatic biodiversity has enormous economic and aesthetic value and is largely responsible for maintaining and supporting overall environmental health. Humans have long depended on aquatic resources for food, medicines, and materials as well as for recreational and commercial purposes such as fishing and tourism. Aquatic organisms also rely upon the great diversity of aquatic habitats and resources for food, materials, and breeding grounds. Factors including overexploitation of species, the introduction of exotic species, pollution from urban, industrial, and agricultural areas, as well as habitat loss and alteration through damming and water diversion all contribute to the declining levels of aquatic biodiversity in both freshwater and marine environments. As a result, valuable aquatic resources are becoming increasingly susceptible to both natural and artificial environmental changes. Thus, conservation strategies to protect and conserve aquatic life are necessary to maintain the balance of nature and support the availability of resources for future generations.

THREATS TO AQUATIC BIODIVERSITY

Human activities are causing species to disappear at an alarming rate. Aquatic species are at a higher risk of extinction than mammals and birds. Losses of this magnitude impact the entire ecosystem, depriving valuable resources used to provide food, medicines, and industrial materials to human beings. Runoff from agricultural and urban areas, the invasion of exotic species, and the creation of dams and water diversion have been identified as the greatest challenges to freshwater environments.

Overexploitation of aquatic organisms for various purposes is the greatest threat to marine environments, thus the need for sustainable exploitation has been identified by the Environmental Defence Fund as the key priority in preserving marine biodiversity. Other threats to aquatic biodiversity include urban development and resource-based industries, such as mining and forestry that destroy or reduce natural habitats. In addition, air and water pollution, sedimentation and erosion, and climate change also pose threats to aquatic biodiversity.

- *Overexploitation of species:* Overexploitation of species affects the loss of genetic diversity and the loss in the relative species abundance of both individual and/or groups of interacting species. The population size gets reduced because of disturbances in age structure and sex composition. Efficient gears remove quick growing larger individuals consequently, the proportion of slow growing ones increases and the average size of individuals in a population decreases. Over-fishing causes change in the genetic structure of fish populations due to loss of some alleles. Thus, genetic diversity gets reduced.
- *Habitat modification:* Physical modification of habitat may lead to species extinction. This is mainly caused due to damming, deforestation, diversion of water for irrigation and conversion of marshy land and small water bodies for other purposes. Construction of dams on river impedes upstream migration of fishes and displaces populations from their normal spawning grounds and separate the population in two smaller groups. Deforestation leads to catchment area degradation due to soil erosion which results into sedimentation and siltation. This not only affect the breeding ground of aquatic organisms but cause gill clogging of small fishes also.
- *Pollution load:* Four forms of pollutants can be distinguished-

- *Poisonous pollutants:* Agrochemicals, metals, acids and phenol cause mortality, if present in a high concentration and affect the reproductive functionality of fish.
- *Suspended solids:* It affects the respiratory processes and secretion of protective mucus making the fish susceptible to infection of various pathogens.
- *Seewage and organic pollutants:* They cause deoxygenation due to eutrophication causing mortality in fishes.
- *Thermal pollution:* It cause increase in ambient temperature and reduce dissolved oxygen concentration leading to death of some sensitive species.

These factors affect the aquatic biodiversity directly or indirectly. Excessive mortality of organisms due to any of these factors may lead to two type of effects:

- Extinction of the species/ populations
- Reduction of population size.

CONSERVATION APPROACHES

Aquatic conservation strategies support sustainable development by protecting biological resources in ways that will preserve habitats and ecosystems. In order for biodiversity conservation to be effective, management measures must be broad based.

- Aquatic areas that have been damaged or suffered habitat loss or degradation can be restored. Even species populations that have suffered a decline can be targeted for restoration (e.g., Pacific Northwest salmon populations).
- An aquatic bio- reserve is a defined space within a water body in which fishing is banned or other restrictions are placed in an effort to protect plants, animals, and habitats, ultimately conserving biodiversity. These bio-reserves can also be used for educational purposes, recreation, and tourism as well as potentially increasing fisheries yields by enhancing the declining fish populations. These bio-reserves are also very similar to marine protected areas, fishery reserves, sanctuaries, and parks.
- Bioregional management is a total ecosystem strategy, which regulates factors affecting aquatic biodiversity by balancing conservation, economic, and social needs within an area. This consists of both small-scale biosphere reserves and larger reserves.
- Watershed management is an important approach towards aquatic diversity conservation. Rivers and streams, regardless of their condition, often go unprotected since they often pass through more than one political jurisdiction, making it difficult to enforce conservation and management of resources. However, in recent years,

the protection of lakes and small portions of watersheds organized by local watershed groups has helped this situation.

- Plantation of trees in the catchment area of water body prevent soil erosion and subsequently reduce the problem of slitation in water body resulting in better survival of aquatic organisms.
- Avoid the establishment of industeries, chemical plants and thermal power plants near the water resources as their discharge affect the ecology of water body resulted in loss of biodiversity.
- The World Resources Institute documents that the designation of a particular species as threatened or endangered has historically been the primary method of protecting the biodiversity.
- Many specialized programs should be instituted to protect biodiversity. For example, the USDA Forest Service started a cooperative state-federal program with a goal to restore the health of riverine systems and associated species.
- Regulatory measures must be taken on wastewater discharge in the water body to conserve biological diversity.
- Increasing public awareness is one of the most important ways to conserve aquatic biodiversity. This can be accomplished through educational programs, incentive programs, and volunteer monitoring programs.
- Various organizations and conferences that research biodiversity and associated conservation strategies help to identify areas of future research, analyse current trends in aquatic biodiversity.

AQUATIC BIODIVERSITY IN FRESWATERR

Freshwater is critical to human society and sustains all terrestrial and aquatic ecosystems. Worldwide, freshwater fishes are the most diverse of all vertebrate groups, but are also the most highly threatened through anthropogenic activities such as river management works, dam building, and land use change in the watersheds. Therefore, studies are being executed to develop tools for freshwater biodiversity conservation, and various methods and strategies have been proposed. The need to protected freshwater habitats, rare or endangered species, and intact waterways have been widely justified.

India has developed a network of 605 protected areas covering approximately 4.74% of the total geographical area of the country in the form of 509 wildlife sanctuaries, 96 National Parks, and three conservation reserves under "Wild life (Protection) Act". The total protected areas have been earmarked for extensive conservation of habitats and ecosystems. However, a review of the protected area network in India reveals a poor representation of freshwater fish biodiversity in that network. Recently, the Ministry of Environment and Forests, Government of India, has prepared National

Biodiversity Action Plan to help conserving biological diversity in both terrestrial and aquatic ecosystems. India has very rich aquatic biodiversity spanning the country. In India there is about 2 319 fish species that have so far been documented, of which about 838 fishes inhabit freshwaters.

The Gerua River originates in the Himalayan mountains, crosses through the Royal Bardi National Park in Nepal, and enters India at the Katerniaghat Wildlife Sanctuary in the Terai region, Bahraich district, Uttar Pradesh. The River is large with a mean annual discharge near 1 500m^3/s. The width of the river channel varies considerably with location, discharge, and the number of channels on a cross section. The presence of protected area on the upper stretch and forest cover on the mid stretch of river tend to have positive impact on its aquatic habitat.

No major degradation of the habitat exists within the sanctuary, except occasional secretive use of insecticides by local fishermen, illegal poaching, fishing by small mesh and large catches of fish by the contractors in the buffer area of the sanctuary. Downstream of the sanctuary, the Girijapuri Barrage diverts some of the water flow of Gerua River for irrigation and then joins the Ghagra River, which is one of the major tributary of River Ganges. In India, studies on freshwater fishes in rivers were primarily focused on the catch data of fishes of commercial value. A review of published literature shows that very few studies on fish diversity have been completed in India. Besides, conservation information on the pattern of fish biodiversity, abundance of threatened and endangered fishes, and threats in the rivers and streams are very limited in India.

Nevertheless, recently, it has been observed to decline rapidly due to environmental degradation like urbanization, damming, abstraction of waters for irrigation and power generation, and pollution. These environmental impacts have induced severe stress on freshwater fish diversity. In view of the worldwide significance of Indian freshwater fishes, effective planning for conservation and management strategies are now important and a pressing challenge. The objectives of this study were to evaluate fish species diversity, composition, individual sizes, habitat, and threats within and outside a river protected area. Our purpose was also to assess whether this protected river area provides benefits to freshwater fishes so that this concept can be used further in India and other developing countries.

MATERIALS AND METHODS

Study site: The studies were made in two section of the river Gerua. One section of 15 km stretch within the Katraniaghat Wildlife Sanctuary (28°20-"12.54" N and 81°06-"57.76" E to 28°22-"11.27" N and 81°12-"96" E). The second section of 85 km stretch belonged to a downstream unprotected area. In the protected area, most of the sampling sites were shaded by dense riparian vegetation. The surrounding land in the sanctuary is protected for conservation purposes and is composed of grasslands, deep forest, terrestrial wild animals and marshy lowlands. The unprotected sites were mainly composed of agricultural land use pattern, rural hamlets with relatively less riparian vegetation.

Field sampling and analysis: The field portion of the study started in April 2000 and continued until March 2004. Field sampling was done annually by the seasons: pre-monsoon (March-May), monsoon (June-September), and winter (October-February). Fishes were collected from nine sampling sites identified as S1 to S9; four sites within sanctuary, and five sites outside this protected area. Fishing is prohibited in all sanctuary sites although some poaching was observed, nevertheless, it was very common outside the sanctuary. The sampling sites (approximately 200 m long) were selected in upstream, midstream, and downstream areas within the sanctuary and outside the protected area. Sites were selected to include multiple representative habitats: mid-channel, shoreline, run and riffle, small channels and wetlands connected ditches, shallow ponds, and pools. Habitat features measured were: water depth (m), water temperature (°C), transparency (cm), water flow (km/hr), conductivity (μ S/cm). Water depth was the average of seven to 10 measurements covering the channel portion of the sampling site at the time of sampling. Water temperature and conductivity were measured by Cyber Scan Waterproof PC 300 multiparameter equipment.

Fishes were collected with gill nets (mesh 2.5 × 2.5 cm; 3 × 3 cm; 7 × 7 cm; length × breadth = 75 × 1.3 m, 50 × 1 m, and 30 × 1 m). At each site, a standard set of gill nets was deployed overnight (17:00-07:00h) and fishing was carried out by using the expertise of local fisher folks. To confirm whether all the species of fish were collected from gill nets, additional data was obtained on number of occasions from the local markets, where fish were sold, and look after the presence of any species that were caught by variety of gears (traps, harpoons, hooked lines, cast nets, among others).

Unless collected directly from the source, the habitat from which the fish originated was not traced, no quantification of fish abundance data was attempted for any of these additional sources, since reliable information was usually not available. All specimens collected were identified using the classification system of Talwar & Jhingran. Any colour, spots, maximum size and other characters of the fishes caught were recorded. Fish samples were preserved in 10% formalin solution. Species relative abundance (RA) in the catch was computed by the following formula:

$$RA = N / S$$

Where:

N=Total number of individuals of specific species

S=Total number of all fishes

The species diversity index was calculated using the Shannon-Weiner Information (Shannon & Wiener 1963):

$$H = \sum (n_i / N) \log_2 (n_i / N)$$

Where:

n_i=Total number of individuals of specific species

N=Total number of individuals of all species

We used Jacquard's index to compare the similarity of the species in the sampling sites within the sanctuary and outside:

$$Sj = j / (x + y - j)$$

Where:

Sj= Is the similarity between any two zones X and Y.

j= The number of species common to both the zones X and Y.

x= The total number of species in zone X and Y.

The similarity in species composition within the sanctuary and unprotected area of the Gerua River was obtained from the Jacquard's index similarity coefficients generated by using the EstimatesS (version 5.0.1) software.

A data matrix was constructed with presence and absence of fish species for each of the sampling sites in the sanctuary and unprotected areas. Analysis of variance was conducted to test the presence and abundance of fish species across the sampling sites. We used the Duncan's new multiple range test to identify the sites that differed, and calculations were performed using SPSS software package. References to conservation status categories within this chapter (endangered species, vulnerable species among others) are based on CAMP and IUCN. Data regarding threats faced by the fish fauna within the sanctuary and unprotected river areas were obtained from direct observations and interactions with local stakeholders and fisherman.

RESULTS

Physicochemical environment: There were physicochemical environmental attributes that showed little variation over season and river reach, and there were some that varied considerably. Table presents mean values and standard deviations across seasons and sites in the sanctuary and unprotected sites. Hydrogen ion concentration (pH) was rather stable across seasons and river sites in the sanctuary and unprotected reaches. The pH across seasons varied slightly (7.0-7.8), and mean pH value in the sanctuary was similar (7.4) to unprotected sites (7.3). Likewise, conductivity recorded in the sanctuary area varied from 201-239 μ S/cm across seasons and was slightly higher in winter. In the unprotected area, conductivity was observed between 201-220μS/cm across seasons and again it was higher in winter. The sanctuary river water was very similar (mean 217 μ S/cm) to the unprotected river water (210 μ S/

cm). Air temperature did not vary much between sites in the sanctuary (28°C) and unprotected sites (26°C). However, through the air temperatures were warm (28-26°C) premonsoon and monsoon seasons and dropped much lower in winter (16-19°C).

Water temperature was warmest during the premonsoon season (25°C), cooled some in monsoon season (21-22°C), and declined during the winter (14-15°C). Turbidity varied by season considerably with the monsoon season with lowest transparency (2-15cm) with downstream sites being the most transparent. Other season had much clearer water (18-50cm) with the clearest water downstream of the sanctuary. There was clearly deeper water in the sanctuary (4-10m) compared to the unprotected sites (1-2m).

Both river sections observed deeper water column during the monsoon season due to greater river flows. Current velocity in the sanctuary area varied from slow (2.93cm/s) to moderately swift (14.9cm/s) across seasons. In the unprotected river velocity were higher than the sanctuary waters, and ranged from 11-13cm/s. However, water depth was much shallower and expected cross section area was much less. That may explain the differences between the sanctuary and unprotected river in terms of current velocity.

FISH COMMUNITIES ACROSS SITES

A total of 6 220 fish were collected from the sanctuary river waters and unprotected river sites, and there were 87 species representing 22 families and 52 genera. The Cyprinidae was the dominant family with 40 species consisting almost half (49%) of all fishes collected. Following, the family Bagridae had seven species and 8% of all fish collected. Then, the Schilbeidae had five species represented 6% of all fish collected. Overall, the most numerous species was a Cyprinid Salmostoma bacaila in terms of relative abundance (RA) with 11% and 1 113 individuals.

The next most numerous were the Schilbeid Eutropiichthys vacha (8%), Notopterid Notopterus notopterus (6%), Schilbeid Clupisoma garua (5%) and Sisorid Bagarius bagarius (5%). While the relative abundance of two Indian major carps, Catla catla and Labeo rohita, was relatively low (<1.5%). However, a bottom dwelling Indian major carp Cirrhinus mrigala was more abundant (2%).

In the collections, there was an abundance of threatened fishes (3 848 individuals, 28 species) like Chitala chitala, Notopterus notopterus, Ompok bimaculatus, Bagarius bagarius, Tor putitora, Tor tor, Eutropiicthys vacha, Pangasius pangasius and others. In this study, 28 endangered and vulnerable species by Indian classifications (National Bureau of Fish Genetic Resources, 1998) constituted almost of third (31%) of the fish collected. Under conservation status as per IUCN (2010) two species (Schizothorax richardsonii and Tor putitora) were categorized as endangered, seven species were near threatened, 10 under least concern, and one species under data deficient. Among minor carps, Labeo bata and Cirrhinus reba showed higher relative abundance. Among exotic species, Cyprinus carpio was recorded inside the sanctuary.

Of the 87 species recorded from the sanctuary river waters, 12 species (14%) that are oriented to coolwater habitat, 46 species (53%) are warmwater and also recognized as food fish, 13 species (15%) as good aquarium fishes, and 4 species (5%) considered as high profile game fishes in India (Tor tor, T. putitora, Aorichthys aorand Bagarius bagarius). The study also showed new record in maximum length of six freshwater fish species:Notopterus notopterus (35 cm), Gudusia chapra (20 cm), Barilius barila (13 cm), Ompok pabo (28 cm), Xenentodon cancila (31.5 cm) and Salmostoma bacaila (20.5 cm).

The Shannon-Weiner diversity index across all sites in the sanctuary ranged between 3.8-5.2 with a mean of 4.4 as compared to 2.9-4.8 with a mean of 4.2 at sites that were in unprotected areas. The Mann-Whitney test of the Shannon-Weiner diversity index values between the sanctuary sites and unprotected sites were non-significant ($p = 0.63$). The Jacqard's similarity index in the sites of protected and unprotected areas showed more similarity across the sites regardless if they were in sanctuary waters or unprotected waters.

Comparison between sanctuary river sites and unprotected sites: In this study, diversity of freshwater fish in the study area showed a considerable difference between the sanctuary and unprotected river sites. Among four sites studied in the sanctuary, site S4 had the maximum diversity in terms of species (65) and genera (45) and site S2 had the lowest (28 species and 21 genera). In the unprotected river, site S7 had the maximum diversity (55 species and 44 genera) and site S9 had the lowest (17 species and 12 genera). Genera like Labeo (8 species) and Puntius (7 species) were dominated in the protected area followed by unprotected area (7 and 6 species).

Within the sanctuary three sites (S1, S2, S4) were distinct in fish composition while for the unprotected sites two were unique (S8, S9).

This study showed a distinct variation ($p < 0.05$) in the relative abundance of 13 threatened fish species in the sanctuary sites and unprotected sites. Many of the threatened fishes were found abundant in the sanctuary areas than the unprotected areas. Fish species like, E. vacha, C. gerua followed by B. bagarius, O. bimaculatus, C. reba, T. tor and T. putitora were the most abundant species in both protected and unprotected areas. In the unprotected areas, species like O. pabda and O. pabo were recorded quite higher in abundance than the protected areas.

In the present study, the diversity of species in the protected and unprotected areas varied considerably from site S1 to S9. On comparison, it was found that all the sites in the protected areas not necessarily showed rich fish diversity than the sites located in the unprotected areas. This variation is of high interest depending on the site specific habitat characteristics. For instance, the river habitat in the sanctuary area at sites S3 and S4 was highly diverse including a network of small channels, wetlands, presence of instream structures (undercut banks, woody logs and debris, overhanging vegetation, among others), and dense riparian vegetation which could be greatly responsible for high species diversity.

Moreover, these sites were located downstream from a barrage forming larger pool habitats supporting fishes to accumulate in certain habitats. Low human disruption and maintenance of ecological integrity in the sanctuary area can be another reason for more diversity and fish abundance. However, sites that did not follow the same pattern might be attributed to fast free flowing river water (site 1 & 2) which prevents fishes for accumulating in certain habitats.

In the present study, water depth was clearly deeper in the sanctuary area as compared to the unprotected sites. In flowing waters the number of fish species were more numerous with increasing water depth. Sarkar & Bain analysed habitat relationships in River Gerua at the sanctuary site, and found three grouped fish species occupying distinct habitats. The deepest habitat had the highest species diversity. Large rivers differ from small streams in several attributes like lack of light penetration to bed, greater depths and velocities on average, greater ratio of volume to margins, and greater physical stability.

Deeper habitats can influence fish species diversity and abundance in the Gerua River. Finally, unprotected sites are known to have higher fishing intensity. When a river is in a sanctuary setting there are a number effects that come with protection of human activities, and it could be hard to pinpoint the cause of increased diversity and abundance.

As described earlier, the diversity of a certain site in the studied river does not directly depends on the protected area boundaries, and can be reflected in rich species diversity of sites S6 and S7 of the unprotected areas; this could be attributed to site specific suitable environmental conditions. In addition, these sites were seen in supporting large group of macrophytes which diversify aquatic habitats and provide support for more fish and greater numbers.

In contrast, the minimum species diversity at S9 might be due to more distance from the protected area, effect of sedimentation, loss of breeding and spawning grounds, and over harvesting. Loss and reduction in diversity and abundance of fishes were consistent with and declining water quality and loss of habitats as a cause of human activities. Fish communities in riverine systems typically follow a pattern of increasing species diversity and abundance from upstream to downstream. We documented a contrasting, inverse relation for species diversity and abundance downstream direction on the river.

The Shannon-Weiner diversity index of the sites within sanctuary and unprotected sites slightly differed from S1 to S8, and considerable difference was observed in S9. This site indicated the maximum effect of protection because it was the furthest from the sanctuary. The Jacqard's index showed the site wise differentiation in species composition in the protected and unprotected areas. The possible reason of low similarity index in the unprotected areas might be the differences of the land use pattern, high sedimentation rate, suburban communal sewage, overfishing, and agricultural runoff. Paller et al. found higher minor variation in Jacqard's index among the undisturbed sites, while Jacqard's index was low at the disturbed site. This pattern is consistent with Paller et al.

Most importantly, the relative abundance of threatened fishes listed as endangered and vulnerable within the sanctuary and unprotected sites differed significantly. This might be due to differences in the land use pattern, intensity of human interference, turbidity, depth, water flow, wind, and differences in fish life histories. However, many factors are involved and real benefit of the protection has many consequences for the habitat and the river. This finding is consistent with the study by McClanahan & Kaunda that also found higher fish abundances inside than outside protected areas.

The composition of 12 endangered freshwater fishes at sites in the sanctuary and unprotected area were relatively heterogeneous. Three of endangered fishes (B. bagarius, C. gerua and E. vacha) were present at all the sites including the unprotected area indicating long range of distribution. The endangered mahseer, T. putitora and T. tor, exhibited a restricted range of distribution in the unprotected sites. A migratory large catfish, P. pangasius, was present at single location (S6) which is located nearest to the sanctuary area, indicating need of special conservation planning for this species. The present study also showed distinct variation in distribution pattern, size classes and abundance of the migratory species like B. bagarius, T. tor, P. pangassius, W. attu, S. silondia and C. chitala. In the sanctuary a unique behaviour was observed with the Feather back, C. chitala, which showed frequent sightings in the river with river dolphins (Platanista gangetica gangetica).

This study recorded maximum length values (TL) that exceeded the earlier records for six freshwater fishes, indicating that the protection of habitat can influence on fish length attainment. This study showed, with more species, greater abundances and larger individuals, and additionally showed higher number and densities of endangered fishes within the sanctuary area. Sarkar et al. 2007 reported a record size (22.5cm TL) of clupeid (Gudusia chapra) from another protected area. In the protected river, there are other studies which recorded higher abundances of fish, greater sizes of individuals and greater biomass.

This study is the first of its kind for the River Gerua in India which showed that a protected area is an effective tool that can benefit freshwater fish conservation, including species with high conservation value. This conclusion is new because the role of protected areas has been shown to benefits river habitats, rare and endangered species, and intact river systems.

In conclusion, it is found that freshwater protected areas commonly result in increased fish abundances for those threatened fishes which are extremely important for biodiversity conservation and management. Our observations also indicated that these areas, within wildlife sanctuaries, can be used as freshwater aquatic sanctuary (FAS), if additional measures are taken to protect these aquatic resources against actual threats.

DEEP SEA BIODIVERSITY

The deep-sea floor is the largest ecosystem on this planet: the average depth of the oceans is around 4.2 kilometres, comprising approximately 65% of the planet's surface. Nevertheless, we know much less about the deep-sea fauna than (for example) the rocky intertidal zone, which makes up a minuscule proportion of the global environment. Of more immediate importance, most of Australia's marine Exclusive Economic Zone is the deep sea (depths exceeding 200 metres) and yet this area is largely unexplored. Although the EEZ sea claim made by Australia is larger than any other country, this nation's basic biological research on the deep-sea floor is limited. My planned research will explore this region to discover how the deep sea can be of broader importance to Australia. This effort is just beginning, and substantial political and funding hurdles must be surmounted before detailed work in Australian waters can begin.

WHY ARE THERE SO MANY SPECIES IN THE DEEP SEA?

In the last 30 years, deep sea floor faunas have been discovered to have incredibly high species diversities, despite consisting of small numbers of individuals. Low population densities owe to the food-poor setting of the deep sea. I evaluate deep-sea faunas on a historical basis, where the processes of speciation, extinction and migration may control the observed levels of animal diversity, a regional pattern. My published work has concerned diversity and biogeography, the processes of speciation, the impact of the disturbance on diversity, and studies of the distribution of species diversity. Reports on NOAA (US National Oceanographic and Atmospheric Administration) funded research has been devoted to the quantitative description of community structure of crustaceans and polychaete worms from the Central Pacific (depths 4500-5100 meters). This research provides baseline data for understanding the impacts of deep-sea mining for manganese nodules.

Current research on deep-sea faunas. A 1992 paper by US colleagues J.F. Grassle and N. Maciolek estimates global diversity of the deep-sea benthos to be at least 10 million species based on extrapolations from work in the Atlantic Ocean; this result was amplified and defended by a letter to Nature. A different extrapolation, using the rate of species change with distance, had surprising consequences: globally isopods are more diverse than polychaete worms, despite local diversities showing the opposite relationship. Isopoda could exceed

400,000 species on a global scale, given the high rates of change observed in the North Pacific. In 1996, a workshop at the NCEAS in Santa Barbara: "Deep-Sea Biodiversity: Pattern and Scale" considered the how processes affecting biodiversity varied at different geographic scales. Research with David Thistle, Florida State University, evaluated the impact of hydrodynamic stress on isopod faunas. Other collaborations with colleagues at the University of Hawaii and The Natural History Museum have evaluated polychaete diversity and its relationship to productivity in the deep sea. Historical influences on isopod diversity in the Atlantic ocean is a continuing interest. More recently, research has evaluated the diversity of isopods in the Gulf of Mexicoat local and regional scales. An unusual fauna of deep-sea isopods on the outer shelf of the Arafura Sea has been discovered in 2005.

SMALL ISLAND BIODIVERSITY

Islands and their surrounding near-shore marine areas constitute unique ecosystems often comprising many plant and animal species that are endemic—found nowhere else on Earth. The legacy of a unique evolutionary history, these ecosystems are irreplaceable treasures. They are also key to the livelihood, economy, well-being and cultural identity of 600 million islanders—one-tenth of world population.

Island species are also unique in their vulnerability: of the 724 recorded animal extinctions in the last 400 years, about half were island species. Over the past century, island biodiversity has been subject to intense pressure from invasive alien species, habitat change and over-exploitation, and, increasingly, from climate change and pollution. This pressure is also keenly felt by island economies. Among the most vulnerable of the developing countries, small island developing States (SIDS) depend on the conservation and sustainable use of island biodiversity for their sustainable development.

The biological diversity and the high degree of endemism of many species on small island developing States is well known. Over 4,00 species of plants and animals are endemic to small island developing States. Because of their small size and the endemic nature of many species, the biological diversity of small island developing States is extremely fragile.

One consequence of the relative isolation of small island developing States is the large incidence of unique biological adaptations flightlessness in birds, gigantism and dwarfism in other groups, and many modifications of form, diet and behaviour.Restrictive habitats and small populations often generate unique features and adaptations to prevailing environmental and climatic conditions, but under such circumstances species often lack the ability to adapt to rapid changes. Small island developing States are not a homogeneous group, although many of them face similar problems with respect to the conservation and management of their natural resources.

CURRENT SITUATION

The current situation on aspects of biodiversity in small island developing States is described below. The focus is on:

- Deforestation and forest degradation,
- Traditional subsistence farming systems,
- In situ and ex situ conservation facilities,
- Coastal and marine biodiversity,
- Freshwater biodiversity, and
- Aquaculture.

Deforestation and Forest Degradation

Deforestation and forest degradation in small island developing States have led to extinction of many animal and plants species, resulting in irreversible losses of genetic resources and ecosystems. Considering their limited land area and their relative fragility, strong winds (e.g., hurricanes, cyclones, typhoons) can cause serious and frequently recurring damage to natural and planted forests. The impacts of human activities are usually even more severe. Deforestation and forest degradation have affected the dynamic interactions of ocean, coral reefs, land formations and vegetation.

Traditional Subsistence Agriculture

This practice, using family labour and a few purchased inputs, remains predominant in over half of all small farms in small island developing States. It has the advantage of being generally environmentally friendly but has the disadvantage of low productivity. Growing population and economic pressures are disrupting ecologically sound farming practices, such as fallow rotation systems. Thus, moderate input systems are becoming more common in small island developing States.

Because of inadequate farm and production management, these inputs are often not used effectively, leading to economic losses and environmental damage, including loss of biodiversity. With the encroachment of agriculture on natural or quasi-natural ecosystems, plant and animal genetic resources are being lost, modern cultivars are replacing local ones and intensive livestock production systems are developing. As a consequence, diseases and pests are increasing in small island developing States, particularly crop pests that are resistant to common pesticides.

Plant and Animal Genetic Resources

These resources are the basis for sustainable agricultural production. It is particularly important for small island developing States to have access to plant genetic resources from countries in the same agro-chemical zone for the diversification of their main crops. The establishment of protected areas —

forest reserves, national parks and wildlife sanctuaries, supported by botanical gardens, herbaria etc. — is essential for conserving small island developing States' biological diversity. Such areas also have the potential to become the basis for ecotourism.

However, establishment of in situ and ex situ conservation systems require both financial and human resources, often not adequately available in small island developing States. Therefore, cooperation among small island developing States is proving desirable as a means of evaluating and conserving small island developing States' genetic resources and safeguarding them for future use.

Fortunately, there appears to be a growing recognition of the dangers posed by the introduction of exotic genetic resources and the need for environmental impact assessments before such introductions take place. Practitioners of sustainable food production need access to reservoirs of crop cultivars, livestock breeds and aquaculture seeds with high productivity potential, good resistance and quality characteristics, and adaptability to local environments.

Marine Ecosystems

Small island developing States' marine ecosystems and biodiversity are especially susceptible to damage, including destruction of coral reefs by fishermen or tourists; pollution, sedimentation and land reclamation; natural disasters; conversion of mangroves and wetlands that result in loss of important nursery areas; use of large-scale pelagic driftnets which impact marine mammals, turtles, birds and non-targeted fish; and overfishing in general.

Coastal fisheries in small island developing States, once abundant, have become scarce owing to overfishing by both artisanal and small-scale commercial fishing. Inadequate monitoring makes it difficult to quantify the overall damage to marine life from such activities.

FRESHWATER BIODIVERSITY

Although it has not been studied as much as marine and coastal biodiversity, some small island developing States have important freshwater biological diversity. They include plants and animals that are rare, endemic, introduced species, including many that are threatened by habitat degradation. Unlike their marine counterparts, which may be dispersed over great distances by ocean currents, inland freshwater aquatic species have more restricted access to dispersion routes. Many native freshwater species are not currently utilized commercially and their status is not well documented.

Aquaculture

Although this industry is becoming more important for some small island developing States, the sector is faced with constraints and sustainability problems. Because of little previous experience in fish farming in many small island developing States, often there are few domesticated native species. Thus, new and/or genetically improved species have been proposed, which may play an important role in future aquaculture development. However, the introduction of non-native species may have environmental impacts and alter traditional ownership rights to land and water resources.

In view of the many factors affecting the biological diversity of small island developing States, the responsibility for conserving this biodiversity is disproportionate to their capabilities. With regard to nature conservation through designation of protected areas, there is considerable variation among small island developing States; many small island developing States have no protected areas, while others, such as Fiji, the Bahamas and Cape Verde, have considerable networks, often incorporating over 10 per cent of their land area.

Upon submission of the first national reports to the fourth meeting of the Conference of the Parties, more information on the current state of biological diversity of small island developing States and the status of implementation of the Convention on Biological Diversity will become available.

EFFORTS OF SMALL ISLAND DEVELOPING STATES AT THE NATIONAL AND REGIONAL LEVELS

Among the 41 States and self-governing territories classified as small island developing States by the United Nations, 32 have ratified the Convention on Biological Diversity; three others have become signatories as of 12 November 1997. Several small island developing States and archipelagoes in developing countries contribute to the international network of biosphere reserves and the United Nations Educational, Scientific and Cultural Organization (UNESCO) World Heritage List. Conservation in such areas is promoted through the exchange of information and experience among insular protected areas, the provision of technical assistance and training opportunities, and the development of comparative studies among groups of islands.

A pilot regional Global Environment Facility (GEF)-funded project in the South Pacific is being coordinated by the South Pacific Regional Environmental Programme (SPREP). It covers 15 island countries in the region, with the goal of protecting their biological diversity by facilitating establishment of conservation areas, with agreed development criteria based on long-term ecological sustainability. At the time of the mid-term review of the project in August 1996, 17 conservation areas were at various stages of planning and establishment.

INTERNATIONAL EFFORTS AIMED AT ASSISTING SMALL ISLAND DEVELOPING STATES IN THE AREA OF BIODIVERSITY

According to the GEF quarterly operational report of June 1997, GEF approved regular projects for the Comoros and Mauritius. It also endorsed enabling activities for 14 small island developing States, including Antigua and Barbuda, the Bahamas, Barbados, Cape Verde, the Comoros, Cuba, Fiji, Maldives, the Marshall Islands, Mauritius, Seychelles, Solomon Islands, Saint Vincent and the Grenadines, and Vanuatu.

In its preamble, the contracting parties to the Convention acknowledged the special conditions of the least developed countries and small island developing States. In its medium-term work programme, the Conference of Parties to the Convention, at its first meeting, agreed that the biological diversity of marine and coastal ecosystems would be the first thematic focus to be addressed at its second meeting.

Subsequently, at its second meeting, the Conference of Parties, in its decision II/10, gave specific guidance to the Executive Secretary regarding the conduct of work in this programme area, and also suggested that the United Nations Secretariat support the work of its Subsidiary Body on Scientific, Technical and Technological Advice in regard to the special conditions of least developed countries and small island developing States. Subsequently, at its third meeting, in its recommendation III/2, the Subsidiary Body recognized the special significance of small island developing States in the global conservation of marine and coastal biological diversity.

In its operational strategy for coastal, marine, and freshwater ecosystems, GEF has agreed that the needs of tropical island ecosystems will receive particular attention. With regard to mountainous ecosystems, it has also indicated that activities in this operational programme will initially address the conservation and sustainable use of biodiversity areas under increasing human pressure and imminent threat of degradation, including tropical islands.

In March 1997, the governing body of the Food and Agriculture Organization of the United Nations (FAO) approved a programme of fisheries assistance for small island developing States, which includes:

- Institutional strengthening and national capacity-building;
- Enhanced conservation and management of exclusive economic zone fisheries;
- Improved post-harvest fish management and marketing;
- A safety-at-sea component;
- Strengthening the economic role of national fisheries industries and privatization of fisheries investments; and
- Aquaculture and inland fisheries conservation, management and development.

During the period under review, FAO prepared a review of the uses and status of trees and forests in land-use systems in Samoa, Tonga, Kiribati and Tuvalu, including an annotated checklist of tree species of agro-forestry. The review contains recommendations for the development of agro-forestry strategies and the conservation of the diversity of species used in traditional agro-forestry systems. FAO previously assisted South Pacific countries in collecting, conserving, characterizing and documenting root and tuber crop genetic resources.

It plans to evaluate the status of existing ex situ collections of plant and animal genetic resources for the South Pacific region, and to propose actions to address related problems. It is also attempting to strengthen national capabilities for the conservation and sustainable utilization of existing genetic resource collections. In the period 1992-1994, it supported an improved seed production project covering 14 countries in the Caribbean region. The objective was to strengthen national capacities to conserve and document existing varieties of crops. Field and laboratory standards were established for seed quality control for sexual seed and vegetative planting materials to facilitate exchange, seed production and further development of seed technology.

The Caribbean Amblyomma Programme, which is supported by the Caribbean Community, is jointly executed by FAO and the Inter-American Institute for Cooperation on Agriculture. It became active in Anguilla and Saint Kitts and Nevis in 1995, and in Montserrat and Saint Lucia in 1996. Dominica, which is surrounded by heavily infested islands, launched its national programme in February 1997, and Antigua and Barbuda and Barbados were expected to join in 1997. The Programme has made notable progress in strengthening veterinary services and helping Governments improve their legislation and quarantine procedures. With funds from Japan, FAO operates the Fiji-based South Pacific Aquaculture Development. Fifteen countries from the region participate in addressing problems of sustainable aquaculture development, including the responsible use of alien species, human resource development in aquaculture, economic viability, and stock enhancement aimed at rehabilitating fisheries and habitats, especially in coral reef areas.

FAO and the United Nations Environment Programme (UNEP) have implemented a field project on stock enhancement of inland waters of Papua New Guinea. The project has facilitated the introduction of international codes of practice on the responsible use of introduced (alien) species intended to increase productivity of inland river systems. FAO's Special Programme on Food Security has also developed a multidisciplinary project in Papua New Guinea, involving water control, the intensification of production systems, diversification, the establishment of demonstration aquaculture farms and the analysis of socio-economic constraints.

The Intergovernmental Oceanographic Commission (IOC) of UNESCO provides a good scientific focus for regional activities in the field of biodiversity, as well as for inputs directly to the Convention secretariat, through the IOC Programme on Ocean Science in Relation to Living Resources (OSLR), the Global Coral Reef Monitoring Network and the IOC Marine Biodiversity Strategy. Over the longer term, the establishment of a Global Ocean Observing

System will provide a basis for systematic monitoring of factors relevant to assessments of marine biodiversity.

OSLR represents a framework programme for IOC member States to share knowledge, develop cooperative actions and address uncertainties about marine living resources and their biodiversity, through joint scientific research, in conjunction with international and intergovernmental initiatives and programmes. The IOC marine biodiversity strategy includes components on monitoring, training and capacity-building.

UNESCO activities relating to biodiversity in small island developing States under the Heritage Convention include promotion of international instruments for protecting biological diversity and the natural heritage; conservation as part of sustainable development; integrated coastal management; and traditional ecological knowledge about small island developing States biodiversity. As part of its medium-term strategy, the Twenty-Eighth UNESCO General Conference adopted a six-year multidisciplinary project, entitled "Environment and development in coastal regions and small islands".

In January 1996, the organization launched an interregional project, entitled "Integrated biodiversity strategies for islands and coastal areas", with the goal of assisting integrated biodiversity strategies for interregional cooperation between coastal States and islands, and promoting implementation of the Convention on Biological Diversity.

There has been heightened interest in traditional approaches to the use and management of natural resources. UNESCO work in this area includes promoting ethnobotany and sustainable use of plant resources within the World Wildlife Foundation (WWF)/UNESCO/Royal Botanic Gardens Kew Initiative on People and Plants; support to the regional non-governmental organization TRAMIL for work on popular use of Caribbean medicinal plants; and regional collaborative research on Caribbean coral reefs and beach stability.

UNEP extended assistance to small island developing States within the framework of the climate change, biological diversity, and desertification conventions. Small island developing States-related work within the UNEP sub-programme "Caring for freshwater, coastal and marine resources", includes facilitation of policy-relevant assessments of the state of small island developing States fresh and marine waters, and their living resources, and development of tools and guidelines for sustainable management and use of small island developing States' fresh and coastal waters and living resources.

Further, the recently adopted Global Programme of Action for the Protection of Marine Environment from Land-based Activities envisions a number of regional workshops, some in small island developing States, to address land-based environmental pollution, including its impacts on biodiversity. Under its sub-programme, "Caring for biological resources", and within the context of the Convention, UNEP is providing assistance to developing countries, including small island developing States, in the formulation of policy instruments for the integrated management of biological resources.

Small island developing States also benefit from UNEP scientific support and technical advisory services in preparation of:

- Biodiversity country studies;
- Biodiversity data management projects;
- The formulation and implementation of national biodiversity strategies and action plans; and
- The preparation of national biosafety frameworks, which also review the biodiversity of small island developing States.

An international coral reef initiative workshop was held in Seychelles in 1996 for the Western Indian Ocean and the East African region, to establish the basis for a strategy and action plan for conserving and managing coral reefs in the region. UNEP continues to assist small island developing States in enhancing their technical and managerial capacities for environmental management, including biodiversity, through its regional advisory services programme, which facilitates the continuous training of national experts.

The organization has secured GEF funds for biodiversity country studies, biodiversity data management, national biodiversity strategy and action plans and/or national biosafety frameworks for a number of small island developing States, including the Bahamas, Barbados, Cuba, Mauritius, Seychelles, Solomon Islands, Saint Lucia and Vanuatu. It has also produced technical documents on the features of biodiversity in small island developing States, using geographic information systems databases, and is currently assisting the Bahamas, Barbados, Seychelles, Solomon Islands and Vanuatu in developing national biodiversity action plans, funded by GEF.

Other activities undertaken by UNEP in support of small island developing States include:

- A workshop on implementation of international conventions related to biological diversity, held in Mozambique in 1997;
- Assistance in developing a draft law on marine conservation areas for Tuvalu; and
- Assistance in preparing a draft framework on environmental law for Kiribati, which contains provisions on conservation of biodiversity.

In 1996, the General Assembly of the World Conservation Union (IUCN) requested its members, commissions and the Director-General to assist small island developing States in the implementation of the Programme of Action for the Sustainable Development of small island developing States. Through its Biodiversity Policy Coordination Division, IUCN convened a workshop on additional funding sources for small island developing States for Convention implementation. Further, it organized a workshop on marine biodiversity jointly with the first meeting of the Conference of parties to the Convention. In collaboration with UNEP, IUCN is developing a global strategy and action plan for alien invasive species, which are a major problem for small island developing States' biodiversity.

IUCN has also engaged in work on management and sustainable use of marine and coastal resources through its regional and country offices. In the context of the Convention on Biological Diversity, IUCN is working to facilitate implementation of the Jakarta Mandate, including key small island developing States issues, such as protected areas, alien species, integrated marine and coastal area management, marine and coastal living resources and mariculture.

In collaboration with the Centre for International Environmental Law and WWF, it has published a document, entitled "Biodiversity in the seas". In collaboration with UNEP and World Resources Institute, it has co-published a document, entitled "National biodiversity planning: guidelines based on early experiences around the world."

The World Conservation Monitoring Centre (WCMC) has developed large databases on small island developing States biodiversity, which contain much information on small island developing States threatened species, habitats and protected areas. These databases have been the foundation for a series of recent WCMC reports on the status of small island developing States biodiversity. It also has undertaken a global review of small island developing States' biodiversity, identifying a number of global threats, including demographic and developmental pressures and natural disasters.

Issues relevant to small island developing States' biodiversity have been addressed in two global WWF publications, World Conservation Strategy, and Caring for the Earth: A Strategy for Sustainable Living. These publications give a comprehensive account of major small island developing States concerns, and provide a solid basis for the consideration of issues covered in the Programme of Action for the Sustainable Development of Small Island Developing States.

Specific actions to be taken by the secretariat of the Convention on International Trade in Endangered Species of Wild Fauna and Flora (CITES) include designation of a small island developing States coordinator for CITES activities; preparation of information packets for non-Party small island developing States; provision of technical assistance to parties and non-parties; and conveyance of specific recommendations to non-party small island developing States encouraging actions to promote compliance with CITES requirements.

ENVIRONMENTAL ASPECTS OF AQUACULTURE

Environmental issues have become of increasing concern for several reasons. The first is recognition of the increasing pressure on resources in some coastal and inland areas. More attention has been given to the impact of aquaculture on the environment in recent years, partly induced by some well publicized "crashes" within the shrimp industry. This has been accompanied by the publicity given to environmental and social issues surrounding aquaculture. This type of exposure has had a profound impact on the public's perception of aquaculture, changing it from the "blue revolution" which would improve the availability of cheap, affordable protein to poorer people, to that of being an environmentally unsound means to produce luxury food items for consumers in developed nations. Australia is adopting integrated farming technologies in a watershed where water is an economic resource and is at a premium (*i.e.* use must be paid for). This situation is in contrast to that encountered in some countries in the region where traditional integrated farming practices are reducing their levels of integration, applying less species diversification and more emphasis on high-value products.

BALANCED RESOURCE USE

There has to be a balanced and more efficient use of resources. Obtaining conflicts. This situation can be particularly acute when the government's capacity to enforce laws and regulations is limited.

It appears likely that economic issues related to resource use and the "polluter pays" principle will be faced, in the next few years, by aquaculturists in some countries of the region.

Minimizing Negative Impacts

Guidance is emerging for the application of practices for improved environmental management, and strategies have been identified for the following:

- Technology and farming systems: In recognizing the importance of appropriate farming technology/system and management of inputs and outputs, special attention is given to the major resources used (*i.e.* feed, water, sediments and seed). At this level, management actions mainly involve farmers and input suppliers.
- Adoption of integrated coastal area management approaches: The importance of integrating aquaculture projects within existing ecological systems in coastal areas is increasingly recognized. This approach requires the consideration of proper site selection and the application of planning and management strategies that allow the allocation of resources among different users.
- Policy and institutional support: It is necessary to have a clear and supportive policy framework for aquaculture. Particular issues include aquaculture legislation, economic incentives and disincentives, actions for the public image of the activity, private-sector and community participation in policy formulation, and increasing the effectiveness of research, extension and information exchange. Policy decisions and their implementation play a strong role in influencing the management possibilities at both the farm and local area levels.

The environmental assessment of an aquaculture area, rather than targeting individual projects, is now being used to identify opportunities for minimizing the impacts of the aquaculture sector within a particular area (*e.g.* in a bay, an estuary or a watershed). Such applicable measures are promoted through a wide range of instruments and should be brought together within the framework of an aquaculture development plan, ideally as part of an Integrated Coastal Management Plan. There is a growing awareness that environmentally friendly aquaculture makes good business sense, and this is particularly true for commercial producers. It is increasingly important when considering the perceptions of target export markets and provides an incentive for both the industry and supporting governments to further advance the adoption of environmentally and socially responsible farming practices through appropriate standards or codes of conduct. Recognized codes of practice require further development and implementation and are becoming particularly important for aquaculture products that are traded internationally.

AQUACULTURE AND ENVIRONMENTAL REHABILITATION

It is well known that inland aquaculture, particularly that within integrated agricultural farming systems, can provide many positive environmental benefits (*e.g.* water storage on small-scale farms, efficient use of farm resources). Perhaps less well known and publicized is the fact that coastal aquaculture, including shrimp culture, can also contribute to environmental improvement. For example, seaweed and mollusc culture can contribute to the removal of nutrients and organic materials from coastal waters. It can also provide an alternative source of income and employment for people involved in environmentally destructive fishing practices and other forms of habitat destruction. Other activities of note include the following:

- Mixed aquaculture-mangrove systems are being used to restore mangrove habitats in some countries (*e.g.* Indonesia's tambaks or earthen ponds, Viet Nam's mixed farming systems);
- Coral reef fish mariculture provides an effective alternative to destructive fishing practices in coral reef areas;
- The rehabilitation of fish populations through stock enhancement; and
- Aquaculture itself is also a technique for the effective monitoring of environmental conditions.

A recent review has suggested that aquaculture within saline inland areas can provide benefits through the use of previously degraded resources, contributing to environmental rehabilitation and poverty alleviation among communities whose livelihoods such damaged resources. Aquaculture is also being integrated as a component of mangrove rehabilitation projects (as in the Mekong Delta) and within coral reef management activities (small-scale cage culture).

BIODIVERSITY, GENETICS AND AQUACULTURE

The potential impact of farmed aquaculture stocks, especially introduced exotic species, on wild fish and fisheries has been highlighted at many international meetings. Among the potential impacts identified are a loss of genetic resources within healthy or undisturbed populations and a dilution of the wild gene pool through interbreeding with escapees from aquaculture facilities. Initiatives have been made recently to breed indigenous species for stock enhancement purposes as well as for aquaculture, but these are new initiatives and results that could guide development work are not available yet.

Aquafeeds and Feeding Strategies

When feeds are used in aquaculture, their contribution to production costs varies following the species and the system employed but can be as high as 50 percent of the total costs. Following concern as to the sustainable supply of fishmeal and fish oils, research has focused on identifying alternative, economically viable sources of proteins and lipids for incorporation into feeds. Improved management for more efficient feed use is another area of interest.

Some research (*i.e.* in Hong Kong) has taken a step further towards the development of environmentally friendly feed. Little information is available on the production of aquaculture feeds in Asia, but a recent estimate indicates a supply of some 1.9 million mt in Asia (excluding China), where 1.2 million mt were for fish and 0.7 million mt for shrimp. An estimate for China (1996) indicated a total aquaculture feed production of 5 million mt, but it was unclear if this figure also included farm-made feeds. The future prospect for fishmeal and oil for animal feeds is that global supply is expected to be stable. A distinction must be drawn between the different feeds manufactured for aquaculture species, *i.e.* whether for high-or low-value species. The higher values of commodity species, or those reared for export, can justify the investment required for the manufacture, distribution, purchase and use of specially formulated feeds. However, much of Asian aquaculture involves smallholder-based production of low-value species, either for personal consumption or for local sale. In the latter case, the input cost of feeds is generally negligible and, if feeds are used, they usually consist of cheap, readily available materials.

AQUACULTURE SYSTEMS AND SUSTAINABILITY DEVELOPMENT

Sustainable development and sustainability are complex issues that are difficult to define and apply to aquaculture. The "systems approach", however, can assist understanding of these issues, as they relate to aquaculture development. The term sustainability has been defined in various ways but perhaps the most widely used is based on the definition of "sustainable development" in the Brundtland report: "sustainable development is development that meets the needs of the present without compromising the ability of future generations to meet their own needs". An even more succinct definition is that of the International Union for the Conservation of Nature (IUCN): "sustainable development improves people's quality of life within the context of the Earth's carrying capacity". These definitions contain two key concepts: meeting the present and future needs of the world's people, and accepting the limitations of the environment to provide resources and to receive wastes for the present and for the future. The Food and Agriculture Organization of the United Nations (FAO), in particular, recognized that increased capacity at the national level is required to achieve sustainable development by including the need for "institutional change" in definitions of sustainable agricultural development. The recognition that institutions are important highlights the need for education and training, effective institutional arrangements and a legal and policy framework to underpin sustainable development of agriculture, and indeed aquaculture.

Sustainability is commonly split into three separate components: social sustainability (SS); economic sustainability (EcS); environmental sustainability (ES). Whilst social sustainability criteria are difficult to define, the definition of economic and environmental sustainability are providing a basis from which management options can evolve in aquaculture projects. To take sustainability to a more practical level requires consideration of environmental, social and economic issues in aquaculture development. Thus, the approach to sustainability implies a systems approach.

There are general guidelines available on the different issues to consider. The Code of Conduct on Responsible Fisheries (CCRF), adopted by the FAO Conference in 1995 in particular, identifies a number of key issues. The Code sets out principles and international standards of behaviour for responsible practices, to ensure effective conservation, management and development of living aquatic resources while, at the same time, recognizing the nutritional, economic, social, environmental and cultural importance of fisheries and aquaculture, and the interests of all those involved in these sectors. The Code's Article on Aquaculture Development contains provisions relating to aquaculture, including culture-based fisheries. Fundamentally, the Code recognizes the importance of activities that support the development of aquaculture at different levels:

- The producer level;
- The local area, *i.e.*, the farm and its integration into local area management and rural development schemes;
- The national institutional and policy environment; and
- Iinternational and transboundary issues.

The Code identifies many key principles in development of management strategies based on an understanding of aquaculture systems-from the farm to national and international levels. It also provides a basis for a systems approach.

THE SYSTEMS APPROACH

Farm Level

There is a lot of information on aquaculture farming systems, and various definitions are available, such as the level of intensity of management and output, and degree of integration with other on-farm activities. A considerable literature exists on integrated (agriculture-aquaculture and vice versa) farming systems (Edwards, 1998 and "Farming species and systems" in these proceedings).

However, there are a wide range of culture species, culture facilities and management practices in use, and thus a very wide range of farming systems. Key factors to be understood in the functioning of a farming system are the technologies of production and social, economic and environmental aspects. At the technology level, feeds, feed additives and fertilizers, water quality, seed quality and availability, chemoterapeutants and other chemicals, disposal of wastes that may adversely affect human health and/or the environment, and food safety of aquaculture products all require consideration. It has also been emphasised that a better understanding of the microbial populations in aquaculture systems, their interaction with the health of the farmed animals, and their role in maintaining a healthy aquatic environment are required.

The systems approach at the farm level can be used to understand and improve the efficiency of use of key natural resources-particularly water, nutrients, land, seed and financial resources. When focussing on the mix of all sustainability criteria through a systems approach, we have to deal with such factors as: appropriate densities, production and husbandry systems geared to the animal's health, maintaining ecological balance within the pond or other growout habitat, and provision of optimal social and economic benefits. It is inevitable that impacts on natural resources will become an increasingly important issue in the new millennium, and the systems approach can be used to analyse and develop the solutions required for more sustainable use of natural resources.

A wide range of management systems is already employed in aquaculture operations with varying degrees of success. Given that aquaculture systems range from small, relatively self-contained farms for subsistence, to large-scale commercial units for trade purposes, variable success is hardly surprising. Thus, a "one-size fits all" approach is unlikely to be successful. A systematic approach to production management, however, allows the farmer to manipulate and control production inputs that will result in more efficient, cost-effective production and minimize excessive outputs with negative environmental impacts. There is a tremendous body of information on site selection, farm construction and design features, aquatic animal health management, broodstock and seed production and care, production techniques, the use of appropriate feeds, feed additives and fertilizers, water and sediment management, including effluent control, and other topics. The challenge is to optimize dissemination and use of such information and experience. The systems approach can also be used for aquatic animal health management. Here the emphasis needs to move more towards management procedures, policies and products that can prevent or effectively eradicate significant pathogens, prevent re-infection through contaminants, and manage diseases in an environmentally sustainable manner. Subasinghe *et al.* (1998) provide a discussion of the role of the systems approach in aquatic animal health management.

The Local Level

The systems approach at the local level recognizes that individual farms cannot be seen in isolation, and that there are many interactions between an aquaculture farm and the external environment-including environmental resources and local communities. Furthermore, there can be significant cumulative effects where there are large numbers of farms crowded in small areas. Environmental interactions with aquaculture arise from a wide range of inter related factors including availability, amount and quality of resources; type of species cultured; size of farm; culture systems management; and environmental characteristics of the farm location. Environmental interactions are not limited to impacts of aquaculture on the environment, but include environmental impacts on aquaculture and impacts of aquaculture on aquaculture. Perhaps less well known or documented are the many ways that aquaculture can contribute to environmental improvement, for example mollusc farms' desedimentation or improved nutrient turnover, or water storage on small-scale freshwater farms.

At the local level, social and institutional interactions are also important and need to be better understood, for example:

- Participation of, and benefits to, rural communities;
- Institutional support through extension services;
- Access to information etc.

A systems approach attempts to understand these linkages and develop management strategies based on such understanding. The future of integration of aquaculture into local ecological and social systems requires more focus on local area development planning. Fortunately, increasing attention is being given to such issues, particularly in coastal areas. Integrated coastal management (ICM) is a process that addresses the use, sustainable development and protection of coastal areas, and according to GESAMP (1996) "comprehensive area-specific marine management and planning is essential for maintaining the long-term ecological integrity and productivity and economic benefit of coastal regions". ICM is made operational through such activities as:

- Land use zoning and buffer zones;
- Regulations, including permitting to undertake different activities;

- Nonregulatory mechanisms;
- Construction of infrastructure;
- Conflict resolution procedures;
- Voluntary monitoring; and
- Impact assessment techniques.

More participatory approaches to planning of aquaculture development will also be given attention with the move towards integrated development planning. Practical experience in implementation ICM for aquaculture is limited, which is in large measure because of the absence of adequate policies and legislation and institutional problems, such as a lack of unitary authorities with sufficiently broad powers and responsibilities, as well as limited training and education of people concerned.

In inland rural areas, increasing attention is being given to integration of aquaculture into rural development and special area management plans. Increasing emphasis is also being given to promotion of aquaculture for poverty alleviation. Such an approach requires emphasis on immediate social needs and people's livelihoods (and how aquaculture might meet these needs and contribute to improved livelihoods) rather than a technology/aquaculture driven approach. The emphasis on aquaculture for development, rather than development of aquaculture may lead to some fundamental changes in the approach to promotion of aquaculture in the coming years.

National, Regional and International Levels

At the national level, government policy, and institutional and human capacity are most important in providing a strong foundation for aquaculture to develop in a su. These issues are covered extensively in the Conference's policy session, except by recognition that most community and farm activities are influenced by national-level policy, legislation and institutional support. For example, the level of aquatic animal disease affecting small-scale producers or enhanced fisheries is related to national policies for quarantine and movement of live aquatic animals, which affect the risk of exposure of small-scale producers to serious aquatic pathogens. Inter-and intra-country trading patterns and movement of aquatic animals also affect these risks. International conventions (*e.g.* the Convention on Biodiversity), trade and consumer preferences, all clearly impact aquaculture development at a local level.

APPLICATION OF THE SYSTEMS APPROACH

Some examples of systems approaches presented during the Conference are given below.

Small-scale Farmer Research and Extension

The systems approach has been widely used in the promotion of sustainable development through small-scale integrated aquaculture. He emphasises that most scientists focus on technical aspects of aquaculture, resulting in the impression that the major constraint facing aquaculture development is a shortage of technical knowledge, overshadowing the developmental and educational constraints.

The most important constraint to aquaculture development is dissemination of existing knowledge, whether derived from research or indigenous technical knowledge of farmers also affect these risks. The limited capacity of developing-country institutions in education, research and development compounds this fundamental failing. Research should follow farming systems research and extension methods in which inter-disciplinary teams work with farmers to evaluate and develop both production systems and extension methods that are appropriate to the local conditions of farmers and their resource base.

A systems approach to management has more potential for success on the typically diverse small-scale farms in developing countries. This approach includes analysis of the resource-base of the farm, and the farmers' perceptions of their needs. The role of institutions in promoting the development of aquaculture can also be analysed using such an approach. The role of national institutions (government and non governmental agencies) as primary facilitators of aquaculture and the necessity of capacity building in order to facilitate the systems development approach should be recognized. International and regional agencies can support such capacity building.

6

Fresh Water Biodiversity

FRESH WATER

DELTAS

Deltas are deposits of sediments (particles of sand, gravel, and silt) at the mouths of rivers that flow into the ocean. The mouth of a river is the end where the body of water flows into the sea. Deltas are shaped by interactions of the river's fresh water with the ocean water, tides, and waves. Throughout history, deltas have been important places for human settlement. They are also vital habitats for many animals and plants. In the fifth century B.C.E, Greek naturalist Herodotus coined the name delta to describe the triangular shape of the sediments deposited at the end of the Nile River. The capital Greek letter delta resembles a triangle. Most deltas are triangular in shape because rivers deposit larger amounts of sediment where they meet the sea, then fan out into the mouth of the sea to deposit the remaining sediment.

FORMATION OF DELTAS

As rivers flow through their beds (a channel occupied by a river), the water breaks up rocks and pebbles, which the river then carries with it as it flows. These pieces of rock, sediment, include sand, pebbles, and silt. When the fast–flowing waters of the river reach the ocean, they push against the ocean water. This decreases the speed of the river water, which spreads out along the coastline. As the speed of the river water slows, the sediments that it carries settle to the bottom of the ocean. Over time, the sediments accumulate and form a delta. In most deltas, the river water is less dense (packed together) than the sea water because it contains less salt. As it flows out into the ocean, it floats on top of the ocean water. This is called hypopycnal flow. (The prefix hypo means "under" or "less" and the root word pycn means "density.") In places where hypopycnal flow occurs, the salt water that lies under the fresh water is called a salt wedge. The bigger sediments, like gravel and pebbles, are deposited at the tip of the salt wedge (nearest to shore) and the smaller sediments, like sand, are deposited farther out along the salt wedge. The smallest silt grains (fine sediment particles) are transported far offshore with the river water.

THE STRUCTURE OF A DELTA

Paths of flowing water called channels run through the sediments of deltas. These channels are called distributaries and they may be large and relatively permanent or small, transient features. The sides of the channels are made up of piles of sediments called *levées*. The areas between the distributaries are called interdistributary areas.

TYPES OF DELTAS

The structure of the distributaries and interdistributary areas depends on where the river empties into the ocean and the forces that affect the river water as it flows into the ocean. There are three major types of deltas: river–dominated, tide-dominated, and wave-dominated. Many deltas are formed by a combination of these forces. River-dominated deltas extend outward from the coast as the river water jets out into the ocean. The sediments deposited by the river tend to form levées that hold channels of water. The aerial view of a river-dominated delta looks like the foot of a bird with several branching channels. River-dominated deltas often have sand bars (long deposits of sand) that are perpendicular to the river. The Mississippi River Delta in Louisiana is an example of a river–dominated delta. Tide-dominated deltas are deltas where the sediments deposited by a river are redistributed by tides. These deltas have channels cut by the river water as well as channels cut by tidal currents. The result is a shoreline that looks like the fringe on the end of a carpet. Tide-dominated deltas may also have coastal features like sand bars and shoals (a sandbank seen at low tide) oriented parallel to the tidal flow.

The Ganges-Brahmaputra Delta in the Bay of Bengal is an example of a tidedominated delta. Wave-dominated deltas occur in places where the sediments deposited by the river water are redistributed by waves. Because waves tend to move sediments along the shoreline, the shape of wave-dominated deltas is usually a smooth coastline. If the waves that affect the delta break parallel to the shoreline, the sediments of the delta will tend to be symmetrical on either side of the river. If the waves break at an angle to the shore, the sediments of the delta will tend to accumulate on one side of the delta. The Nile River Delta in Egypt is an example of a delta that is wave–dominated.

HUMANS AND DELTAS

Deltas play an important role in human life both historically and in present times. The land on deltas is typically very good for agriculture. On the parts of the delta close to the river, the soil is fertilized each year by nutrient-containing floodwaters when the river floods. In addition, by building aqueducts (canals or pipelines used to transport water) from the wetter lands near the river to the dryer lands farther from the water source, it is easy to expand the amount of land available for farming. This practice is called land reclamation. In particular, the fertile Nile delta has supported much of the agriculture in Egypt for thousands of years. Trade is another reason why humans have lived on deltas. Because of the many distributaries and the access to a major river and the ocean, deltas are regions where goods can be easily exported both inland and overseas. Many of the world's largest ports are in deltas. Also, communication is relatively easy in deltas. Because the land is usually flat, it is easy to build roads. The many waterways make boat travel easy as well.

LIFE ON DELTAS

The interdistributary areas of deltas can support a variety of different habitats, depending on whether they are closer to the freshwater of the river or the salt water of the ocean. If the interdistributary areas are close to the river and affected by annual floods, they are called floodplains. Other interdistributary areas near the river water may be freshwater marshes, freshwater swamps, or lakes. The interdistributary areas that are closer to the ocean are likely influenced by the tides. They may be tidal flats (flat, barren, muddy areas periodically covered by tidal waters), mangrove (a tree that grows in saltwater) swamps, salt marshes, or marine embayments (an indentation in the shoreline of the sea that forms a bay). Because deltas have such a broad range of environments, they host a diversity of species.

Many different species of plants flourish on deltas, from saltwater trees called mangroves, to sea grasses, swamp sedges, and shrubs. Deltas also serve as nursery grounds for many species of fish and invertebrates (animals without a spine) and many land animals such as snakes and birds. The marine (sea) areas are important habitats for burrowing worms and mollusks, crustaceans (sea animals with hard outer shells) that hunt for food along the seafloor, as well as a variety of different species of fish. Deltas also serve important environmental roles. They remove harmful chemicals that are deposited in them by river pollution. These chemicals are absorbed by sediments and trapped as new sediments settle on top. Over time, bacteria break down many of these harmful substances and release chemicals that are not dangerous to the health of humans or other animals. Deltas are also known as nutrient recharge zones. When animals and plants die, they are buried in sediments and bacteria digest them. This converts the chemicals in their bodies into the raw materials needed for plants to grow.

FRESHWATER LIFE

The animals and plants that live in freshwater are called aquatic life. The water that they live in is fresh, which means that it is less salty than the ocean.

The terrestrial (land) environment that surrounds the freshwater environment has a large impact on the animals and plants that live there. Some factors that influence the freshwater environment include climate, soil composition, and the terrestrial animals and plants in the area. Just as on land, aquatic plants require carbon dioxide, nutrients (substances such as phosphate and nitrogen needed for growth) and light for photosynthesis, the process where plants make their food from sunlight, water, and carbon dioxide. Aquatic animals need to breathe in oxygen and consume food.

The physical conditions surrounding the body of water or wetland (lands that are covered in water often enough so that it controls the development of the soil) control the availability of these resources. For example, the concentrations of nutrients, oxygen, and carbon dioxide in the water depend on how much air gets into the water and on the chemical composition of the land nearby. The sediments (particles of sand, gravel, and silt) in the water influence how much light reaches the bottom of the lake or river. The temperature of the water affects how quickly animals and plants grow. The characteristics of the bottom of the body of water (sand, mud, rocks) and the speed of the currents (horizontal movement of water) control what kinds of plants and animals can live and reproduce in an area.

In general, freshwater environments are divided into two major categories: lentic waters and lotic waters. Lentic waters are those that are moving, as in rivers and streams. Lotic waters are those that are stationary, as in lakes and ponds. Sometimes, however, rivers and streams flow into lakes and ponds and the two different habitats merge together. Some wetlands may also contain many characteristics of freshwater environments.

LIFE IN RIVERS AND STREAMS

Rivers and streams are characterized by several physical features. They are generally comprised of freshwater that flows in one direction. The flow of water is most often from an area of high altitude (like a mountain range) to an area of low altitude (like an ocean). Usually, the water flows quickly initially and slows as it moves downstream. Streams often join rivers, so there is more water at the end of a river than at the beginning. As rivers flow, they erode (wear away) rocks and pick up sediments, making rivers often murkier at the end. Because rivers and streams change so much from their beginnings to their ends, there are many different types of habitats for animals. As a result, the number of animal species that live in rivers and streams is greater than the number of species that live in lakes and ponds.

Plant Life in Rivers and Streams

A major challenge facing plants that live in rivers and streams is staying in place, especially in swift currents. Plants have several different techniques to overcome the drag (the pull) of the water. Diatoms are a type of algae. Algae are marine organisms that range in size from microscopic phytoplankton to giant kelp and that contain chlorophyll, the same pigment used by land plants to perform photosynthesis. Diatoms avoid currents by using their small size.

They grow in a single layer on the surfaces of rocks. Because of the friction between the rock surface and the water, the water flow slows nearly to a stop within about a tenth of an inch (one–quarter of a centimeter) from the rock's surface. This region is called the boundary layer, and it provides the diatoms with protection from the forces of the current that would otherwise drag them downstream. Typical large river plants include algae, mosses, and liverworts.

These plants overcome the drag of the water by using special adaptations to grip rocks. Large algae often attach themselves to rocks with root-like structures called holdfasts. In addition, plants often anchor themselves in nooks between rocks or where waters pool, to avoid the drag of the river water. River plants that live within the currents have developed techniques to withstand the forces of the water. These forces would quickly snap any plant with rigid stems or leaves. As a result, plants that live in rivers are very flexible so that they can easily bend and move with the currents.

Animal Life in Rivers and Streams

Animals that live in rivers and streams also face the challenge of staying where they are. Many animals have hooks and suckers that they use to attach themselves to rocks. Blackfly larvae that live in streams in the northern United States and Southern Canada have suction cups that they use to stick to rocks in streams. Mayfly larvae have hooks that they use to fasten themselves to the algae growing on rocks. Other animals have streamlined shapes that minimize drag by presenting little resistance to water. Trout, which are extremely common in oxygen-rich fast–flowing waters, are shaped like torpedoes. Limpets are flattened molluscs that cling to the surfaces of rocks.

Their flat shape decreases the currents' drag on them. Animals that live in streams and rivers have developed interesting ways of gathering food in the fast-flowing waters. Snails, limpets, and caddis fly larvae scrape algae from rocks using special mouthparts. Many different insect larvae, as well as freshwater clams, filter the water for small bits of food. They have specialized mouthparts that look like brushes or combs that they use to strain the water and extract the edible plankton (animals and plants that float with currents) that float into their reach.

Rivers and streams are homes to a large number of fish. Perch, smallmouth bass, largemouth bass, bullhead, carp, pike, and sunfish prefer the parts of rivers where waters slow. These fish tend to be large, visual predators (animal that hunts another animal for food) that hunt in pools for smaller fish and invertebrates (animals without a backbone). Sculpins and darters prefer the faster moving sections of the river where waters are highly oxygenated. They use the swift current to bring food to them rather than hunting for their prey. Trout are also found in these faster moving parts of the river.

LIFE IN LAKES AND PONDS

Large lakes are often divided into zones. The near-shore area is called the littoral zone. This is the part of the lake that is shallow enough for aquatic plants to grow. The limnetic zone, also called the epilimnion, is the surface water of the lake away from the shore. (The prefix epi means "on the surface" and the root word limn means "lake.") It extends down as deep as sunlight penetrates. The majority of the plant life in this zone is phytoplankton (microscopic plants that float in currents). The deep part of the lake is called the profundal zone or the hypolimnion. (The prefix hypo means "under.") No plant life exists in this zone because of the absence of light. Most of the biological activity is that of bacteria decomposing dead animals and plants.

Seasonal Changes in Lakes

Lakes and ponds are greatly influenced by the temperature changes throughout the seasons. The description below is typical for a lake in a temperate (moderate) climate, which experience seasonal temperature changes. Tropical lakes (those in hot and humid areas) will have less dramatic fluctuations in temperatures. In the summer, the Sun warms the epilimnion. Warmer water is less dense than colder water, so it floats on top of the cooler water in the hypolimnion.

The region between the warm surface waters and the cold deeper waters is a transition zone where the water changes temperature very quickly with depth. This region is called the thermocline. The thermocline acts as a kind of barrier between the surface and the deep waters. In the early summer, the epilimnion is full of life. Phytoplankton can grow quickly because they have plenty of light and nutrients and the water temperature is warm. In turn, zooplankton (animals like crustaceans and small fish that float in the waters) feed on the phytoplankton. These zooplankton are food for larger fish and birds.

As summer progresses, the phytoplankton use up the nutrients in the epilimnion. They begin to die and sink to the bottom of the lake. There, decomposers, like fungi and bacteria, break up the dead phytoplankton and animals and convert them into the nutrients that phytoplankton need to grow. Because the thermocline acts as a barrier between the bottom and the top of the lake, these nutrients are unavailable to the phytoplankton in the epilimnion. Phytoplankton cannot grow in the hypolimnion, where there are nutrients, because there is no light. In the fall, the air temperature cools, which cools the surface of the lake. Eventually the temperature in the epilimnion becomes the same temperature as that of the hypolimnion. The thermocline disappears and the nutrient–rich waters from the hypolimnion mix with the waters in the surface of the lake. This is called the fall turnover. At this time, the nutrients from the bottom of the lake are mixed throughout the lake. However, because the amount of sunlight decreases in the fall and into the winter, the phytoplankton in the surface cannot grow very quickly.

During the winter, the surface of the lake continues to cool. Freshwater is densest at 39°F (4°C). Ice, with a temperature of 32°F (0°C), is less dense than the deeper waters and so it forms on the surface of the lake. This provides fish and other invertebrates room to live under ice-covered lakes. The ice also acts an blanket–like insulation that helps keep the water underneath from freezing. In the springtime, the temperatures warm so the ice melts. Eventually the whole lake becomes 39°F (4°C) and so the waters from the bottom mix with the waters from the surface. This is called the spring turnover. As summer begins, the surface waters warm and the thermocline again separates the epilimnion from the hypolimnion. Because of the fall and spring turnovers, the nutrients from the bottom of the lake are available to the phytoplankton in the surface waters. This sets the lake up for the summer's rapid growth of phytoplankton and all the animals that depend on them.

Plant Life in Lakes and Ponds

Some of the most plants in lakes and ponds are the smallest. These phytoplankton are usually single-celled plants grouped with the algae. Sometimes they connect themselves together into long strings called colonies. Common phytoplankton in lakes and ponds are diatoms, which have beautiful shells made of silica (the same material that comprises sand); dinoflagellates, which move by snapping their flagella (long whip–like cell extensions that can propel an organism; and cyanobacteria, which are bacteria that perform photosynthesis. The larger plants in ponds and lakes include large algae and mosses, cattails, reeds, water lilies, bladderworts, willows, and button bush. These plants often grow in mud where the gases that they need to grow-such as oxygen and carbon dioxide-are scarce.

Many larger plants have stems that are spongy and they pull gases from the air down into their roots. Plants on land use their roots to gather water and nutrients, however aquatic plants are surrounded by water, and nutrients are dissolved in the water. Some aquatic plants have given up their roots. For example, duckweed (or water lentil) and watermeal are small pea-sized plants that float on the surface of lakes and ponds in the spring and summer. They absorb nutrients from the water and produce a lot of starches. By the fall, they are so heavy with nutrients that they sink to the bottom of the lake. They live out the winter in the mud at the bottom of the lake, existing on their stores of starch.

By spring, they have used up so much of the starch, they are light enough to float again. They pop to the surface just in time to use the strong light of spring and summer for photosynthesis and they begin to use their starch stores once again. Other large plants, like milfoil, water soldier, and water hyacinths also float on the surface of lakes and ponds. The edges of lakes are often divided into four zones based on the physical environment and the types of plants found there. Beginning farthest from the water, the swamp plant zone contains plants that have roots in the shallow water.

At times the water can recede from this zone, leaving the plants roots exposed to the air. Typical plants in the swamp plant zone are rushes and sedges (a type of plant that looks like a stiff grass). The next zone is called the floating-leaf and emergent zone.

Here the water never dries up, but the lake is shallow enough that the tops of plants emerge out of the water. A typical plant that lives in this zone is the water lily, which has special gas filled chambers in its leaves that allow it to stay floating on the surface of the water. In the submerged plant zone, plants live entirely underwater. Canadian waterweed and many types of mosses live in this zone. The freefloating plant zone takes up the center of the lake. Here plants without roots, like duckweed and water soldier, float freely on the surface.

Animal Life in Lakes and Ponds

Zooplankton float in the epilimnion of lakes and eat phytoplankton and other zooplankton. Usually, these animals are nearly transparent, in order to avoid being seen by their predators. Typical zooplankton in lakes include the water flea, Daphnia, which can reproduce without mating. Under normal conditions all of its offspring are female. However, when the animals are stressed, by lack of food for example, they will produce males. This mixes up the gene pool of the population and creates individuals that are likely to withstand environmental changes. Another typical freshwater zooplankton is the rotifer, which has bristles on top of its head that it whirls like propellers in order to move through the water and capture prey.

Many insects have juvenile stages that are aquatic. Mayflies, caddis flies, mosquitoes, and dragonflies all live for some period underwater in lakes and ponds. They swim among the rocks and plants in the lake bottom for a season or several years. Then they metamorphose (change in appearance) into their adult form and fly away from the water. The bottom of the lake is also home to many different worms, mussels, and crustaceans. These animals feed on the remains of plants and animals that drop to the bottom of the lake from above. Larger animals live in lakes and ponds.

In particular, fish, birds, and amphibians prey on the invertebrates that live in the lakes. Fish such as bluegills eat juvenile insects that swim in the bottom of the lake, while crappies eat zooplankton near the surface. Birds like flycatchers and warblers fly near the surface of lakes, preying on insects that are hatching from their juvenile stage. Frogs also hunt for insects that live near the pond. Still other birds and fish prey on smaller fish. Bass, salmon, osprey, loons, and heron hunt for fish by using their keen eyesight. Beavers and muskrats are mammals that depend on water for their homes. They build dams and lodges, which provide them with protection from predators.

GROUNDWATER FORMATION

Groundwater is fresh water in the rock and soil layers beneath Earth's land surface. Some of the precipitation (rain, snow, sleet, and hail) that falls on the land soaks into Earth's surface and becomes groundwater. Water-bearing rock layers called aquifers are saturated (soaked) with groundwater that moves, often very slowly, through small openings and spaces.

This groundwater then returns to lakes, streams, and marshes (wet, low-lying land with grassy plants) on the land surface via springs and seeps (small springs or pools where groundwater slowly oozes to the surface). Groundwater makes up more than one-fifth (22%) of Earth's total fresh water supply, and it plays a number of critical hydrological (water-related), geological and biological roles on the continents.

Soil and rock layers in groundwater recharge zones (a entry point where water enters an aquifer) reduce flooding by absorbing excess run-off after heavy rains and spring snowmelts. Aquifers store water through dry seasons and dry weather, and groundwater flow carries water beneath arid (dry) deserts and semi-arid grasslands. Groundwater discharge replenishes streams, lakes, and wetlands on the land surface and is especially important in arid regions that receive limited rainfall. Flowing groundwater interacts with rocks and minerals in aquifers, and carries dissolved rock-building chemicals and biological nutrients. Vibrant communities of plants and animals (ecosystems) live in and around groundwater springs and seeps.

Almost all of the fresh liquid water that is readily available for human use comes from underground. (The bulk of Earth's fresh water is frozen in ice in the North and South Pole regions. Water in streams, rivers, lakes, wetlands, the atmosphere, and within living organisms makes up only a tiny portion of Earth's fresh water.) For thousands of years, humans have used groundwater from springs and shallow wells to fill drinking water reservoirs, and water livestock and crops. Today, human water needs far exceed surface water supplies in many regions, and Earth's rapidly-growing human population relies heavily upon groundwater to meet its ever larger demand for clean, fresh water.

FISH (SALTWATER)

There are over thirty thousand different species of fish, and they are the most numerous vertebrates. Vertebrates are animals that have a bony spine that contains a nerve (spinal) chord. Vertebrates usually have an internal skeleton that provides support and protection for internal organs. This spine and skeleton allow vertebrates to move quickly and to have great strength. Fish usually live surrounded entirely by water. They all have gills for breathing and fins for swimming. Most fish are ectotherms, which means that their bodies are nearly the same temperature as the water in which they live. About 60% of all fish live entirely in saltwater, while the rest live in freshwater or both freshwater and saltwater.

One of the most remarkable things about fish is their diversity. Fish can be very large, like the whale sharks that can reach 90,000 pounds (41,000 kilograms) or very small, like gobies that can weigh as little as 0.0004 ounce (0.1 gram). They have a variety of diets, including plants, other fish, invertebrates (animals without a backbone) and microscopic plankton (freefloating plants and animals). Fish find their food in many different ways, including hunting with their eyes, grazing, scraping the sea floor, and digging. Some fish have special organs that bioluminesce (create light) and lure prey towards their mouths, while others use this glow to make themselves blend in with coral or other ocean features, disguising themselves from predators. Fish come in a variety of colours that can serve to scare predators away or blend into their environment. Some fish live their entire lives within one bay or cove, while others may migrate thousands of miles (kilometers) across the ocean. Fish can live alone or they may swim with many other fish, called schools, to protect against predators. Fish are classified into three groups. The jawless fish belong to the class Agnatha. They include hagfish and lampreys. The rays and sharks belong to the class Chondrichthyes. They have skeletons that are made of a tough material called cartilage. The bony fish belong to the class Osteichthyes. This largest group includes many familiar fish like tuna, halibut, anchovy, and cod.

CLASS AGNATHA

There are about fifty species of Agnathans and they are divided into two groups: hagfish and lampreys. These fish have no jaws and their fins are not evenly matched across their bodies, so they are not efficient swimmers. They have mouths that look like suckers with small teeth that are used for grasping on to prey. Organs for smelling and sensing surround their mouths and help them identify prey. These fish have very poor vision. Hagfish live in colonies (groups) on the sea floor. They dig in sediments (sand, gravel, and silt) for worms to eat. If approached by predators, hagfish emit large quantities of foulsmelling slime from glands along the sides of their bodies. This usually discourages or confuses predators.

After the danger has passed, the hagfish will tie itself in a knot, which it slides along the length of its body to scrape off the slime. The mouth of the lamprey is called an oral disc. It is cone-shaped and contains sharp teeth that it uses to bore a hole into the side of another animal. The lamprey then attaches its oral disk to the live animal's wound and feeds off the blood and tissue of its host. Lampreys usually detach after some period of time without killing their host. Lampreys are usually most often found attached to bony fish, but they also have been seen on whales and dolphins.

CLASS CHONDRICHTHYES

The class Chondrichthyes includes about 700 species of sharks and rays. They are an extremely old group, having been in existence for approximately 280 million years. Nearly all members of this class are marine (live in seawater). These fish have skeletons made out of cartilage, which is a tough but flexible tissue. It is the same material that is found in human ears and noses. Sharks and rays do not have gas bladders (internal sacs that fill with gas to help the fish rise or fall in the water so that it does not waste energy by continually swimming). Sharks and rays must continually swim in order to prevent themselves from sinking.

However, the liver of sharks is large and it contains a lot of oily materials. As oil is less dense than water, this special liver helps the shark stay afloat. Cartilaginous fish have several rows of teeth that fall out as they age. They are then replaced with new teeth that grow in from behind.

Sharks do not have scales; instead they have rough plates called dentricles embedded in their skin. These dentricles make the skin feel abrasive, like sandpaper. Many sharks have electroreceptors on their heads. These specialized organs allow the shark to sense the electrical currents generated by fish as they swim through the water.

The shark's well–developed nervous system, including a large brain, also helps it locate its prey (animals that are food). Rays have a more flattened shape than sharks. Their fins are attached to their bodies so that they look like triangular or semicircular wings.

Large rays, like the manta ray, can measure 22 feet (7 meters) from fin to fin. These huge animals feed on plankton. Other rays, like the stingray, have sharp barbs attached to the base of their tail. These are used as defence against predators. Another family of rays can actually produce an electric current, which they can use to stun prey.

CLASS OSTEICHTHYES

The class Osteichthyes, or the bony fish, make up the majority of fish, with almost 28,000 different species. These fish all have a strong, but lightweight, skeleton that supports their organs. They have gas bladders that help them maintain buoyancy.

The teeth of bony fish are fused to their jawbone and do not fall out as do the teeth of the cartilinagous fish. Osteichthyes are found in every type of marine environment from near-shore tidepools and coral reefs to the very bottom of the deep ocean. Nearly 90% of all the bony fish are categorized into one order, Teleostei. These fish include many common fish, like the cod, tuna, seabass, and perch. Teleostei also includes unusual fish like the mola, which floats near the surface of the ocean in warm currents; the angler fish, which lives on the seafloor and lures its prey using a worm-shaped appendage; and the football fish, which permanently fuses with its mate.

Movement

Many teleost fish have bodies that are shaped to allow them to move easily through water. In particular, fast or constantly swimming fish have body shapes that minimize drag, or resistance to movement. The less surface area comes into contact with water in the forward direction, the less drag the fish will have. A torpedo shape, with a body that tapers towards the rear, is one of the most effective shapes for minimizing drag, and many fast–swimming fish have this sort of shape. An example of a fish with a shape that minimizes drag is the swordfish, which can reach speeds up to 75 miles per hour (120 kilometers per hour) in short bursts.

Some fish, like eels, wave their entire bodies back and forth in an "S" shape to move through the water. This is not a very efficient way of swimming as it requires a lot of energy and it increases the surface that confronts the water as the fish swims forward. More advanced swimmers have a stiff body, with a tail that bends back and forth behind the fish. This allows the fish to conserve energy because it only moves its tail, not its entire body. It also minimizes the area that confronts the water to the head of the fish.

Water and Gas Balance

Just like humans, metabolism (the process of cells burning food to produce energy) in fish requires oxygen and produces carbon dioxide as a waste product. Fish breathe through gills, which are found underneath flaps on both sides of the head. Water containing dissolved oxygen is brought in through the mouth and pumped over the gills. The gills are packed with blood vessels that absorb the oxygen from the water and produce the carbon dioxide waste that is generated within the fish. The body fluids of saltwater teleost fish are about one-third as salty as ocean water.

Another way of thinking about this is that the concentration of water inside these fish is greater than the concentration of water in the ocean. Because of osmosis (the passage of a liquid from a weak solution to a more con centrate solution), the water from the inside of the fish is constantly diffusing (moving outward) from the fish. As a result, saltwater fish must regularly drink seawater to replenish the water that is lost by osmosis. The fish have special salt glands in their gills that remove the excess salt that comes inside the fish with the seawater.

LATERAL LINE

Teleost fish have developed an interesting sensory organ that helps to sense vibrations in the water. Along both sides of most teleost fish is a long row of canals (tubes) that are packed with nerve cells. These nerves detect movements of currents, changes in water pressure, and even noises.

AQUIFERS: FRESH WATER UNDERGROUND

An aquifer is a body of rock or soil that yields water for human use. Most aquifers are water-saturated layers of rock or loose sediment. With the exception of a few aquifers that have water-filled caves within them, aquifers are not underground lakes or holding tanks, but rather rock "sponges" that hold groundwater in tiny cracks, cavities, and pores (tiny openings in which a liquid can pass) between mineral grains (rocks are made of minerals).

The total amount of empty pore space in the rock material, called its porosity, determines the amount of groundwater the aquifer can hold. Materials like sand and gravel have high porosity, meaning that they can absorb a high amount of water. Rocks like granite, marble, and limestone have low porosity, and make poor groundwater reservoirs. Aquifers must have high permeability in addition to high porosity. Permeability is the ability of the rock or other material to allow water to pass through it. The pore space in permeable materials is interconnected throughout the rock or sediment, allowing groundwater to move freely through it.

Some high–porosity materials, like mud and clay, have very low permeability. They soak up and hold water, but don't release it easily to wells or other groundwater discharge points, so they are not good aquifer materials. Sandstone, limestone, fractured granite, glacial sediment, loose sand, and gravel are examples of materials that make good aquifers. Water enters aquifers by seeping into the land surface at entry points called recharge zones and leaves at exit points called discharge zones. (Some aquifers discharge into the ocean.)

Influent or "water-losing" streams, ponds, or lakes are bodies of surface water in recharge zones that contribute groundwater from their water supply. Groundwater flows into effluent or "watergaining" streams and ponds in discharge zones. For the water level in an aquifer to remain constant, the amount of water entering at recharge zones must equal the amount leaving at discharge zones. (Imagine a bucket punched with holes under a dripping faucet. If water drips in at the same rate that it drips out, the water level stays the same.)

If water discharges or is pumped from an aquifer more quickly than it recharges, the groundwater level will fall. The time an average water molecule spends within an aquifer is called its residence time. Water in some fast–flowing aquifers spends only a few days underground, while other rock layers can hold water for ten thousand years. Average aquifers have residence times of about two hundred years.

The Water Table and Unconfined Aquifers

Water enters aquifers by moving slowly down through a layer of surface rocks and soil whose pore spaces are partially filled with air (zone of infiltration). The water continues moving downwards until it reaches a level where all the pore spaces are completely filled with water (zone of saturation).

The top of the zone of saturation is called the water table. In some wet, lowland regions, southern Florida for example, the water may be only a few feet (meters) below the surface. In others, like the American Southwest, water–saturated rocks may be hundreds of feet below the land surface. Groundwater reservoirs that have uniform rock or soil properties (porosity and permeability) throughout are called unconfined aquifers. The water table forms the upper surface of an unconfined aquifer. The shape of the water table in an unconfined aquifer mirrors the shape of the land surface, but its slopes are gentler.

In temperate (moderate) climates that receive moderate amounts of groundwater–replenishing rainfall, water infiltrates into unconfined aquifers in hilltop recharge zones and discharges into effluent streams and ponds in low areas where the water table intersects the land surface. Water will only rise to the level of the water table in a well, so a pump or bucket is required to extract water from an unconfined aquifer.

Confined Aquifers and Artesian Flow

Confined aquifers are pressurized groundwater reservoirs that lie beneath layers of non-permeable rock (granite, shale) or sediment (clay). Groundwater enters a confined aquifer in recharge zones beyond the uphill edges of the confining layer and discharges beyond the downhill edges. Groundwater trapped beneath an impermeable barrier cannot rise to the height of the water table, so pressure builds up in confined aquifers. Artesian wells are wells drilled in confined aquifers where the pressure is great enough to make water flow at the surface.

Lakes are large inland bodies of fresh or saline (salty) water. Lakes form in places where water collects in low areas or behind natural or man-made dams (barriers constructed to contain the flow of water). Some lakes are fed by streams (natural bodies of flowing freshwater), and some form where groundwater (water flowing in rock layers beneath the land surface) discharges onto the land surface. Water leaves lakes by flowing into outlet streams, infiltrating (soaking in) into groundwater reservoirs called aquifers, and by evaporating into the atmosphere (mass of air surrounding Earth). Lakes vary in size from large lakes such as the Great Lakes of North America, to small mountain lakes. Lakes are larger than ponds, which are small bodies of fresh water that are shallow enough for rooted plants to grow.

The study of ecology (relationships between living organisms and their environment) in lakes, inland seas, and wetlands is called limnology. Lakes store only a tiny percentage of Earth's fresh water. They are, however, an extremely important water resource for humans. Freshwater lakes provide water for agricultural irrigation (watering), industrial processes, municipal uses, and residential water supplies. People who live in the continental interiors use lakes for fishing and recreation, and very large lakes have important shipping and transportation routes. Humans also construct artificial lakes, called reservoirs, by building dams across rivers. In addition to providing the benefits of natural lakes, reservoirs also store water for specific communities, control floods, and generate hydroelectricity (electricity generated from water power).

HOW LAKES FORM AND DISAPPEAR

Earth scientists who study water on the continents (hydrologists and hydrogeologists) see lakes as temporary reservoirs within stream and groundwater systems. All water that falls as precipitation (rain, snow, sleet, hail) on the land surface of continents eventually makes its way to the ocean or evaporates back into the atmosphere. Water collects in lakes because it enters more rapidly than it escapes, but it is never permanently trapped there. As in a tub with a running faucet and an open drain, individual water molecules (smallest unit of water, each containing two hydrogen atoms and one oxygen atom) are constantly entering and escaping.

After arriving in a lake, an average water molecule spends about one hundred years before moving to a new reservoir. (The time that an average water molecule spends in a reservoir is called it residence time. Water resides for about two weeks in rivers, forty years in glaciers (slow moving mass of ice), and between two hundred and ten thousand years in groundwater reservoirs.) To geologists (Earth scientists), lakes are temporary features. Stream-fed lakes within stream systems are destined for destruction. Every stream seeks to create a constant slope, called a graded profile, between where the stream's waters begin and ends by eroding (wearing away) and depositing sediment (particles of sand, silt, and clay) along its course.

When a natural or man-made obstruction blocks a stream, such as a river, streams deposit sediment in the lake or reservoir behind the obstruction and erode away in front of it. Eventually, the dam will collapse, and the lake will empty. Lakes that fill depressions and have no outlets fill when the regional climate becomes wetter or when warm periods melt mountain snows and glacial ice. They evaporate away during periods of dry weather and dryer climate. It may take thousands, or even tens of thousands of years, but lakes eventually drain, collapse, or dry up.

LAKE LAYERS AND OVERTURNS

Contrary to their common image as evenly mixed pools of unmoving water, lakes are complex, dynamic bodies of moving surface water. Lake water varies within the lake; its temperature, chemical content, light infiltration, and biological habitats vary from top to bottom and side to side.

Furthermore, the vertical layering (stratification), horizontal variations, and circulation patterns within lakes change over time. Waves, currents (a moving mass of water), and even tides affect circulation of water within lakes. Lakes are thermally stratified (layered just as to temperature); they have layers of warm and cool water that are separated by layers where the temperature changes (thermoclines). Like the oceans, many lakes have a thin layer of warm surface water, and a thicker layer of cool deep water that is separated by a thermocline layer. Wind generates currents on lake surfaces and creates some mixing. Unlike the oceans, however, many lakes have seasonal overturns that mix their waters. water is denser than solid water (ice). Water reaches its maximum density at 39°F (4°C). Because of this odd property, the warm less-dense water rises, the cool denser water sinks, ice floats, and lakes overturn.

Lakes that are ice-covered for part of the year undergo overturns that partially or completely mix their waters. Many lakes in temperate (moderate temperatures) regions like the northern United States overturn and mix completely twice a year (dimictic lakes). During the warm summer months, these lakes have a usual temperature profile with warm surface waters, a thermocline, and cool bottom water. In the fall, when the surface water cools down to 39°F (4°C), it becomes denser than the water underneath it and the surface layer sinks to the bottom.

The bottom water rises to the surface, and the lake overturns. Over the winter, the bottom water is the warmest, and the frozen surface water is the coldest. (Plants and animals survive the winter on the lake floor in the chilly, but not frozen, bottom water.) In the spring, when the ice melts, and the water warms to 39°F (4°C), it sinks, and the lake overturns again. Limnologists classify lakes just as to the number of mixing events they undergo each year.

Lake type classifications have the root term mictic, meaning "to mix," and include:

- *Oligomictic*: Warm, ice-free lakes that rarely mix. They are warmest at the top, and coolest at the bottom. Tropical, oligomictic lakes have warm bottom water and very warm surface water. They rarely overturn because their water does not near 39°F (4°C).
- *Meromictic*: Warm, ice–free lakes that mix incompletely. These are deep lakes that are warmest at the top, and coolest at the bottom.
- *Monomictic/dimictic/polymictic*: Lakes with seasonal ice covers that overturn and mix completely once (monomictic), twice (dimictic), or many times per year (polymictic). Lake overturns are the norm in temperate regions, but local conditions affect the timing and number of overturns in specific lakes.
- *Amictic*: Lakes that never overturn because they are icecovered

throughout the year. These lakes exist near the North and South Poles and atop very high mountains. They have cold bottom water that hovers near 39°F (4°C) and frozen surface water.

LAKE CHEMISTRY: SALINE LAKES

Many of Earth's largest and most important lakes contain salt water. All surface water contains some dissolved chemicals, called salts. Groundwater, streams, and freshwater lakes all contain the chemical components of rocks and minerals. Humans can drink fresh water because our bodies can use or at least tolerate the types and concentrations of dissolved chemicals it contains. Salt water, on the other hand, has a very high concentration of dissolved salts, and is undrinkable. The Dead Sea, on the border between Israel and Jordan, is Earth's saltiest body of water. It is truly a dead sea because it is too salty to support life. Saline lakes generally form in arid (dry) regions where surface water evaporates quickly.

When water evaporates, the salts stay behind. Over time, the lake water becomes saltier. Some saline lakes, such as the Great Salt Lake, are all that remains of a much larger fresh water lake that has evaporated over time. Others, like the Caspian Sea in central Asia, began as saltwaterfilled ocean basins that have since closed. Saline lakes are often temporary features that fill during periods of wetter climate and then dry up when stream flow or groundwater discharge slows. Playa lakes are flat desert basins that occasionally fill with water. Desert oases (watering holes) form and disappear with such regularity that thirsty travellers think they imagined them. The Great Salt Lake, Caspian Sea, Aral Sea, and Dead Sea are all presently evaporating. Over time, the dissolved chemicals become so concentrated in drying lakes that they bond together and form solid salt crystals. Thick layers of salt cover dry lake beds.

LAKE BIOLOGY

Lakes support rich communities of plants and animals (ecosystems) that have adapted to live within ever-changing conditions on lake beds, within the water column (water running from the surface to the lake floor, often showing differences in temperature, nutrients, etc.), and along lake shores. Lakes, like islands, are often closed systems that only rarely gain new species or individuals from other lakes.

Many lakes host groups of rare species that have evolved (changed over time) together in their specific lake. These ecosystems are rich and unique, but fragile. They have little defence against foreign predators or diseases. Human alterations and water pollution have threatened many lake species. Environmental groups and government agencies are presently attempting to protect and revive threatened lake species such as cichlids (rare doublejawed fish) that inhabit the lakes of the Great Rift Valley in east Africa. Lake organisms live in zones that are determined by the physical structure of their lakes such as the amount of available light, water depth, and distribution of nutrients. Most lake plants and animals live in shallow, well-lit surface waters called the euphotic zone.

Most plants depend on the Sun's energy to produce food by the chemical process of photosynthesis, and they cannot grow in water that is too deep or too cloudy for light to penetrate. Lake animals such as fish need oxygen that plants give off during photosynthesis, so they live mostly in the euphotic zone as well. Plants with roots grow in shallow water along edges of lakes where light reaches the lake floor (littoral zone) and floating plants perform photosynthesis in the open surface waters (limnetic zone). Oxygen-consuming bacteria inhabit the deepest, darkest parts of lakes (benthic zone) where dead plant and animal materials accumulate. Limnologists also classify lakes by the balance of organisms and nutrients in their waters.

Types include lakes that are described as oligotrophic, eutrophic, and mesotrophic:

- *Oligotrophic*: Nutrient poor lakes that support very few plants and animals. Oligotrophic lakes are typically cool, deep, and have very clear water. Very little organic (relating to or from living organisms) mud accumulates in oligotropic lakes, and they often have sand and gravel beds.
- *Eutrophic*: Lakes rich in plant nutrients that support abundant plant life in their surface waters. Their water is often clouded by microscopic plants, and their beds covered with thick layers of decaying plant material. Bacteria that live on the organic mud use up oxygen, and eutrophic lakes often have oxygen–poor deep water. Plants and bacteria eventually take over eutrophic lakes. They become oxygen-poor bogs and marshes where fish cannot live. Some chemicals that humans use, including fertilizers and detergents, cause a process called eutrophication when they run off into lakes, which causes the population of plants to increase to such an extent that eventually oxygen-starved fish die.
- *Mesotrophic*: Lakes with moderate amounts of nutrients and healthy, balanced communities of plants, animals and bacteria. Mesotrophic lakes receive adequate amounts of fresh water and nutrients, and seasonal overturns allow nutrient–poor and nutrient–rich layers to mix. Mesotrophic lakes are intermediate between crystal-clear, lifeless oligothrophic lakes and cloudy, muddy eutrophic lakes.

WHERE LAKES FORM: LAKE BASINS

Lakes form where water collects in depressions, or basins. Many lakes fill low areas created by plate tectonic movements (tectonic basins) and volcanic activity. (Plate tectonics is the movement of large, rigid pieces of Earth's outer rock shell called the lithosphere.) Retreating glaciers and ice sheets leave behind large basins and small depressions that fill with meltwater. Though flowing streams and rivers generally act to fill in and drain lake basins, other sedimentary processes can create landscape depressions and natural dams that confine water in lakes.

Lakes in Tectonic Basins

Rift valley lakes fill long, linear valleys within rift zones. (Rifts are areas where the continental lithosphere is stretching and beginning to break into pieces. They are the precursors of ocean basins.) A chain of large lakes including Tanganyika, Naivete, and Malawi follows the Great Rift Valley through eastern Africa. Lake Victoria, the world's second-largest lake, lies between two branches of the rift valley. The Red Sea, Sea of Galilee, Dead Sea, and Gulf of Ababa fill the northern branches of the rift where it crosses the Arabian Peninsula in the Middle East. Russia's Lake Baikal, the world's deepest lake, fills an ancient, inactive rift valley in central Asia. Lakes also form in places where continents are moving towards each other.

The Black Sea, Caspian Sea, and Mediterranean Seas fill a closing ocean basin between Africa and Europe. When continents collide, water fills depressions in the landscape over folded and broken (faulted) rock layers that were caught between the land masses. Slopes that are too steep collapse and block rivers with natural dams. Blocks of uplifted, erosion-resistant rock form bedrock that holds back mountain lakes.

Volcanic Lakes

Volcanoes are mountains that form from eruptions of molten rock (lava) on the land surface. When a volcanic peak collapses into its emptied magma chamber (a pool or room of magna held under tremendous pressure within a volcano prior to a volcanic eruption), it forms a large circular basin called a caldera. (Craters, the small basins near the top of active volcanoes, sometimes also contain small lakes, but most significant volcanic lakes, including inaccurately-named Crater Lake, fill calderas.) Yellowstone Lake in Wyoming and Crater Lake in Oregon are examples of caldera lakes. Volcanic ash, mud, and lava flows also create natural dams in river valleys. A dam of volcanic rock confines Lake Tahoe in a high valley of the Sierra Mountains on the California–Nevada border.

Glacial Lakes

The thick continental ice sheets that covered northern North America, Europe, and Asia during the Pleistocene ice ages (a division of geologic time that lasted from two million to ten thousand years ago) left behind thousands of lake and ponds when they retreated about twenty thousand years ago. The weight of the ice sheets pushed down on the continents, leaving broad basins that filled with melt water when they retreated. The Great Lakes of North America (Superior, Huron, Michigan, Erie, and Ontario) formed this way. Hundreds of lakes, such as the Winnipeg, Athabasca, Great Slave, and Great Bear cover the central and eastern provinces and territories of Canada that are still rebounding from their heavy ice load. Advancing glaciers also pile tall ridges of sediment, called moraines, at their toes (the end of extensions of glaciers along the ground). When glaciers retreat, moraines hold back meltwater. Small lakes and ponds also form in glacial depressions called kettles that form when blocks of ice buried in glacial sediment melt. Melting mountain glaciers feed many mountain lakes and glacial sediment traps streams and meltwater.

Groundwater Discharge Lakes

Water moving through pore spaces in rock and soil layers discharges on the land surface in places where the water table (level below which pore spaces are saturated with water) intersects the land surface. In regions with wet climates, the water table is near the land surface and ground water discharges in low spots. Groundwater chemically erodes limestone and other rocks and creates caves, cavities, sink holes, and collapse basins called karst features. Florida's many lakes, including Lake Okeechobee, are groundwater filled karst features.

LIFE IN AND AROUND PONDS

Ponds are havens for plants. Because the sunlight is abundant all through the water, plant can grow from every location in a pond. (Plants need sunlight to live as they convert the Sun's energy into food in a process called photosynthesis.) Often, the surface of a pond will be almost entirely covered with pond-loving plants such as the water lily and other plants that need higher levels of sunlight or that need direct exposure to air. Ponds also support various species of animal life, both in and surrounding its waters. Often, the bottom of a pond will be muddy, rather than rocky, and the mud hosts a variety of living creatures, such as crayfish. The still pond water and muddy bottom are also favourable conditions for the eggs of insects and creatures such as frogs to develop (often attached to the stems or leaves of plants). For microscopic life such as bacteria and algae, a pond offers plenty of food and the sunlit water provides a suitable temperature for the microscopic cells to grow and divide.

Animals such as deer often use natural ponds as a source of drinking water. Birds feed upon fish that live in ponds. Beavers find the still pond waters a good place to build their lodge. Ponds are often a source of relaxation and recreation. In warmer times of the year, a pond's edge can be a place where people picnic or rest outdoors. In the cold winter season of northern climates, ponds can freeze solid and host winter sports such as ice skating.

THE FATE OF PONDS

The flow of water into and out of a pond can be slow. This feature, along with its shallow depth, makes a pond vulnerable to contamination. If chemicals that upset the natural composition of the pond are introduced, then the water quality necessary to sustain life can be destroyed. Ponds that form in arid (dry) regions where rainfall is briefly heavy then sparse throughout the rest of the year continue a cycle of filling up, then slowly drying. These ponds attract animal life only when water is abundant, which can sometimes cause conflicts with humans. In some regions of Africa, crocodiles return during the rainy season to newly filled ponds that form near populated villages. Hungry after hibernating (being in an inactive state) the rest of the year, the crocodiles pose a threat to livestock that also drink from the pond, and the people who tend the livestock. As time passes, the vast majority of ponds will naturally fill in, as sediment (particles of gravel, sand, and silt) and other debris collect in the shallow water. After about one hundred years, what was once a pond often becomes a field, and the water source of the pond is diverted by the changing landscape or by changes in rainfall amounts.

PONDS

A pond is a depression in the ground that is filled with water that remains year round. Ponds range in size from the artificial backyard projects about the size of a bathtub to bodies of water that are about the size of a football field. Ponds support a variety of animal and plant life, and are also used as recreational sites by people. The difference between a pond and a lake involves size and water depth. A lake is big enough to have at least one beach (sand or rock that slopes down to the water) and contains enough water to generate waves from the wind that blows across the surface of the water. In contrast, a pond is usually too small for waves of any size to form. At the center of a lake, the water can reach depths of many hundreds, even thousands of feet (meters). A pond, however, is a shallow and still body of water where sunlight can usually reach down to the bottom.

HOW PONDS FORM

Natural ponds form in shallow depressions where rainwater (including run-off from nearby higher areas) collects. Water from an underground source such as an underground spring can also collect into a pond. Ponds that people enjoy in their backyard are often artificial, created by preparing the hole and adding water and plants to create a backyard oasis. These ponds can provide relaxation and a habitat for attracting insects, birds, and amphibians (such as frogs and salamanders), even in backyards located in a bustling city. Other artificial ponds are workhorses. One example is a sewage treatment pond. This type of pond keeps the sewage in a place where the growth of microorganisms can occur in the shallow and warm water. As microorganisms such as algae grow, they can use some of the materials in the sewage as food. This helps clean the water, and is an example of bioremediation, the process of using natural substances such as bacteria, to clean a contaminated natural resource, such as water.

RIVERS SURFACE WATER

Rivers are bodies of flowing surface water driven by gravity. Hydrologists, scientists who study the flow of water, refer to all bodies of flowing water as streams. In common language, it is accepted to refer to rivers as larger than streams. Water flowing in rivers is only a very small portion of Earth's fresh water. The oceans contain about 96% of the water on Earth, and most fresh water is bound up in glacial ice near the North and South Poles. Rivers shape the landscape and are integral to the hydrologic cycle (circulation of water on and around Earth) on the continents. Rivers shape the lands as they erode (wear away) and deposit sediment (particles of gravel, sand, and silt) along their courses. Running river water acts to level the continents.

When geologic forces slowly raise (uplift) mountain ranges, rivers wear them away. The streams that form the Ganges River of India (headwater streams), for example, are presently tearing down the Himalayas almost as quickly as they are uplifted by the movements of Earth's crustal plates (plate tectonics). When geologic forces create depressions or low areas on the continents, rivers act to fill them. River sediment replenishes floodplain (Flat land next to rivers that are subject to flooding) soils and coastal sands. Earth's major rivers, including the Nile, Amazon, Yangtze, and Mississippi, drain the waters of vast continental areas and set down (deposit) huge deposits of sediments at the ends of rivers that flow into the ocean (for example, in deltas at the end of many rivers) Rivers host vibrant communities of plants and animals, and refill groundwater reservoirs and wetlands that support biological life far beyond their banks. Rivers are a main focus of human interaction with the natural environment.

Human agriculture, industry, and biology require fresh, accessible water from rivers. Ancient human civilizations first arose in the fertile valleys of the world's great rivers: the Yangtze and Yellow Rivers in China, the Tigris and Euphrates Rivers in the Middle East, and the Nile River in Egypt. The distribution of Earth's rivers and systems of rivers has influenced human population patterns, commerce, and conquest since ancient times. Rivers flow through the great cities of the world, and the imagery of rivers is deeply embedded in our language, culture, and history. Today, billions of people depend directly and indirectly on rivers for food and water, transportation and recreation, and spiritual and religious inspiration. Almost all major rivers are today confined by man–made dams and levees (walls along the banks) that provide people with the means to generate electricity and protection from floods. These alterations to rivers have come at an environmental cost. When floodwaters are contained by levees or other flood–control dams, they no longer supply nutrients and sediment to floodplain soils that support agriculture. Furthermore, dams and levees that upset a river's natural path and profile (side view) cause changes to the patterns of erosion and deposition (depositing sediments) throughout the entire river system. Dams have contributed to beach erosion on many coastlines because dams trap sediment in reservoirs.

Agricultural and urban development along riverbanks has threatened many species of plants and animals that live in riverside wetlands. Also, the very dams and levees that prevent frequent small floods create an increased risk of infrequent, disastrous flooding. The city of New Orleans, for example, lies at a lower elevation than the bed of the Mississippi River that runs through the center of the city in an artificial channel behind massive levees. If the levees failed, a flash flood would engulf the city and potentially threaten the lives of its residents.

MAJOR RIVERS

Earth's largest river systems define the natural and human environment within their watersheds. A watershed is the land area that drains water into a river or other body of water. A list of the world's major rivers is also a list of the major natural and cultural geographic regions on six continents. (The continent Antarctica is too cold for liquid water. Its fresh water is bound up in large masses of moving ice called glaciers.)

- *Africa*: The Nile is, by most measurements, the world's longest river. (River lengths are difficult to measure because rivers constantly shift their courses and change length. There is also disagreement about which branches of water (tributaries) are considered part of the main river. By some measurements, the Amazon River in South America is actually slightly longer than the Nile.)

  The Nile has sustained life in the inhospitable Sahara desert of eastern Africa for thousands of years. Its headwater (uphill end) streams flow from lakes in Ethiopia and Uganda and feed two branches, the White Nile and the Blue Nile, which meet in the Sudanese city of Khartoum. From there, the Nile cuts a green–bordered lifeline through the Egyptian desert. It flows through Cairo, the bustling capital of modern–day Egypt, past the pyramids of Giza and the ancient Egyptian capital of Thebes, to its outlet in the Mediterranean Sea. The Congo River (called the Zaire River from 1971 to 1997) makes a long loop through the equatorial rainforests and war-torn nations of central western Africa.

  The Congo is the main trade and travel route into the African interior, and it is the setting for Joseph Conrad's famous novel Heart of

Darkness. The Limpopo, Okavango, Ubangi, and Zambezi are other major African rivers.

- *Asia*: Huge rivers drain water from the massive Asian continent into the Pacific, Indian, and Arctic oceans. In China, the Yangtze (Chang Jiang), Yellow (Huang He) and Pearl Rivers carry flowing waters (run-off) from the northern slope of the Himalayan Mountains and western China to the East China Sea. Hundreds of millions of Chinese people depend on these rivers for their electricity, food, and livelihoods. Water moving south from the Himalayas flows into the rivers of India and South Asia, including the Ganges–Bramaputra system and the Mekong River.

  The Ganges River of northern India is sacred in the Hindu religion. Hindus travel to its banks to meditate and wash away their sins. Upon death, cremated remains are placed into the Ganges in hopes of improving the deceased's fortunes in the afterlife. The Ob, Ikysh, Amur and Lena Rivers run across the northern forests and wind-swept tundra (treeless arctic plains) of Siberia (the Asian portion of Russia) into the icy Arctic Ocean. In the Middle East, rivers play an important role in the history and mythology of western civilization.

  The ancient civilizations of Sumeria and Mesopotamia arose in the "fertile crescent" between the Tigris and Euphrates Rivers (Shat-al-Arab) in what is today Iraq. Along with the Jordan River, they play major roles in Jewish, Christian, and Islamic history.

- *Australia*: The island continent of Australia has only a few major rivers, and its central desert, the outback, is extremely dry. The Murray River and its major tributary (major branch), the Darling, make up Australia's largest river system. The Murray drains water from the southeastern states of Victoria, New South Wales and southern Queensland and its floodplains are Australia's most productive farmlands. ü Europe: Rivers are intertwined in the history, culture, and geography of Europe.

  The capital cities of Europe are synonymous with their rivers (London and Thames, Paris and Seine, Vienna, Budapest and Danube). By their very names, the Rhone (France), Rhine (Germany), Volga (Russia), Oder and Elbe (Germany, Poland, Czech Republic), Po and Tiber (Italy), and Ebro (Spain) conjure images of great art and fine wine, desperate battles and bloody conquests, grand castles and ancient hamlets.

- *North America*: The Mississippi and its major tributaries, the Missouri, Ohio, and Arkansas Rivers, collect water from a huge drainage basin that spans the central plains of North America between the Rocky Mountains and the Appalachians.

Canada's Mackenzie and Churchill Rivers empty into the Arctic Ocean, and the St. Lawrence River empties the Great Lakes into the Atlantic Ocean. The mighty Yukon River of northern Canada and Alaska carried prospectors to mines and mills during the Alaskan gold rush (1898–99). Many of the great ports of the Atlantic seaboard and Gulf of Mexico lie near river mouths (the end of a river where the river empties into a larger body of water): New York (Hudson), Philadelphia and Washington, D.C. (Potomac, Susquahana), Norfolk (Delaware), New Orleans (Mississippi), and Houston (Brazos). Rivers, including the Mississippi, Missouri, Colorado, Rio Grande, and Columbia, played central roles in European exploration and settlement of the American West. Today, the rivers that carried explorers Meriwether Lewis, William Clark, John Wesley Powell, and other legendary frontiersmen across the continent are used for agricultural irrigation, drinking water, recreation, and power generation. Their water is a valuable and heavily-sought resource.

- *South America*: The Amazon is the largest river in the world. It flows from the Andes Mountains of Peru, across the Brazil and empties into Atlantic Ocean on the northeast coast of Brazil. The Amazon has more than 1,100 tributaries, 17 of which are longer than 1,000 miles (1,609 kilometers) long. The main river runs from west to east just a few degrees south of the equator, and its massive watershed lies entirely within the warm, wet tropical zone. The central Amazon contains Earth's lushest, wettest, most biologically diverse rainforest. The Orinoco (Venezuela), Sao Francisco (Brazil), Parana (Argentina, Paraguay) and Uruguay (Uruguay, Brazil) rivers are other major waterways of South America.

STREAM SYSTEMS

Streams are any size body of moving surface fresh water driven towards sea level by gravity (force of attraction between two masses). Water scientists refer to all bodies of flowing sur-face water as streams regardless of size, yet in common language, streams are considered smaller than rivers. Stream systems are networks that collect fresh water run-off from the land and carry it to the ocean. Together, tree-shaped systems of small branch streams drain vast areas of the continents into large rivers. Stream systems of all sizes erode (wear down) sediment (particles of gravel, sand, and silt) along their courses and carve complex patterns into the landscape.

They wear down slowrising mountains and fill valleys and lowlands (low and level lands) with layers of sediment. Stream systems change character along their courses. Steep mountain streams feed shallow elevated streams that in turn flow into meandering rivers that snake across broad floodplains (flat, low-lying land near a stream that is covered with water when the stream overflows its banks). Deposits of sediment form at river mouths, the area where fresh river water enters the ocean.

If a rubber duck was dropped into a mountain stream on Pike's Peak in Colorado, it might tumble down the mountainside in whitewater rapids to Cripple Creek. From there, the duck would rush over gravel beds where Colorado miners once panned for gold, and then float serenely across Kansas, Oklahoma, and Arkansas on the Arkansas River. It would pause to drift across huge man-made reservoirs, and then plunge through the spillways of dams before entering the swift, muddy waters of the Mississippi River. A few weeks or months later, you might spot the duck heading out to sea amid barges and river boats in New Orleans.

WATERSHEDS AND DRAINAGE PATTERNS

The land area that drains water into a stream is called a watershed or a drainage basin. A basin is a natural depression in the surface of the land. Watersheds can be as small as a hillside that feeds a wet-weather creek, and as large as a drainage system like the Amazon Basin that carries the run-off from most of a continent. Large watersheds are composed of many smaller drainage basins. The boundaries between watersheds, called drainage divides, are ridge lines or high points where water flows down and away in all directions.

A divide can be limited, like a ridge between two mountain gullies (deep ditches or channels cut in the earth by running water, usually after a rainstorm), or extensive, like the North American Continental Divide along the spine of the Rocky Mountains. Water that falls east of the Continental Divide eventually flows into the Atlantic Ocean, and water that falls west of the Rockies ends up in the Pacific Ocean. Streams are arranged within watersheds in networks that feed water into larger and larger streams. Tree-shaped (dendritic) systems composed of small branch tributaries (small streams that flow into larger streams) that join and flow into large trunk streams are the most common type of stream drainage pattern. Less common drainage patterns develop where rock layers and geologic features affect the paths of streams. Drainage patterns shaped like cross-hatched garden trellises develop in hilly areas where there are ridges and valleys, and streams flow out from round volcanic mountains in radial patterns like spokes on wheels.

VALLEY AND CHANNELS

Streams cut down into the land surface and create valleys. A stream valley includes the entire area between hills on either side of a stream. The water–filled path of the stream at a specific point in time is called a channel. Over time, channels migrate back and forth and fill stream valleys with thick layers of river sediment. Some streams, particularly those in steep, mountainous terrain have narrow, V-shaped valleys and channels that fill most of the valley floor. Others, including most streams in gently–sloping basins and coastal lowlands have narrow channels that snake across wide sediment filled valleys. For example, the Mississippi River has carved a valley more than 100 miles (161 kilometers) wide and filled it with sediment hundreds of feet (meters) thick over thousands of years.

CHANNEL PATTERNS

Stream channels assume different patterns within their valleys: straight, braided and meandering. While many channels have straight segments between meanders or braids, truly straight channels are quite rare. They develop in steep, mountainous areas where geologic forces are slowing lifting up the land surface.

Water flowing rapidly downhill from mountains saws straight channels down into solid rock. Braided streams have many intertwined channels and islands of loose gravel that constantly shift across gravel-filled valley floors. They are common in streams that receive large pulses of water and course-grained sediment.

The sediment-choked streams that carry water from the toes of melting glaciers are typically braided. Streams that bend and curve across gently sloping valleys and coastal plains are called meandering streams.

(Individual loops and bends are called meanders.) During normal weather conditions, water flows in a narrow channel that snakes across broad plains of soft sediment. During floods, muddy water overflows the banks of the channel and deposits layers of mud and silt on the surrounding floodplains. River floodplains are typically fertile farmlands that have been replenished by floodwaters.

The coarser grained sediment settles out of flood waters closer to the channel builds natural levees (walls along the banks of a stream channel) along its banks. The path of a meandering channel changes over time.

Meanders grow from slight bends into nearly-circular loops. At a river bend, fast-flowing water erodes the outer channel bank and sediment accumulates on the inside of the curve in a deposit called a point bar. Eventually, the bends at the neck of the meander grow so close that the water bypasses the loop.

This process strands crescent-shaped segments of the former channel and round point bar deposits called oxbows on the floodplain. Oxbow lakes are abandoned meanders that contain water. Channel patterns change down the course of a stream system between headwater streams and lowland trunk rivers.

They also change over time as streams adjust to changing conditions of water flow, land incline, and amounts of sediment. Stream waters continuously erode and deposit sediment over time, and stream channels constantly shift across valley floors.

STREAM WATER FLOW

Water flows downhill due to Earth's gravity (force of attraction between two masses) pulling it. Streams, like rivers, are gravity-driven bodies of moving surface water that drain water from the continents. Water scientists, called hydrologists, refer to all bodies of running water as streams, no matter their size so, in one sense, rivers are large, well-established streams). In everyday communication, it is common to refer to streams as smaller than rivers. Streams transfer water that falls on the land as precipitation (rain, snow, sleet, and hail) to the oceans. Streams, again like rivers, constantly shift their courses and change length.

The stream is carried along a defined path, called a channel. Water flowing in stream channels is a powerful sculptor that carves landscapes and molds sediment (particles of rock, sand, and silt). It wears down mountain ranges and cuts deep canyons through solid rock. Stream waters support vibrant communities of plants and animals, and they have been the lifeblood of human civilization for thousands of years. Streams shape the land and are also integral to the hydrologic cycle (circulation of water on and around Earth).

EROSION AND DEPOSITION

Streams are the main agent of erosion (wearing away) on land. Water in fast-moving streams is usually turbulent. The flowing water is filled with swirls and small localized whirlpools of swirling water called eddies. Turbulent water picks up particles of sediment that have weathered from rock and soil and carries them downstream. (Weathering is the breaking up of rocks by physical and chemical processes, such as being exposed to the actions of water, ice, chemicals, and changing temperature.) Faster-moving water can carry more sediment in the water, and can push larger stones along the bottom of the channel.

Some mountain streams move huge boulders, while sluggish lowland (low country and level) streams carry only fine grains of silt and mud. The sand grains and larger rock fragments that slide and bounce along stream beds wear away solid rock. In a straight stream, the fastest-moving water and area of greatest erosion is generally in the middle of the channel. Where a stream bends, the strongest current (a moving mass of water) is on the outside of the curve. When water slows down, it drops its sediment load, causing sedimentary deposits to form along stream courses in areas of slowmoving water.

The slower the current, the finer the sediment it deposits. In straight channels, stream water lays down sediment along the stream banks. In bending channels, sedimentary deposits called point bars form on the inside of the bends. Individual sediment grains travel downstream like hitchhikers. Sometimes the grains are picked up by a strong current or flood that moves them far downstream, but usually they don't go very far in a single trip.

The grains of sand on a beach each made a long trip with many stops before they arrived at the ocean. Whether an individual grain of sediment moves depends on the speed of water currents that vary as the amount of water moving through a stream changes. As water currents become faster they can move larger grains. Stream waters also erode rocks by dissolving its minerals, which causes them to crumble. Chemical weathering, also called dissolution, occurs when the slightly acidic water chemically alters the minerals in rocks, which causes them to break down. Clear stream water carries the chemical components (parts) of the rock's minerals called ions (electrically charged particles).

When conditions in the water change (the water slows or cools), the ions recombine into solid mineral crystals. This form of sedimentary deposition is called precipitation. Limestone, salt, and gypsum form by precipitating from water. Ocean animals like corals and shellfish take in ions and use them to build their shells. Some types of rocks, including chalk and flint (also known as chert) form from the remains of organisms.

GRADED STREAMS AND BASE LEVEL

All streams strive to reach a constant slope (incline) called a graded profile by eroding and depositing sediment. The profile (side-view) of a graded stream (a stream with a graded profile) is steep near the uphill end and gently sloping near the point at the end where a stream pours its water into a larger body of water. The position of the downstream end of the profile is determined by the water level at the outlet, called base level. Streams cannot erode below base level. Almost all stream systems run to the sea, so sea level is the ultimate base level for most streams. Conditions change constantly in all streams, and the process of readjustment by erosion and deposition is ongoing.

As conditions change along its course, a stream will readjust its profile by eroding sediment in some places and depositing it in others. If base level falls, stream waters cut down into the land surface. If it rises, they deposit more sediment. If the movements of the underlying plates of Earth's crust rise (geologic uplift) to steepen the upper part of a stream, it will erode down to regain its graded profile. Streams also attempt to level out obstructions along their path. They work to tear down dams, both natural and man-made, by erosion and filling the reservoir behind it with sediment. Lakes are, therefore, only temporary features of stream systems, and dams are interrupting the natural flow of a stream's water.

7

Taxonomy and Biodiversity

TAXONOMY

Taxonomy was once only the science of classifying living organisms (alpha taxonomy)

Taxonomy, sometimes alpha taxonomy, is the science of finding, describing and naming organisms, thus giving rise to taxa.

- In today's usage, Taxonomy (as a science) deals with finding, describing and naming organisms. This science is supported by institutions holding collections of these organisms, with relevant data, carefully curated: such institutes include Natural History Museums, Herbaria and Botanical Gardens.
- Systematics (as a science) deals with the relationships between taxa, especially at the higher levels. These days systematics is greatly influenced by data derived from DNA from mitochondria and chloroplasts. This is sometimes known as molecular systematics and is doing well, likely at the expense of taxonomy.

Cladistics is a branch of biology that determines the evolutionary relationships between organisms based on derived similarities. It is the most prominent of several forms of phylogenetic systematics, which study the evolutionary relationships between organisms. Cladistics is a method of rigorous analysis, using "shared derived traits" of the organisms being studied. Cladistic analysis forms the basis for most modern systems of biological classification, which seek to group organisms by evolutionary relationships. In contrast, phenetics groups organisms based on their overall similarity, while approaches that are more traditional tend to rely on key characters (morphology).

As the end result of a cladistic analysis, treelike relationship-diagrams called "cladograms" are drawn up to show different hypotheses of relationships. A cladistic analysis can be based on as much or as little information as the researcher selects. Modern systematic research is likely to be based on a wide variety of information, including DNA-sequences (so called "molecular data"), biochemical data and morphological data.

In a cladogram, all organisms lie at the leaves, and each inner node is ideally binary (two-way). The two taxa on either side of a split are called sister taxa or sister groups. Each subtree, whether it contains one item or a hundred thousand items, is called a clade. A natural group has all the organisms contained in any one clade that share a unique ancestor (one which they do not share with any other organisms on the diagram) for that clade. Each clade is set off by a series of characteristics that appear in its members, but not in the other forms from which it diverged. These identifying characteristics of a clade are called synapomorphies (shared, derived characters). For instance, hardened front wings (elytra) are a synapomorphy of beetles, while circinate vernation, or the unrolling of new fronds, is a synapomorphy of ferns. Willi Hennig (1913-1976) is widely regarded as the founder of cladistics.

DEFINITIONS

A character state that is present in both the outgroups (the nearest relatives of the group, that are not part of the group itself) and in the ancestors is called a plesiomorphy (meaning "close form", also called ancestral state). A character state that occurs only in later descendants is called an apomorphy (meaning "separate form", also called the "derived" state) for that group. The adjectives plesiomorphic and apomorphic are used instead of "primitive" and "advanced" to avoid placing value-judgments on the evolution of the character states, since both may be advantageous in different circumstances. It is not uncommon to informally refer to a collective set of plesiomorphies as a ground plan for the clade or clades they refer to.

Several more terms are defined for the description of cladograms and the positions of items within them. A species or clade is basal to another clade if it holds more plesiomorphic characters than that other clade. Usually a basal group is very species-poor as compared to a more derived group. It is not a requirement that a basal group is present. For example when considering birds and mammals together, neither is basal to the other: both have many derived characters. A clade or species located within another clade can be described as nested within that clade.

CLADISTIC METHODS

A cladistic analysis is applied to a certain set of information. To organize this information a distinction is made between characters, and character states. Consider the color of feathers, this may be blue in one species but red in another. Thus, "red feathers" and "blue feathers" are two character states of the character "feather-color." The researcher decides which character states were present before the last common ancestor of the species group (plesiomorphies) and which were present in the last common ancestor (synapomorphies) by considering one or more outgroups. An outgroup is an organism that is considered not to be part of the group in question, but is closely related to the group. This makes the choice of an outgroup an important task, since this choice can profoundly change the topology of a tree. Note that only synapomorphies are of use in characterising clades. Next, different possible cladograms are drawn up and evaluated. Clades ideally have many "agreeing" synapomorphies. Ideally there is a sufficient number of true synapomorphies to overwhelm homoplasies caused by convergent evolution (i.e. characters that resemble each other because of environmental conditions or function, not because of common ancestry). A well-known example of homoplasy due to convergent evolution is the character wings. Though the wings of birds and insects may superficially resemble one another and serve the same function, each evolved independently. If a bird and an insect are both accidentally scored "POSITIVE" for the character "presence of wings", a homoplasy would be introduced into the dataset, and this gives a false picture of evolution.

Many cladograms are possible for any given set of taxa, but one is chosen based on the principle of parsimony: the most compact arrangement, that is, with the fewest character state changes (synapomorphies), is the hypothesis of relationship we tentatively accept. Though at one time this analysis was done by hand, computers are now used to evaluate much larger data sets. Sophisticated software packages such as PAUP* allow the statistical evaluation of the confidence we have in the veracity of the nodes of a cladogram.As DNA sequencing has become cheaper and easier, molecular systematics has become a more and more popular way to reconstruct phylogenies. Using a parsimony criterion is only one of several methods to infer a phylogeny from molecular data; maximum likelihood and Bayesian inference, which incorporate explicit models of sequence evolution, are non-Hennigian ways to evaluate sequence data. Another powerful method of reconstructing phylogenies is the use of genomic retrotransposon markers, which are thought to be less prone to the reversion and convergence that plagues sequence data.

Ideally, morphological, molecular and possibly other (behavioral etc.) phylogenies should be combined: none of the methods is "superior", but all have different intrinsic sources of error. For example, character convergence (homoplasy) is much more common in morphological data than in molecular sequence data, but character reversions are more common in the latter. Cladistics does not assume any particular theory of evolution, only the background knowledge of descent with modification. Thus, cladistic methods can be, and recently have been, usefully applied to non-biological systems, including determining language families in historical linguistics and the filiation of manuscripts in textual criticism.

Cladistic Classification

Three ways to define a clade for use in a cladistic taxonomy.Node-based: the most recent common ancestor of A and B and all its descendants.Stem-based: all descendants of the oldest common ancestor of A and B that is not also an ancestor of Z.Apomorphy-based: the most recent common ancestor of A and B possessing a certain apomorphy (derived character), and all its descendants.

Three ways to define a clade for use in a cladistic taxonomy. *Node-based:* the most recent common ancestor of A and B and all its descendants. *Stem-based:* all descendants of the oldest common ancestor of A and B that is not also an ancestor of Z. *Apomorphy-based:* the most recent common ancestor of A and B possessing a certain apomorphy (derived character), and all its descendants. A recent trend in biology since the 1960s, called cladism or cladistic taxonomy, requires taxa to be clades. In other words, cladists argue that the classification system should be reformed to eliminate all non-clades. In contrast, other taxonomists insist that groups reflect phylogenies and often make use of cladistic techniques, but allow both monophyletic and paraphyletic groups as taxa.

A monophyletic group is a clade, comprising an ancestral form and all of its descendants, and so forming one (and only one) evolutionary group. A paraphyletic group is similar, but excludes some of the descendants that have undergone significant changes. For instance, the traditional class Reptilia excludes birds even though they evolved from the ancestral reptile. Similarly, the traditional Invertebrates are paraphyletic because Vertebrates are excluded, although the latter evolved from an Invertebrate.

A group with members from separate evolutionary lines is called polyphyletic. For instance, the once-recognized Pachydermata was found to be polyphyletic because elephants and rhinoceroses arose from non-pachyderms separately. Evolutionary taxonomists consider polyphyletic groups to be errors in classification, often occurring because convergence or other homoplasy was misinterpreted as homology.

Following Hennig, cladists argue that paraphyly is as harmful as polyphyly. The idea is that monophyletic groups can be defined objectively, in terms of common ancestors or the presence of synapomorphies. In contrast, paraphyletic and polyphyletic groups are both defined based on key characters, and the decision of which characters are of taxonomic import is inherently subjective. Many argue that they lead to "gradistic" thinking, where groups advance from "lowly" grades to "advanced" grades, which can in turn lead to teleology. In evolutionary studies, teleology is usually avoided because it implies a plan that cannot be empirically demonstrated.

Going further, some cladists argue that ranks for groups above species are too subjective to present any meaningful information, and so argue that they should be abandoned. Thus they have moved away from Linnaean taxonomy towards a simple hierarchy of clades. The validity of this argument hinges crucially on how often in evolution gradualist near-equilibria are punctuated. A quasi-stable state will result in phylogenies, which may be all but unmappable onto the Linnaean hierarchy, whereas a punctuation event that balances a taxon out of its ecological equilibrium is likely to lead to a split between clades that occurs in comparatively short time and thus lends itself readily for classification according to the Linnaean system.

Other evolutionary systematists argue that all taxa are inherently subjective, even when they reflect evolutionary relationships, since living things form an essentially continuous tree. Any dividing line is artificial, and creates both a monophyletic section above and a paraphyletic. Paraphyletic taxa are necessary for classifying, the tree–for instance, the early vertebrates that would someday evolve into the family Hominidae cannot be placed in any other monophyletic family. They also argue that paraphyletic taxa provide information about significant changes in organisms' morphology, ecology, or life history–in short, that both taxa and clades are valuable but distinct notions, with separate purposes. Many use the term monophyly in its older sense, where it includes paraphyly, and use the alternate term holophyly to describe clades (monophyly in Hennig's sense).

xAs an unscientific rule of thumb, if a distinct lineage that renders the containing clade paraphyletic has undergone marked adaptive radiation and collected many synapomorphies-especially ones that are radical and/or unprecedented -, the paraphyly is usually not considered a sufficient argument to prevent recognition of the lineage as distinct under the Linnaean system (but it is by definition sufficient in phylogenetic nomenclature). For example, as touched upon briefly above, the Sauropsida ("reptiles") and the Aves (birds) are both ranked as a Linnaean class, although the latter are a highly derived offshoot of some forms of the former which themselves were already quite advanced.

A formal code of phylogenetic nomenclature, the PhyloCode, is currently under development for cladistic taxonomy. It is intended for use by both those who would like to abandon Linnaean taxonomy and those who would like to use taxa and clades side by side. In several instances it has been employed to clarify uncertainties in Linnaean systematics so that in combination they yield a taxonomy that is unambiguously placing the group in the evolutionary tree in a way that is consistent with current knowledge.

A taxon is usually assigned to a rank in a hierarchy. The basic rank is that of species, and if an organism is named it most often will receive a species name. The next most important rank is that of genus: if an organism is given a species name it will at the same time be assigned to a genus, as the genus name is part of the species name. Of the botanical names used by Linnaeus only names of genera, species and varieties are still used. The third-most important rank, although it was not used by Linnaeus, is that of family.

Thus, taxonomy is a branch of biology which deals with collection of organisms, their identification, nomenclature and systematic grouping or classification into various categories. This is done on the basis of similarities and differences of their morphological, anatomical, cytological, genetical, physiological, biochemical, developmental and other characteristics.

The similarities of characteristics between species or groups of species indicate their relationship. This is also gives us some idea about their phylogeny (i.e. their evolutionary history). The classification of plants into various groups is called plant taxonomy or systematic botany. Similarly, classification of animals is called animal taxonomy or systematic zoology.

The original concept of species has undergone a considerable change during the progress of taxonomy. John Ray (1627-1705) was the first to distinguish genus and species. However, the clear morphological concept of species was first given by Linnaeus (1707-1778). Later on, Darwin proposed the biological concept of species. The concept was further modified by Ernst Meyr.

Morphological concept of species by Linnaeus. A species is the group of individual which resemble each other in most major morphological (vegetative and reproductive) characteristics. Biological concept of species by Darwin. In addition to morphology, the biological concept also takes into consideration ecology, geography, cytology, physiology, behavior, etc.

According to the biological concept, a species is a group of individuals which resemble each other in morphological, physiological, biochemical, and behavioral characteristics. These individuals are capable of breeding with each other under natural conditions, but are unable to breed successfully with members of other species. Thus, species is a group of fertile organisms that can interbreed and produce fertile offspring only among themselves. The recent trend is to consider species as the groups of actually or potentially interbreeding natural populations of closely resembling individuals (Ernst Meyr).

A species is considered to be the smallest, most basic unit of classification in most of the systems. It was thought to be an indivisible, stable and static unit (taxon). However, in modern taxonomy, sub-divisions of species, such as sub-species and populations, have been created which aid in our understanding through classification.

Taxa and Categories

- Taxa (Singular: Taxon): A taxon is the taxonomic group of any rank in the system of classification. For example, in plant kingdom, each one of the following such as, angiosperms, dicotyledons, polypetalae, Malvaceae, Hibiscus esculentus, etc. represents a taxonomic group i.e. a taxon. A taxon may be a very large group such as a Division (e.g. angiosperms), or it can be a very small group such as a species (e.g. Hibiscus esculentus).
- Categories (Singular: Category): In the system of classification, the various taxa are assigned definite ranks or positions according to their

taxonomic status. Each such taxonomic rank is called the taxonomic category. The various major categories in the classification of plant kingdom are Kingdom, Division (Phylum), Class, Series, Order, Family, Genus and Species.

The difference between the taxon and the category should be clearly understood. For example, when we say "Division- Angiosperms", 'Division' represents the taxonomic category while 'angiosperms' represents the taxon. Thus, a taxon is a group of organisms (living beings), whereas a category only indicates the rank or status of the taxon in the systematic hierarchy. Arranging various taxonomic categories in their proper order on the basis of their taxonomic ranks is called taxonomic hierarchy (systematic hierarchy). In this hierarchy, the kingdom represents the category of highest rank while the species is the category of the basic rank.

Following is an example of the taxonomic hierarchy representing the methodology of classifying a plant and an animal in a scientific manner.

A broad scheme of ranks in hierarchical order:

- Domain
- Kingdom
- Phylum (animals or plants) or Division (plants)
- Class
- Order
- Family
- Genus
- Species
- Subspecies

The prefix super-Indicates a rank above, the prefix sub- indicates a rank below. In zoology the prefix infra- indicates a rank below sub-

BINOMIAL NOMENCLATURE

The system of giving a scientific name to each properly identified plant or animal is called nomenclature. A system of nomenclature of plants and animals in which each scientific name consists of two parts or sub-names is called the system of binomial nomenclature. Thus according to this system the scientific name of sunflower is Helianthus annuus and that of man is Homo sapiens. In the above names, the first part of the name (i.e. Helianthus or Homo) represents the name of the genus (generic name). The second part of the name (i.e. annuus or sapiens) represents the name of the species (specific name).

This system of binomial nomenclature was introduced by Carolus Linnaeus in 1753 in his book Species Plantarum.

The system follows certain rules, such as:

- The scientific name must be in Greek or Latin language.
- Genetic name should come first and must begin with a capital letter.

- The same name should not be used for two or more species under the same genus.
- The scientific name must be either underlined or written in italics.
- The name of the author who first described the species should be written after the specific name (e.g. Homo sapiens Linnaeus).

PRINCIPLES OF CLASSIFICATION

While developing a system of classification of organisms, certain basic principles are observed. Some of these are as follows:

Morphological Criteria

Morphology forms the primary basis for classifying organisms into various taxonomic groups or taxa. In earlier artificial systems, only one or a few morphological characters were taken into consideration (e.g. plants were classified into herbs, shrubs, trees, climbers, etc. on the basis of their habit). The sexual system proposed by Linnaeus was based mainly on the characteristics of stamens and carpels. Later on, in the natural systems of classification (e.g. Bentham and Hooker's system of classification of plants), a large number of morphological characters were taken into consideration. As a result, classification of plant groups was more satisfactory and their arrangement was showing natural relationships with each other.

The similarities in the morphological characters are used for grouping the plants together. Because, these similarities indicate their relationships. On the other hand, differences or dissimilarities of characters are used for separating the plant groups from each other. Plant groups with greater differences are considered to be unrelated or distantly related. For example, all flowering plants with ovules enclosed in an ovary cavity are grouped together as Division-Angiosperms whereas, the angiosperms are further classified into two classes: Dicotyledons and Monocotyledons, on the basis of differences of the characters of root system, leaf venation, flower symmetry and number of cotyledons in the embryo.

Phylogenetic Considerations

In the more recent systems of classification of plants, a greater emphasis is given on the phylogenetic arrangement of plant groups, an arrangement which is based on the evolutionary sequence of the plant groups. These systems also reflect on the genetic similarities of the plants. Some of the phylogenetic systems of classification of plants are the ones proposed by Engler and Prantle (1887-1899), Bessey (1915), Hutchinson (1926 and 1934), etc. However, none of these or any other systems is a perfect phylogenetic system. This is because, our present knowledge of the evolutionary history of plant groups is very fragmentary and incomplete. At best, the present day systems can be described as the judicious combination of both natural and phylogenetic systems. Modern taxonomy takes into consideration data available from all disciplines of botany for classification of plants. This helps immensely in establishing inter-relationships of various plant groups. As a result, taxonomic arrangement becomes more authentic and convincing.

Chemical Taxonomy or Chemotaxonomy

It is a comparatively recent discipline. Chemotaxonomy is the application of phyto-chemical data to the problems of systematic botany. The presence and distribution of various chemical compounds in plants serve as taxonomic evidences. Nearly 33 different groups of chemical compounds have been found to be of taxonomic significance.

Numerical Taxonomy

Application of numerical methods (data) in the classification of taxonomic units is called numerical taxonomy. Edgar Anderson (1949) was the first to make use of numerical taxonomy in the classification of flowering plants. It involves exhaustive quantitative estimation of taxonomic characters from all parts of the plant as well as from all stages in the life cycle. The numerical data thus collected for various plant groups is tabulated systematically. Computers are used for this purpose. The main objective of numerical taxonomy is to clarify and illustrate degrees of relationship or similarity in an objective manner.

CLSSIFICATION OF TAXONOMY

The effect of Theory of evoluion on smaller level of species, genus and family is very little but it have large effect on taxonomic theory. Systemation started developing classification based on theory of evolution. It was reasoned that such classifications should designed to reflect revolutionary lineages and should reflect the course of evolution. Since the start of the twentieth century, numerous taxonomists have attempted to produce systems of classification reflection the phylogeny of the flowering plans. This is very tough because of the lack of fossil records and partly to the complexity and diversity of the plants themselves.

NUMERIAL PHENETICS

Phenetics in taxonomy deals with collection and organisation of data on similarity basis for classification purpose, workers are called pheneticists. They attempt to develop system having high predictive values that allow the maximum number of generalizations to be made form the schemes.

They give importance in using a large number of character, at least 60 but preferabley 80 to 100, that can be correlated on the basis of similarity. Phenetic systems are based on date derived from the phenotypes of organisems–hence the desiganation phenetic systems. The ides of phenetic system classification was proposed by Michel Adanson. With availability of computers and new technology, many taxonomists began developing methods for grouping of taxo using quantitative methods. One assumption was that it should theoretically be possible to develop methodology that would quantitatively allow a person to arrive at basically the same classification. Sneath and Sokal summarized their findings in 1963 and in a revised edition in 1973. They believed in equal weighting of characters and made claims about the value of numerical phenetics. However, their refusal to weight characters annd their attitude to ignore phyletic information had negative impact on the validity and acceptance of their method in the beginning. Modifications of these extreme viewpoints by later pheneticists has brought about wider acceptance of the methodology. In an attempt to be objective, pheneticists have replaced generic names with Operational Taxonomic Units (OTU's) the term given to the lowest-ranking taxon studied in the investigation.

A problem is also the definition of characters for example, can leaf shape be defined as a single character or is it actually condensation of several independent characters that determine the shape. Phenetic methodology requires that character states be determined for each OTU. A data matrix (OTUs versus character states) is prepared and the data are codified for computer processing. In cluster analysis, the computer sorts out (clusters) the OTUs according to their overall similarity.The available computer programmes produce a phenogram that is a dendrogram of phenetic relations in which less and less similar OTUs are successively linked together with the contourlike lines encircling different numbers of OTUs or by linking OTUs with straight lines of varying lenghts. These computer methods attempt to demonstrate a classification graphically but can be comprehanded by the taxonomist. Some workers have attempted to delimit subjectively taxa by phenetic method especially be equating different levels of similarity on the phenogram but, thus far, this has not been successful or practical. The main aim of phenetics is to lend objectively to somewhat subjective methods employed by traditional taxonomist. Younger generation of numerical taxonomists seem to appreciate the difference in information content of characters and is willing to weight characters besed on a posteriori evidence. Numerical method appear to be useful in difficult situations where other methods have not proved satisfactory.Most workers today agree that classifications developed by numerical methods are probably no better or no worse than classifications produced by the traditional taxonomists. One by advantage of this method is that it had compelled taxonomists to define characters.

CLADISTICS

Cladistic analyst uses a branching diagram called cladogram to summarize similarities among organisms. The cladogram is a graph which shows the sequence of branching points in the phylogeny. Workers in this system of classifications are known as cladists. The main aim of cladistics is productions of objective and repeatable branching diagreams showing hypothetical evolutionary histories, thus eliminating arbitrariness from classification. The cladistic school of classification was founded by the German entomologist Willi Henning around 1950. Many systematic zoologists such as J.S. Farris, N.I. Platnick, E. O. Wiley, and many others quickly adopted Henning's principles.

Botanists have started realizing the potential for applying the methods of cladistics to their problems. There are many viewpoints among cladists; however, most will agree that natural monophyletic groups are recognized by uniquely derived characters termed synapomorphies, and that only natural groups defined by synapomorphies are to be included in classification schemes. A shared primitive character-state is known as a sympleisomorephy.

Cladists attempt to adjvst similarities on a branching diagram according to the hierarchical level at which they are postulated to be a synapomorphy. Cladists reasoned that branching points are determined by backward tracing of uniquely derived characters since derived characters are to be found only among the descendants of the ancestor. In general, two comparative methods are used for evaluating characters-ontogenetic criterion and a outgroup criterion. Both methods attempt to establish the partitioning of characters into ancestral or derived.

An ontogenetic criterion uses data from comparative embryology to assess the direction of character transformation. As most of the times development information is not available for plant so outgroup comparison is use. In this method, if a shared character in a group being studied is also found outside the group in closely related taxa, then that character is considered too general. This requires that the systematist know the relationship of one group to other related groups in order to analyse critically the relationship within the group.

Cladists attempt to maximize the adjustment of characters while attempting to account for their distribution. In theory, cladistic analysis has value as a method for delimiting monophyletic groups. It forces the careful analysis of all characters; it introduces a new concept of character weighting, that is, the possession of uniquely derived characters. Determination of direction of evolutionary changes and determination of primitive characters possess further difficulties and also some groups can not be analysed easily because of the absence of easily detectable synapomorphies. Continuous characters and variation within a taxon present difficulties when using cladistic methods. The importance of hybridzation, may not be great in animal taxa but in plant taxa it is of quite importance as many plants are produced by hybridization. Many times two cladistic hypotheses can equally explain character distributions. At such time it depends on systematist to judge and select. To this problem, cladists apply the principle of parsimony by selecting the hypothesis that can best be defended by the investigator as having the fewest character state changes. It is common knowledge that monocotyledons are derived from dicotyledons. So monocotyledons can't be grouped like dictotyledons and cannot have same rank in classification. Using this reasoning, the monocyledons should have rank within the dicotyledons. From this example, it is clear that the application of cladistics to botanical classification will have an impact on the practice of developing classification.

TAXONOMY ON THE WEB: EVOLUTION, REVOLUTION OR INVOLUTION

Technology has enormously progressed during the last decades, and the birth of the internet has also affected all aspects of science. Today, we can hardly imagine how the previous generations of scientists were able to man-age without computers or e-mail! Today, we can easily electronically access a plethora of databases, journals contents, huge amounts of information, we have developed modern apparatus and can use complex statistical models and programmes to process our data, our scientific colleagues all over the world are available at the click of a mouse. Nobody can deny that the tech-nological progress has contributed positively to science. But... can the web be a tool to improve the traditional 'bad concept' of taxonomy? Godfray was one of the first in proposing that all new taxonomic revisions should be placed on the web, available and accessible to all.

He comments that taxonomy is made for the web: it is information-rich and often requires copious illustrations. According to Godfray, taxonomy and systematics have an image problem among funding bodies and the community in general and it is time for a change: taxonomy needs to reinvent itself if it is to survive and flourish. He advocates a unitary taxonomy; all taxonomic information about each group (descriptions, photographs, illustrations, keys) would be on the web and new information could be added, each group being under the administration of an authoritarian body. In this way the information would be more attractive to financial support than taxonomy as presently practised. Major government and private research funders would consider construction and maintenance of a unitary taxonomy. Godfray (2002a) consider that it might also attract new sources of funding 'it surely isn't impossible that a major company might sponsor the web revision of, say, the Lepidoptera, and if it wants to put its logo on the site, then why not?'

Several subsequent papers after Godfray (2002a) have followed this logic. Bisby et al. (2002) supported the creative ideas for modernizing taxonomic practices and endorse Godfray's suggestion that species descriptions, images and a platform for publication and debate should be provided on the web. Wilson (2003) imagines an electronic page for each species of organism on Earth, available everywhere by single access on command; the page contains the scientific name of the species, a pictorial or genomic representation of the primary type specimen and a summary of its diagnosis, photographs, pictures, etc, since the page is indefinitely expansible and its contents are continuously peer reviewed and updated. All the pages together for an en-cyclopedia, the encyclopedia of life. Wheeler et al. (2004) insist on the need of taxonomy to accommodate to the new technologies, 'it is time to approach taxonomy as large-scale international science. Gewin (2002) compiled some opinions and views of the pioneers who are trying to turn this vision into reality.

On the other hand, several other scientists did not receive the suggestion by Godfray (2002a) with such optimism. Knapp et al. (2002) pointed out that some changes are clearly necessary, but science cannot be replaced by informatics. These authors consider that working within the current enabling conventions is more positive and practical than throwing them out and be-ginning again; web-based taxonomy is clearly the way of the future, but the technologies needed for this to operate successfully on the scale required are only starting to be available, and quality control is something that also must be addressed. Thiele & Yeates (2002) also showed some objections to the Godfray's model, highlighting that a taxon is an hypoth-esis, not an observation or fact so the proposed solutions for a change cannot be so easily transferred to the domain of taxonomy. According to Scotland et al. (2003a), advanced technology does not necessarily result in increased taxonomic productivity, and web-based taxonomy as a technical solution (together with the molecular approach) may provide a tantalising mirage for politicians concerned about conservation of biodiversity, but, in practice, these ideas are largely a red herring, they do little to address the real problem.

NEW PROJECTS AND FUNDING INITIATIVES FOR SUPPORTING TAXONOMY

Although we are not sure yet if the advances of the web will be the solution or not for the taxonomical crisis, we must admit, at least, that the Internet has facilitated various web-based ambitious projects and initiatives. Many Internet taxonomy initiatives exist, perhaps too many, as pointed out by Mallet & Willmott (2003). We have already mentioned previously the importance of the Global Taxonomy Initiative (GTI) and the Convention on Biological Diversity (CBD). Several more exists, Species 2000 and the Integrated Taxonomic Information System, are two major players in creating an electronic global framework for tax-onomy, which joined forces last 2001 in the Catalogue of Life consortium and are now making rapid progress with a catalogue of all known organ-isms. The 2002 Catalogue of Life now lists 260000 species on CD-ROM and on the Web. The Global Biodiversity Information Facility is also a vital step toward accessible spe-cies-level information. The Catalogue of Life and the GBIF are each funded at about US$3 million a year and we should see real progress over the next years. The All Species Founda-tion was launched, with the goal of cataloguing every species on Earth in 25 years. The Tree of Life project is another ambitious project, from a phylogenetic perspective. These are among the largest projects, but more than 50 other web-based projects exist worldwide.

There are also several other interesting initiatives with more discrete and realistic goals focused on restricted geographic areas rather than globally. For example three major European programmes funded by the European Commission:

- Fauna Europaea, which started in 2000 and has the objective of producing a web-based checklist of all European land animals.
- In 1997 the European Commission funded also another project to compile a taxonomic checklist of marine organisms, and today the European Register of Marine Species (ERMS) is now complete in its first edition, available both on the web and in printed format.
- The EuroMed PlantBase project has a similar objective for plants. A further objective is to link these European treatments with other systems around the world through the Species 2000 framework.

In several countries, separately funded national inventories could be the key to significant acceleration of the biodiversity census. We can use Sweden and the Iberian Peninsula, as exam-ples. The Iberian Peninsula constitutes without doubt, the richest and most diversified region in Western Europe. Ten years ago, a nationally funded project, Fauna Iberica, was launched, and this project is, without doubt, the most ambitious taxonomic project than ever existed in the Iberian Peninsula and the only one capable at present of bringing together all the necessary resources to produce an inventory of the animal diversity in this area. Seventy-two monographs on animal groups belong-ing to 11 phyla are already edited, in press or in preparation, representing approximately 20% of the estimated total number of animal species in the Iberian Peninsula and Balearic Islands but, at this rate, more than 75 years will be needed to complete it. More taxonomists and more funding are required to save time.

The so-called Swedish Taxonomy Initiative (STD) provides other example. It was launched in January 2002 and aims to complete an inventory of Sweden's fauna and flora of multicel-lular organisms within 20 years. Following the tradition of its most famous taxonomist, Sweden aims to be the first country to complete an inventory and pictorial guide to its biodiversity.

If most of scientists agree on something concerning taxonomy, it is the lack of funding. Funds are needed to train new taxonomists and to provide facilities and resources to the taxonomists that already exist. We have also two nice examples to show that, probably, we are moving now into the right direction: The NSF Partnerships for Enhancing Expertise in Taxonomy (Peet) developed in the Usa, and the Synthesis of Systematic Resources (Synthesys) project supported by the European Community. These two funding initiatives, together with others such as PBI (Planetary Biodiversity Initiative) of the Us National Science Foundation, the Edit programme (European Distributed Institute of Taxonomy), the UK-Nerc funded Cate (Creating a Taxonomic e-Science) project could serve as ap-propriate models for future organisations.

The dismissal of taxonomy worldwide possibly originated in the USA and some measures are now being implemented to correct it. The National Science Foundation has realized that taxonomy is dying and that the USA cannot have a scientific community which is deprived of taxonomists. This led to the launch of the PEET initiative, which has been training new generations of taxonomists since 1995. This program includes substantial budgets to fund projects, enabling intensive training, targeting poorly known groups of organisms for revisionary or monographic research. The principal investigators of the projects are prestigious taxonomists who train young people from different countries. As shown by Rodman & Cody (2003) many of the PEET trainees have secured employment in the USA and abroad in academic, museums, or government agency positions relevant to systematics.

Synthesis of Systematic Resources (Synthesys) is an initiative launched by the Consortium of European Taxonomic Facilities. In 2004, 20 European natural history museums and botanic gardens were successful in securing this integrated infrastructure initiative grant. This programme has two parts, the first one enables European researchers to access the collections comprising more than half of the world's natural history specimens, world-class libraries, facilities for microscopy, physical, chemical and molecular analysis and experienced hosts and trainers at 20 European institutions.

The second part is related to networking activities focused on creating a single museum service, an integrated European resource bringing together the collections of the major natural history museums and other institutions in Europe. SYNTHESYS integrates in the same programme some previous similar programmes conducted in each country, such as Sys-Ressource (Great Britain), Colparsyst (France), Cobice (Denmark), Hight-Lat (Sweden), Bioiberia (Spain) and ABC (Belgium).

Natural history museums must play an important role in supporting taxonomy, and they are not only about pure science, but also about educating the public. It is important to bring taxonomy to the general public, as society must know the importance of our work. Information from natural history collections about the diversity, taxonomy and historical distributions of species worldwide is becoming increasingly available over the internet. Computerization of collections and development of electronic catalogues are providing new capabilities for curating collections. Funding programmes should be also addressed to support this type of collection management and to provide more educational activities between scientists and, for example, children, since some of them may be-long to the next generation of taxonomists.

PROGRESS IN PHYLOGENETICS

The main concepts related to phylogeny were already defined above, since phylogenetic ideas are really implicit in the concepts of taxonomy and systematics. In fact, "Nothing in Biology makes sense except in the light of Evolution", following the famous essay by C. T. Dobzhansky. Consequently, it seems more than reasonable and justified the importance of Phylogenetic Systematics, since classifications should reflect the relationships among taxa in an evolutionary framework. Furthermore, phylogenies are fundamental to comparative biology; there is no doing it without taking them into account. Phylogenies provide new ways to measure biodiversity, to assess conservation priorities, and to quantify the evolutionary history in any set of species. If evolution is the unifying theme of biology, then the tree of life is the framework from which it hangs and reconstructing it should be one of the great scientific goals of the new century.

Shortly after the publication of Darwin's The Origin of Species, biolo-gists were enamoured with the concept of phylogeny. In 1866, E. Haeckel (who coined the term "phylogeny") published a collection of detailed phylogenetic trees that depicted much of what was known about the evolutionary history of life. By the 1940s and 1960s the study of phy-logeny greatly diminished, but, fortunately, the reemphasis on phylogenetic perspectives in biology began in the 1960s and 1970s, with the accumulation of new phylogenetic data (especially from molecular biology), the develop-ment of explicit and objective methods for phylogenetic inference, and the construction of computer hardware and software sufficient to the task of applying the new methods to the new data. Because no per-son was present to observe directly the evolution of a group of organisms, biologists must infer phylogenies from the characters of living and fossil taxa.

Schools of Phylogenetic Inference

Presently there are three main schools dealing with the phylogenetic inference: Evolutionary taxonomy, numerical taxonomy or phenetics, and Phylogenetic Systematics (or Cladistics sensu lato). Evolutionary taxonomy is an early school of phylogenetic inference, which recognises that similarity among species could arise either because species were closely related or because of convergent or parallel evolution. The term "evolutionary taxonomy" could be confusing since it seems that only this school explicitly uses the term evolution in its classification system. For this reason, some authors, such as Brusca & Brusca (1990), suggested the name of orthodox taxonomy. This school emerged during the middle of the XIX century with G.G. Simpson, E. Mayr, W.J. Bock and P.D. Ashlock-and being the main advocates of this school. The traditional classification is mainly based on the evolutionary taxonomy, which accepts paraphyletic groups. Criticism of this school is primarily based on the lack of an explicit and objective methodology. Evolutionary taxonomy is essentially traditional taxonomy with evolution taken into account.

In the early 1960s, a group of statisticians and biologists introduced a new approach, known as numerical taxonomy or phenetics, initially developed by Sokal and Sneath (1963). This school considers that organisms should be grouped on the basis of overall similarity, independently of whether these groupings represent phylogeny.

Many characters are analysed and taxa are arranged using clustering methods based on overall similarity. Numerical taxonomy was popular, however, its use has declined since the clusters do not necessarily reflect true phylogenetic relationships. Some authors talk today about phenetic cladistics, in which the treatment of individual characters as units that measure similarity, without evaluation of character quality, is retained.

In 1966, W. Hennig published the book Phylogenetic Systematics, giv-ing rise to the highly influential school of systematics known as cladistics. This school has common objectives with the pheneticists bringing in ob-jectiveness, but focusses on common ancestry, homologies, synapomorphies (shared derived characters) and monophyly.

Only monophyletic groups are considered valid. Phylogenetic systematics is strictly founded on the logic of scientific argumentation in the sense of Karl Popper; hypotheses on homologies and on monophyly can be substantiated and falsified with intersubjectively verifiable criteria. Using the logic of an-cestral and derived traits, cladistics construct sets of monophyletic groups to construct phylogenetic trees thought to identify true ancestral-descend-ant relationships. Hennig's method has been developed for morphological characters, but it can also be applied to other discrete characters including DNA sequences.

According to Pagel (2002), by 1980s, statistical and model-based methods for inferring phylogenetic trees began to provide an alternative perspective to cladism. Many authors tend to consider that this new statis-tical approaches are the basis of a different school, the statistical school.

There is presently a lot of controversy and different points of view about the exact relationships and interplay between the terms Phylogenetic Systematics, Cladistics, Maximum Parsimony, etc., and numerous questions can be asked: "Is Cladistics really a synonymous with Phylogenetic Systematics?"

Cladistics, as defined by Wägele (2005), is the construction of dendrograms from character/taxa datasets using the maximum parsimony method (one of the several available methods). According to this, if we are using other methods such as Maximum likelihood or the Bayesian approach to infer phylogenetics, it seems that we are not doing cladistics, but aren't we doing Phylogenetic Systematics?

Methods to Infer Phylogenetics the Statistical Domains?

One of the main problems in inferring phylogenies is that the number of possible phylogenies grows very fast, as the number of taxa increases: for three species there are three choices, but for 10 there are over 34 mil-lion, and for 20 there are over 8.2×10^{21}. Phylogenetic trees describe the pattern of descent amongst a group of species. With the rapid accumulation of DNA sequence data, more and more phylogenies are being constructed based upon sequence comparisons. The combination of these phylogenies with powerful new statistical approaches for the analysis of biological evolution is challenging widely held beliefs about the history and evolution of life on Earth.

The three principal methods of phylogenetic inference are parsimony, distance methods, and maximum likelihood, and all draw upon ideas that emerged from the debates among the different schools of phylogenetic inference. Parsimony methods are closely linked to cladism, and the maximum likelihood methods arose from the statistical school.

Distance-based methods of inference share a number of features with the phenetic school. Beside this, there is also a recent ap-plication of Bayesian methods; these methods are not new in the field of statistics but are being used recently with phylogenetic purposes. Maximum likelihood and Bayesian have their base in statistics and these probabilistic techniques represent a parametric approach, while maximum parsimony can be considered nonparametric.

Parsimony methods seek, out of all the evolution-ary trees that could possibly describe the relationships among a group of organisms, the tree that implies the fewest evolutionary changes in the characters being examined, the simplest tree. This most parsimonious tree is taken to be the best estimate of the unknown true tree.

The principle of parsimony is based on the ideas of Occam, a fourteenth-century philosopher, who advocated the view that when alternative explanations for an observed phenomenon exist, the simplest (or most parsimonious) explanation is to be preferred. This principle is known as Occam's Razor. One of the main problems of parsimony methods is that they do not use all the available information. To use the information in a more efficient way and have the possibility to choose statistically the best tree Felsenstein (1985) used the new approach initiated by Edwards, the maximum likelihood.

Distance methods find phylogenetic trees whose branch lengths most closely reflect the actual "distances" that are observed among all possible pairs of species. These methods are based on the transformation of discrete characters (as the presence or absence of a morphological character, or the identity of a nucleotide in a homologue region of a gene) in a distance value. Distance methods are also usually more used for molecular data and are based on the comparison of pairs of aligned sequences. These methods constitute the last remnant of phenetics in systematics (they have received much criticism because of this phenetic component), and consequently, the assumptions of these methods are valid in absence of homoplasy.

If convergent and parallel changes are rare, then the observed distance between any species will reflect evolutionary events that have occurred since the two species separated from their common ancestor. Dendrograms based on distance can be obtained by (1) searching clusters of most similar sequences based on pairwise distances between sequences (clustering methods such as UPGMA or Neighbour-joining), and (2) seek-ing the tree whose sum of branch lengths is minimized (minimum evolution methods)

Maximum likelihood estimation of phylogenetic trees was first intro-duced by Edwards and Cavalli-Sforza in the early 1960s and Felsenstein implemented the method for DNA sequence data. According to this method, the best tree is the most probable tree. These methods have been developed mainly for molecular data instead of morphological data and most recent advances have focused on the analysis of DNA sequences. One of the drawbacks of these methods is that they requires powerful computers and are very slow to run, and consequently, it is difficult to analyse large amounts of data as exhaustedly as with parsimony methods.

Bayesian analysis. An important recent advance in phylogenetic infer-ences is the application of Bayesian Markov Chain Monte Carlo (MCMC) methods. Bayesian methods are also based on a statistical approach, similar to the Maximum Likelihood, but in this case a tree is found with maximum posterior probability, evaluating features in common among the sampled trees.

Using Bayesian algorithms one searches the tree or set of trees that maximize the probability of the tree for the given data and the selected substitution model. These methods, by virtue of collecting a random sample of trees from the universe of possible trees, allow one to estimate aspects of the phylogeny. Bayesian inference of phylogeny brings a new perspec-tive to a number of outstanding issues in evolutionary biology, including the analysis of large phylogenetic trees and complex evolutionary models and the detection of the footprint of natural selection in DNA sequences.

Once the trees have been obtained, there are several a posteriori crite-ria that measure the fit between data and topology, such as bootstrapping, Bremer's index and Jacknife percentages.

For the past two decades, there has been an ongoing debate within the phylogenetics community over whether model-based approaches for mo-lecular systematics (such as maximum likelihood) should be preferred over the more traditional 'maximum parsimony' approach. Some authors recommend that those who infer and make use of trees should adopt a pluralistic and critical approach, using both maximum parsimony and maximum likelihood and evaluating the results of both methods cautiously, in the light of the understanding of the strengths and weaknesses of each technique. In fact, as shown by Crisp & Cook (2005) different methods using the same data can give different results. These authors summarised the main advantages and disadvantages of each method; parsimony fails in indicating probability of estimates, while the statisticals maximum likelihood and Bayesian inference can fail if the model is unrealistic.

Although many authors tend to consider that maximum likelihood and Bayesian methods are really suplanting parsimony methods, Kolaczkowski & Thornton (2004) have recently shown that maximum likeli-hood and Bayesian approaches can become strongly biased and statistically inconsistent when the rates at which sequence sites evolve change non-identically over time. Maximum parsimony performs substantially better than current parametrics methods over a wide range of conditions, specially when evolution is heterogeneous.

Anyway, as pointed by Crisp & Cook (2005), phylogenetic trees are often misinterpreted, so we must be cautious inferring phylogenies. Fur-thermore, in spite of the progress in statistical models and methodological approaches, there is no magic pill for the phylogenetic error. We still have a long way to try to get the final Tree of Life.

Morphology Versus Genetics to Address Phylogeny

Misof et al. (2005) discuss the recent proposals of DNA taxonomy and summarises some advantages and disadvantages of the molecular and morphological approaches in taxonomy. They do not intend to dismiss DNA based taxonomy, but they emphasise that molecular characters pose completely new problems to taxonomy. DNA taxonomy is currently promoted because of its potential for automation. These authors show that species identifica-tion can not be entirely automated since the result of a species description in taxonomy is equivalent to the formulation of a valid hypothesis.

Several papers have been recently published advocating the incorpora-tion of molecular techniques into taxonomic protocols. Tautz et al. (2002, 2003) made a plea for DNA taxonomy and indicated that DNA is pointing the way ahead in taxonomy. According to these authors, it's time for DNA's unique contribution to take a central role. DNA sequences are much used in phylogenetic analysis because of the many potential combinations in only a few hundred base pairs. In this sense the genes with the broadest taxonomic coverage currently available are those encoding the ribosomal small subunit sequences, both of nuclear and mitochondrial origin.

As considered by Misof et al. (2005), without any doubt, the incorporation of as many different character sets as possible into a species taxon description will improve the fit between species taxa and real evolutionary units. An issue in recent articles supportive of DNA taxonomy is the emphasis on molecular techniques over morphological approaches; it seems that the new techniques should replace the "old" ones to 'solve' the lack of adequate classifications and effective identification tools. In this sense, after the initial contributions of Tautz et al. (2002, 2003) supporting the molecular approach, several critiques appeared fast in the literature. According to these authors, to relegate taxonomy, rich in theory and knowledge, to a high-tech services industry would be a decisive step backwards for science. Molecular data certainly contribute, but when nothing is known about organisms except their DNA, there are no evolutionarily interesting patterns to explain. According to these authors, there is no credible reason to give DNA characters greater stature than any other character type.

Indeed, they pointed out several problems of the molecular approach:

- Difficulties of aligning sequences of dif-ferent length,
- Problems of distinguishing paralogs from orthologs,
- Difficulty of selecting appropriate genes for any particular taxonomic study,
- This new expensive technology would add to the North-South divide in taxonomy, since only the more developed countries would be able to use the new technology and many taxonomists with limited access to sequencing technology would be excluded.

From a phylogenetic point of view Scotland et al. (2003b) considered that the increased use of DNA sequence data, relative to morphology, for phylogeny reconstruction is inevitable and well founded. However, curiously, in other article of the same first author, Scotland et al. (2003a) reflected a totally different, much more conservative, point of view, and indicated that the lack of taxonomic progress will not be solved by DNA, and that promoting DNA sequences as the central and essential scaffold for all taxonomy would be an extremely inefficient and retrograde step for most groups.

In fact they pointed out methodological problems and pitfalls of the DNA approach similar as those reported above by Lipscomb et al. (2003) and Seberg et al. (2003) The article by Scotland et al. (2003b) purporting to examine the value of morphological data in phylogeny reconstruction has been received critically by several systematics.

Two of them seem to us especially interesting: Wiens (2004), who explained why we still need to collect more morphological data, and Smith and Turner (2005) who, as a paleontologist, provided a unique perspective in the debate. Wiens (2004) analysed the important role of morphological data in phylogeny reconstruction and reported many reasons to continue to do morphologi-cal phylogenetics, in spite of the advances in molecular systematics:

- Morphological data are necessary to solve the phylogenetic relationships of fossil taxa and their relationships to living taxa;
- For many extant rare taxa, there are no specimens available for molecular studies; indeed many species remain known from a single

specimen that was collected decades ago and frequently this scarce material was fixed in formalin, which difficult the DNA extraction;

- There are many factors that may cause molecular analyses to reconstruct clades that, although statistically well-supported, they are incorrect (i.e. long-branch attraction, deviations between gene and species trees, and even contamination and misidentification of specimens),
- We are very far from describing all the living species on earth, much less sequencing them. Smith & Turner (2005) supported the ideas of Wiens (2004) and strength the importance of the use of morphological characters in paleontology. In fact, molecular data cannot reconstruct the phylogenetic relationships of extinct taxa, except for rare cases involving recently extinct forms

Summarising, Mallet and Willmott (2003) pointed out that we might be only one tenth of the way through describing the world's species, questioning when it is sensible to add an extra requirement to the already slow process of describing new taxa, even if funds became available for DNA taxonomy.

These authors doubt DNA taxonomy will catch on as a mandatory step for species description in all organisms, and believe that most biologists will prefer to see DNA sequence information as a supplement rather than a required replacement for morphological data. In our opinion, as concurred by Lee (1999), homoplasy, for example, can contaminate both types of data (morphological and molecular), suggesting that morphological and molecular systematics might have more in common than previously assumed. Both approaches must be used to properly address phylogeny. And not only morphology and genetics must contribute, but also behavioural, ecological, biochemical and physiological data should be also considered. As many tools we will be able to use and integrate, a closer knowledge of the real world we will have. On the other hand, naturalists and molecular biologists often share questions, methods and explanations. In any case, it seems reasonable, that morphology should continue playing a major role in taxonomy: as Dunn (2003) pointed out, it is hard to understand how taxonomy will be taught to students, volunteers, parataxonomists, etc. without starting first with morphology. Quick and accurate identification of species in the field and laboratory based on morphological characters is also critical to many other areas of biology besides systematics (e.g. ecol-ogy, behaviour, physiology).

MEASUREMENT OF BIODIVERSITY

A variety of objective measures have been created in order to empirically measure biodiversity. Each measure of biodiversity relates to a particular use of the data. For practical conservationists, measurements should include a quantification of values that are commonly shared among locally affected organisms, including humans. For others, a more economically defensible definition should allow the ensuring of continued possibilities for both adaptation and future use by humans, assuring environmental sustainability.

Fig. Polar bears on the sea ice of the Arctic Ocean, near the north pole.

As a consequence, biologists argue that this measure is likely to be associated with the variety of genes. Since it cannot always be said which genes are more likely to prove beneficial, the best choice for conservation is to assure the persistence of as many genes as possible. For ecologists, this latter approach is sometimes considered too restrictive, as it prohibitsecological succession.

TAXONOMIC DIVERSITY

Biodiversity is usually plotted as taxonomic richness of a geographic area, with some reference to a temporal scale. Whittaker described three common metrics used to measure species-level biodiversity, encompassing attention to species richness or species evenness:

- Species richness - the least sophisticated of the indices available.
- Simpson index
- Shannon-Wiener index

Recently, another new index has been invented called the Mean Species Abundance Index (MSA); this index calculates the trend in population size of a cross section of the species. It does this in line with the CBD 2010 indicator for species abundance.

Other Measures of Diversity

Alternatively, other types of diversity may be plotted against a temporal timescale:

- Species diversity
- Ecological diversity
- Morphological diversity
- Genetic diversity

A few studies have attempted to quantitatively clarify the relationship between different types of diversity. For example, Sarda Sahney a researcher at the University of Bristol has found a close link between vertebrate taxonomic and ecological diversity.

Scale

Diversity may be measured at different scales. These are three indices used by ecologists:

- Alpha diversity refers to diversity within a particular area, community or ecosystem, and is measured by counting the number of taxa within the ecosystem (usually species)
- Beta diversity is species diversity between ecosystems; this involves comparing the number of taxa that are unique to each of the ecosystems.
- Gamma diversity is a measurement of the overall diversity for different ecosystems within a region.

FACTORS AFFECTING BIODIVERSITY

Factors that affect biodiversity in an ecosystem include area, climate, diversity of niches, and keystone species.

The factors that affect biodiversity are:

- Habitat loss- occurs when human activities result in conversion of natural ecosystem to humandominated systems.
- Overexploitation - occurs when humans harvest faster than the organisms are able to reproduce.
- Introduction of exotic species- can also have a significant effect on biodiversity.
- Persecution of pest organisms- many large carnivores were hunted to extinction because of their threat to humans and livestock.
- Natural events such as hurricanes, fires, floods, or volcanic eruption may have destroyed the original vegetation, resulting in patches of early successional stages that contribute much to the diversity of organisms present.

IUCN CRITERIA OF ENDANGERMENT

Biodiversity loss is continuing at an unprecedented rate, with many species declining to critical levels and significant numbers going extinct. The IUCN Red List is the most comprehensive information source on the status of wild species and their links to livelihoods. It is the clarion call for fighting the extinction crisis. The overall aim of the Red List is to convey the urgency and scale of conservation problems to the public and policy makers, and to motivate the global community to work together to reduce species extinctions.

The IUCN Red List assesses the extinction risk of species. Assessments of all mammals, birds, amphibians, sharks, reef-building corals, cycads and conifers have been completed. Efforts are underway to assess all reptiles, fishes and selected groups of plants and invertebrates. This sample indicates how life on Earth is faring, how little is known, and how urgent the need is to assess more species. In this way, The IUCN Red List is becoming The Barometer of Life.

How Does the IUCN Red List Help Save Species?

The IUCN Red List has many uses in conservation including:

- Conservation Planning – informing species-based conservation actions and identifying globally important sites for conservation including Important Plant Areas, Important Bird Areas, Key Biodiversity Areas and Alliance for Zero Extinction sites.
- Decision-making – Influencing conservation decisions at multiple scales, from environmental impact assessments to international multilateral environmental agreements.
- Monitoring – Indicating the current status of species and revealing trends in their extinction risk over time, to track progress towards biodiversity targets.

The IUCN Red List is used by government agencies, wildlife departments, conservation-related non-governmental organisations (NGOs), natural resource planners, educational organisations, and many others interested in reversing, or at least halting the decline in biodiversity

IUCN Red List Criteria for Endangered

The IUCN Species Programme working with the IUCN Species Survival Commission (SSC) has for more than four decades been assessing the conservation status of species, subspecies, varieties, and even selected subpopulations on a global scale in order to highlight taxa threatened with extinction, and therefore promote their conservation.

The IUCN Red List of Threatened Species provides taxonomic, conservation status and distribution information on plants and animals that have been globally evaluated using the IUCN Red List Categories and Criteria. This system is designed to determine the relative risk of extinction, and the main purpose of the IUCN Red List is to catalogue and highlight those plants and animals that are facing a higher risk of global extinction (*i.e.* those listed as Critically Endangered, Endangered and Vulnerable). The IUCN Red List also includes information on plants and animals that are categorised as Extinct or Extinct in the Wild; on taxa that cannot be evaluated because of insufficient information (*i.e.*, are Data Deficient); and on plants and animals that are either close to meeting the threatened thresholds or that would be threatened were it not for an ongoing taxon-specific conservation programme (*i.e.*, are Near Threatened). The IUCN Red List Categories and Criteria are intended to be an easily and widely understood system for classifying species at high risk of global extinction. The general aim of the system is to provide an explicit, objective framework for the classification of the broadest range of species according to their extinction risk. However, while the Red List may focus attention on those taxa at the highest risk, it is not the sole means of setting priorities for conservation measures for their protection.

The IUCN Red List Categories and Criteria have several specific aims:

- To provide a system that can be applied consistently by different people;
- To improve objectivity by providing users with clear guidance on how to evaluate different factors which affect the risk of extinction;
- To provide a system which will facilitate comparisons across widely different taxa;
- To give people using threatened species lists a better understanding of how individual species were classified.

The current version of the Categories and Criteria (3.2) in 2001 following a meeting of Criteria Review Working Group, in February 2000. The criteria can be applied to any taxonomic unit at or below the species level. In the following information, definitions and criteria the term 'taxon' is used for convenience, and may represent species or lower taxonomic levels, including forms that are not yet formally described. The the process and categorisation of taxa is illustrated in the following diagram:

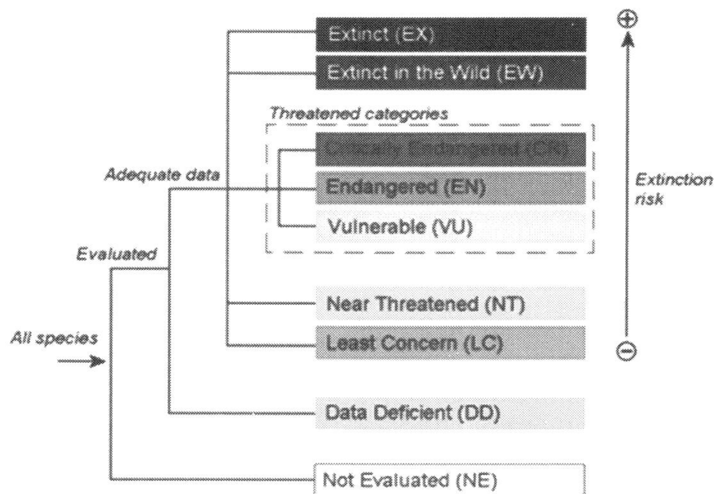

Fig. Structure of Categories

For listing as Critically Endangered, Endangered or Vulnerable there is a range of quantitative criteria; meeting any one of these criteria qualifies a taxon for listing at that level of threat. Each taxon should be evaluated against all the criteria. Even though some criteria will be inappropriate for certain taxa (some taxa will never qualify under these however close to extinction they come), there should be criteria appropriate for assessing threat levels for any taxon. The relevant factor is whether any one criterion is met, not whether all are appropriate or all are met. Because it will never be clear in advance which criteria are appropriate for a particular taxon, each taxon should be evaluated against all the criteria, and all criteria met at the highest threat category must be listed.

RED DATA BOOKS

A Red Data Book contains lists of species whose continued existence is threatened. Species are classified into different categories of perceived risk. Each Red Data Book usually deals with a specific group of animals or plants (e. reptiles, insects, mosses). They are now being published in many different countries and provide useful information on the threat status of the species. The red-listing assessment is a simple logical process to determine the status of threat to a species based on available information. In 2004 a training workshop was conducted in Lao PDR on redlisting assessment methodologies.

The MWBP red-listing process will be conducted on selected species groups in addition to the flagship species. Of the four Flagship Species of the MWBP, three have been categorised as Critically Endangered (Cr). One of the key outputs of this process is production of a Regional Red Data Book for the Lower Mekong Basin in five languages.

The process:

- Determine through a consultative process which groups of wetland species to carry out red listing process
- Carry out initial red-listing process for flagship and associated species, as part of action planning process
- Organise a training workshop on the Red-listing process
- National expert groups trained government counterparts collect information on species selected, and carry out process at national level
- Regional expert groups meet to collate the data and go through red listing process
- Send results to Species Survival Commission for ratification

RED DATA BOOKS OF KARELIA AND EASTERN FENNOSCANDIA

The first attempt at the publication of the Red Data Book of Karelia was made by the Karelian Branch of the USSR Academy of Sciences in 1985. The Book included 160 vascular plant species, 22 fungal species, 9 mammal species, 21 bird species, 2 amphibian species, 4 reptile species, 31 insect species. The first edition of the Red Data Book of Karelia became a cornerstone of nature conservation while it draw attention to endangered and threatened species. Data on rare plant and animal species were taken into account in the organisation of the PA network.

The second edition of the Red Data Book of Karelia prepared by a large team of specialists from Karelian Research Centre of RAS and Petrozavodsk State University was published in 1995.

The materials represented in the volume are much more comprehensive than in the first edition. Of particular value are distribution maps of rare plant and animal species. The Red Data Book was supplemented with information about rare species of lichens and mosses, fish, molluscs and higher crustaceans. The Red Data Book was published with active participation of the Republic of Karelia Ministry of Ecology and Natural Resources and Finnish Ministry of the Environment.

Other important publications:

- State report of the natural environment in Republic of Karelia - published annually since 1992 by the State Committee for Environmental Protection in Republic of Karelia.

Some species included in the Red Data Book of Karelia:

Vascular Plants

Karelian flora numbers 1200 wild vascular plant species (33 Pteridophyta, 8 Equisetales, 12 Lycopsida, 6 Gymnospermae and over 1100 Phanerogamae). Over 300 species are classed as rare and to various degrees endangered. More than 205 of them require some form of protection.

Mosses

The species composition of mosses in Karelia is insufficiently studied with no clear idea even of the approximate number of species in the republic. The highest diversity is recorded for cormophytic mosses - 426 species in Karelia of which 86 are included in the Red Data Book.

Lichens

This numerous group of living organisms is little studied in Karelia. 77 species are included in the Red Data Book.

Fungi

There are 272 species and forms of fungi recorded today in Karelia of which 23 species are included in the Red Data Book.

Mammals

Of the 56 animal species found in Karelia 26 are rare and endangered.

Birds

Among the 280 bird species found in Karelia 130 can be classed as rare and scarce.

Reptiles and Amphibians

Karelia is inhabited by 3 species of lizards and 2 snake species of which 3 are included in the Red Data Book.

Fish

At present Karelian water bodies are inhabited by 57 fish species; 28 of the species are rare and endangered.

Insects

The species composition of insects in Karelia is not sufficiently studied. 255 species are in need of protection.

Molluscs and Higher Crustaceans

Karelian water bodies are inhabited by ca. 100 species and forms of molluscs and 10 species of higher crustaceans. Four of the species are protected.

SOURCES OR TAXONOMIC EVIDENCE

Taxonomic evidence needed for the purpose of classifications and phylogenies is acocumulated and gathered from a variety of sources. Because all parts of a plant at all stages of its developemnt can provide taxonomic characters, data must be assembled from many different fields. The use of information can be gethered from comporative anatomy, embryology phlynology, cytogenetices, chemistry, and so on.

CATEGORIES OF TAXONOMY

A hierarchy is an orderly arrangement of a series of levels called categories. These categories are ranks to which taxa are assigned. Examples are species, genus, family, etc. The higher categories, such as family or order, are more broader than the lower categories. The other categories are regularly used to reflect evolutionary relationships in very large and complex group. The categories are defined by the Code according to their position and relation to other categories. The Code does not define or explain what is meant, in a biological sense, by a species, a genus, or a family. The classifications used in botany are based on the fundamental category of the species.

MONOPHYLETIC

When we derive a taxon from one ancestral population, it is called "monophyletic taxon" and when it is derived from 2 or more ancestral population it is called "polyphyletic taxon". If assemblages or plants looks like polypohyletic, they should be divided into monophyletic groups. As a result, monophyletic categories at all ranks in the hierarchy express evolutionary lineages and relationships. Because of the lack of fossil recored, it is often difficult to determine the origin of the higher ranks in the angiosperms, the main reason for differences in contemporary classification systems.

TYPOLOGICAL CONCEPTS OF TAXA

Systematists, when they want to remember essential characters of a family, genus or species think of thier salient features. This is what we mean by typological concept of taxa.Taxonomists who are familiar with the flora of an area or when travelling in new are as unconociously use typologial methods in recognision of the hundreds of species, genera and families of the flora. Taxonomist also recognises plant material by comparing them with previously identified specimens or by using keys, figures and descriptions. Since these tools represent only a sample of the populations, a taxonomist build up typological pictures of species, geners, and families from experience with the plants.

SPECIES

A "species" is defined as a group of individual organism that are fundamentally alike. A species should be different from other closely related species. This is necessary in order to have a practical classification that can be used by other. However, it is sometimes difficult define and limit a species precisely. If each minor change in plant populations were to be made the basis of species distinction, there would be no end to the number of species. A species is a concept which connot be defined within some limit and also it is not absolute and inelasfic. Disagreement over the definition of species is vague. Specimen of same species are genetically related individuals which can reproduce with each other. Different species have different strategies of reproductive isolation which prevents interbreeding. For example, many species are sexual, a few are asexual; some have arisen by polypoidy, changes in chromosome number and taxonomists other mechanisms. Since species represent lineages, system of living populations may be found at various stages or levels of morphological divergence from one another in reproductive isolation.

The biological species concept regards the species as distinct population system of central importance in nature and in evolutionary biology. According to this concept, biological species are kept isolated from closely related species living in the same region by reproductive isolating mechanisms that prevent gentic exchange between them. Two such reproductively isolated species can coexist in the same general area with their own identity. In actual taxonomic practice, the biological species concept is difficult to apply to plant species.

Sometimes plant species come in contact, and in some cases are able to exchange genes and may produce fertile hybrids. First-generation hybrids produced between plant species may be completely fertile or partially or completely sterile.

Genera

Genera is also a concept like species Genus is a broad category containing group like species. Some can say that genera are aggregatias of closely related species. The function of the genus concept is to put together species in a phylogenetic manner by placing the closest related species together within the generic classification. The genera in most families, for example, Magnoliaces, show striking differences in feature, whereas genera in other families, for instance, compositae, are separated on less obvious characters. There is no standard size for a genus. A genus having one species is called *monotypic example. Leitneria* (Leitneriaceae), consisting of a single species, *L. floridana.* Some genera, such as *Senecio* of the Compositae family, contain 2000 to 3000 species. It we cannot clearly separate species of two genera, then it is convenient to have bigger unitas of classification. When a genus is divided into two or more genera, nomenclatural changes are needed for certain species. We can avoid such changes by using infrageneric ranks as they do not cause changes in binomials. The present trend in plant taxonomy is to have a broad, conservative view of generic concepts and limits. This is expressed in a relatively stable generic nomenclature.

Families

Concept of species and genus also work well with the concept or family. Ideally, families should be monophyletic. Both reproductive and vagetative features are used to characterize families. The parts of the flower and fruits of flowering plants are numerous and varied, and they are not subject to environmental modifications. So they can provide more characters for the definition of families than do vegetative features. The category of family is more inclusive than either the genus or the species.

Ovary position, kinds of pistils and stamens, carpel number, fruit type, symmetry of the flower, leaf arrangement, leaf morphology, and habit are some of the features used for characterization. Certain characters used for defining families can also be used to define taxa at higher or lower ranks. Size and uniformily of families differ very much. Some families such as Cruciferae, are relatively homogeneous. Some such as Leguminosae are large but not homogeneous. Taxonomist differ in classification of Leguminosae family. Some divide it into three families, while other divide this family into three subfamilies. The Compositae, is one of the largest families of flowering plants It has around 22,000 species; whereas Leitneriaceae has a single genus and one species. The size and diversity of plant families depends on their evolutionary history, age, and radiation. Taxonomic schemes are subjective; the whole aim of taxonomy is to convey information through a derived scheme. When distinction are clear among closely related families, there is no problem in defining the family. But problem arises when there is no clear cut disfinction, so assigenment of genera to pooly defined family is matter of educated judgement.

Order

Order is an inclusive category having one or more families. The monophyletic requirement needed for establishing an order should not be too restrictive when developing a workable classification. As order this are a bigger category and more inclusive much more difficult to define and delimit than families. Orders and other higher categories are characterized by an aggregate of characters. Most taxonomists agree that orders should be monophyletic, but there is no agree- ment on their sizes or to the familises that should be included. Bessey defined many orders containing only few families, whereas Hutchinson defined orders containing large number of families. Here flowering plants are arranged into 83 orders.

8

Remote Sensing and Biodiversity

REMOTE SENSING

Remote sensing is the acquisition of information about an object or phenomenon, without making physical contact with the object. In modern usage, the term generally refers to the use of aerial sensor technologies to detect and classify objects on Earth (both on the surface, and in the atmosphere and oceans) by means of propagated signals (*i.e.,* electromagnetic radiation emitted from aircraft or satellites).

There are two main types of remote sensing: passive remote sensing and active remote sensing. Passive sensors detect natural radiation that is emitted or reflected by the object or surrounding area being observed. Reflected sunlight is the most common source of radiation measured by passive sensors. Examples of passive remote sensors include film photography, infrared, charge-coupled devices, and radiometers. Active collection, on the other hand, emits energy in order to scan objects and areas whereupon a sensor then detects and measures the radiation that is reflected or backscattered from the target. RADAR is an example of active remote sensing where the time delay between emission and return is measured, establishing the location, height, speed and direction of an object.

Remote sensing makes it possible to collect data on dangerous or inaccessible areas. Remote sensing applications include monitoring deforestation in areas such as the Amazon Basin, glacial features in Arctic and Antarctic regions, and depth sounding of coastal and ocean depths. Military collection during the cold war made use of stand-off collection of data about dangerous border areas. Remote sensing also replaces costly and slow data collection on the ground, ensuring in the process that areas or objects are not disturbed.

Orbital platforms collect and transmit data from different parts of the electromagnetic spectrum, which in conjunction with larger scale aerial or ground-based sensing and analysis, provides researchers with enough information to monitor trends such as El Niño and other natural long and short term phenomena.

Other uses include different areas of the earth sciences such as natural resource management, agricultural fields such as land usage and conservation, and national security and overhead, ground–based and stand–off collection on border areas. By satellite, aircraft, spacecraft, buoy, ship, and helicopter images, data is created to Analyse and compare things like vegetation rates, erosion, pollution, forestry, weather, and land use. These things can be mapped, imaged, tracked and observed. The process of remote sensing is also helpful for city planning, archaeological investigations, military observation and geomorphological surveying.

DATA ACQUISITION TECHNIQUES

The basis for multispectral collection and analysis is that of examined areas or objects that reflect or emit radiation that stand out from surrounding areas.

Applications of Remote Sensing Data

- Conventional radar is mostly associated with aerial traffic control, early warning, and certain large scale meteorological data. Doppler radar is used by local law enforcements' monitoring of speed limits and in enhanced meteorological collection such as wind speed and direction within weather systems. Other types of active collection includes plasmas in the ionosphere. Interferometric synthetic aperture radar is used to produce precise digital elevation models of large scale terrain.
- Laser and radar altimeters on satellites have provided a wide range of data. By measuring the bulges of water caused by gravity, they map features on the seafloor to a resolution of a mile or so. By measuring the height and wave–length of ocean waves, the altimeters measure wind speeds and direction, and surface ocean currents and directions.
- Light detection and ranging (LIDAR) is well known in examples of weapon ranging, laser illuminated homing of projectiles. LIDAR is used to detect and measure the concentration of various chemicals in the atmosphere, while airborne LIDAR can be used to measure heights of objects and features on the ground more accurately than with radar technology. Vegetation remote sensing is a principal application of LIDAR.
- Radiometers and photometers are the most common instrument in use, collecting reflected and emitted radiation in a wide range of frequencies. The most common are visible and infrared sensors, followed by microwave, gamma ray and rarely, ultraviolet. They may also be used to detect the emission spectra of various chemicals, providing data on chemical concentrations in the atmosphere.
- Stereographic pairs of aerial photographs have often been used to

make topographic maps by imagery and terrain analysts in trafficability and highway departments for potential routes.

- Simultaneous multi–spectral platforms such as Landsat have been in use since the 70's. These thematic mappers take images in multiple wavelengths of electro–magnetic radiation (multi–spectral) and are usually found on Earth observation satellites, including (for example) the Landsat programme or the IKONOS satellite. Maps of land cover and land use from thematic mapping can be used to prospect for minerals, detect or monitor land usage, deforestation, and examine the health of indigenous plants and crops, including entire farming regions or forests.
- Hyperspectral imaging produces an image where each pixel has full spectral information with imaging narrow spectral bands over a contiguous spectral range. Hyperspectral imagers are used in various applications including mineralogy, biology, defence, and environmental measurements.
- Within the scope of the combat against desertification, remote sensing allows to follow-up and monitor risk areas in the long term, to determine desertification factors, to support decision–makers in defining relevant measures of environmental management, and to assess their impacts.

Geodetic

- Overhead geodetic collection was first used in aerial submarine detection and gravitational data used in military maps. This data revealed minute perturbations in the Earth's gravitational field (geodesy) that may be used to determine changes in the mass distribution of the Earth, which in turn may be used for geological studies.

Acoustic and Near–acoustic

- *Sonar*: Passive sonar, listening for the sound made by another object (a vessel, a whale etc); active sonar, emitting pulses of sounds and listening for echoes, used for detecting, ranging and measurements of underwater objects and terrain.
- Seismograms taken at different locations can locate and measure earthquakes (after they occur) by comparing the relative intensity and precise timing.

To coordinate a series of large–scale observations, most sensing systems depend on the following: platform location, what time it is, and the rotation and orientation of the sensor. High–end instruments now often use positional information from satellite navigation systems. The rotation and orientation is often provided within a degree or two with electronic compasses. Compasses can measure not just azimuth (*i.e.,* degrees to magnetic north), but also altitude (degrees above the horizon), since the magnetic field curves into the Earth at different angles at different latitudes. More exact orientations require gyroscopic-aided orientation, periodically realigned by different methods including navigation from stars or known benchmarks.

Resolution impacts collection and is best explained with the following relationship: less resolution=less detail and larger coverage, More resolution=more detail, less coverage. The skilled management of collection results in cost–effective collection and avoid situations such as the use of multiple high resolution data which tends to clog transmission and storage infrastructure.

DATA PROCESSING

Generally speaking, remote sensing works on the principle of the inverse problem. While the object or phenomenon of interest (the state) may not be directly measured, there exists some other variable that can be detected and measured (the observation), which may be related to the object of interest through the use of a data–derived computer model. The common analogy given to describe this is trying to determine the type of animal from its footprints. For example, while it is impossible to directly measure temperatures in the upper atmosphere, it is possible to measure the spectral emissions from a known chemical species (such as carbon dioxide) in that region. The frequency of the emission may then be related to the temperature in that region via various thermodynamic relations.

The quality of remote sensing data consists of its spatial, spectral, radiometric and temporal resolutions:

- *Spatial Resolution*: The size of a pixel that is recorded in a raster image–typically pixels may correspond to square areas ranging in side length from 1 to 1,000 metres (3.3 to 3,300 ft).
- *Spectral Resolution*: The wavelength width of the different frequency bands recorded–usually, this is related to the number of frequency bands recorded by the platform. Current Landsat collection is that of seven bands, including several in the infra-red spectrum, ranging from a spectral resolution of 0.07 to 2.1 mm. The Hyperion sensor on Earth Observing–1 resolves 220 bands from 0.4 to 2.5 mm, with a spectral resolution of 0.10 to 0.11 mm per band.
- *Radiometric Resolution*: The number of different intensities of radiation the sensor is able to distinguish. Typically, this ranges from 8 to 14 bits, corresponding to 256 levels of the gray scale and up to 16,384 intensities or "shades" of colour, in each band. It also depends on the instrument noise.

- *Temporal Resolution*: The frequency of flyovers by the satellite or plane, and is only relevant in time–series studies or those requiring an averaged or mosaic image as in deforesting monitoring. This was first used by the intelligence community where repeated coverage revealed changes in infrastructure, the deployment of units or the modification/introduction of equipment. Cloud cover over a given area or object makes it necessary to repeat the collection of said location.

In order to create sensor–based maps, most remote sensing systems expect to extrapolate sensor data in relation to a reference point including distances between known points on the ground. This depends on the type of sensor used. For example, in conventional photographs, distances are accurate in the center of the image, with the distortion of measurements increasing the farther you get from the center. Another factor is that of the platen against which the film is pressed can cause severe errors when photographs are used to measure ground distances.

The step in which this problem is resolved is called georeferencing, and involves computer–aided matching up of points in the image (typically 30 or more points per image) which is extrapolated with the use of an established benchmark, "warping" the image to produce accurate spatial data. As of the early 1990s, most satellite images are sold fully georeferenced.

In addition, images may need to be radiometrically and atmospherically corrected:

- *Radiometric Correction*: Radiometric correction gives a scale to the pixel values, *i.e.*, the monochromatic scale of 0 to 255 will be converted to actual radiance values.
- *Atmospheric Correction*: Atmospheric correction eliminates atmospheric haze by rescaling each frequency band so that its minimum value (usually realised in water bodies) corresponds to a pixel value of 0. The digitizing of data also make possible to manipulate the data by changing gray–scale values.

Interpretation is the critical process of making sense of the data. The first application was that of aerial photographic collection which used the following process; spatial measurement through the use of a light table in both conventional single or stereographic coverage, added skills such as the use of photogrammetry, the use of photomosaics, repeat coverage, Making use of object's known dimensions in order to detect modifications. Image Analysis is the recently developed automated computer-aided application which is in increasing use. Object–Based Image Analysis (OBIA) is a sub-discipline of GIScience devoted to partitioning remote sensing (₹) imagery into meaningful image-objects, and assessing their characteristics through spatial, spectral and temporal scale. Old data from remote sensing is often valuable because it may provide the only long–term data for a large extent of geography. At the same time, the data is often complex to interpret, and bulky to store. Modern systems tend to store the data digitally, often with lossless compression. The difficulty with this approach is that the data is fragile, the format may be archaic, and the data may be easy to falsify. One of the best systems for archiving data series is as computer–generated machine–readable ultrafiche, usually in typefonts such as OCR–B, or as digitized half–tone images. Ultrafiches survive well in standard libraries, with lifetimes of several centuries. They can be created, copied, filed and retrieved by automated systems. They are about as compact as archival magnetic media, and yet can be read by human beings with minimal, standardized equipment.

Data Processing Levels

To facilitate the discussion of data processing in practice, several processing "levels" were first defined in 1986 by NASA as part of its Earth Observing System and steadily adopted since then, both internally at NASA and elsewhere; these definitions are:

Level	Description
0	Reconstructed, unprocessed instrument and payload data at full resolution, with any and all communications artifacts (*i.e.*, synchronization frames, communications headers, duplicate data) removed.
1a	Reconstructed, unprocessed instrument data at full resolution, time-referenced, and annotated with ancillary information, including radiometric and geometric calibration coefficients and georeferencing parameters (*i.e.*, platform ephemeris) computed and appended but not applied to the Level 0 data (or if applied, in a manner that level 0 is fully recoverable from level 1a data).
1b	Level 1a data that have been processed to sensor units (*i.e.*, radar backscatter cross section, brightness temperature, etc.); not all instruments have Level 1b data; level 0 data is not recoverable from level 1b data.
2	Derived geophysical variables (*i.e.*, ocean wave height, soil moisture, ice concentration) at the same resolution and location as Level 1 source data.
3	Variables mapped on uniform spacetime grid scales, usually with some completeness and consistency (*i.e.*, missing points interpolated, complete regions mosaicked together from multiple orbits, etc).
4	Model output or results from analyses of lower level data (*i.e.*,variables that were not measured by the instruments but instead are derived from these measurements).

A Level 1 data record is the most fundamental *(i.e.,* highest reversible level) data record that has significant scientific utility, and is the foundation upon which all subsequent data sets are produced. Level 2 is the first level that is directly usable for most scientific applications; its value is much greater than the lower levels.

Level 2 data sets tend to be less voluminous than Level 1 data because they have been reduced temporally, spatially, or spectrally. Level 3 data sets are generally smaller than lower level data sets and thus can be dealt with without incurring a great deal of data handling overhead. These data tend to be generally more useful for many applications. The regular spatial and temporal organization of Level 3 datasets makes it feasible to readily combine data from different sources.

HISTORY

The modern discipline of remote sensing arose with the development of flight. The balloonist G. Tournachon (alias Nadar) made photographs of Paris from his balloon in 1858. Messenger pigeons, kites, rockets and unmanned balloons were also used for early images. With the exception of balloons, these first, individual images were not particularly useful for map making or for scientific purposes.

Systematic aerial photography was developed for military surveillance and reconnaissance purposes beginning in World War I and reaching a climax during the Cold War with the use of modified combat aircraft such as the P–51, P–38, RB–66 and the F–4C, or specifically designed collection platforms such as the U2/TR-1, SR-71, A–5 and the OV–1 series both in overhead and stand-off collection. A more recent development is that of increasingly smaller sensor pods such as those used by law enforcement and the military, in both manned and unmanned platforms.

The advantage of this approach is that this requires minimal modification to a given airframe. Later imaging technologies would include Infra–red, conventional, doppler and synthetic aperture radar.

The development of artificial satellites in the latter half of the 20th century allowed remote sensing to progress to a global scale as of the end of the Cold War. Instrumentation aboard various Earth observing and weather satellites such as Landsat, the Nimbus and more recent missions such as *RADARSAT* and *UARS* provided global measurements of various data for civil, research, and military purposes.

Space probes to other planets have also provided the opportunity to conduct remote sensing studies in extraterrestrial environments, synthetic aperture radar aboard the Magellan spacecraft provided detailed topographic maps of Venus, while instruments aboard SOHO allowed studies to be performed on the Sun and the solar wind, just to name a few examples.

Recent developments include, beginning in the 1960s and 1970s with the development of image processing of satellite imagery. Several research groups in Silicon Valley including NASA Ames Research Center, GTE and ESL Inc. developed Fourier transform techniques leading to the first notable enhancement of imagery data.

REMOTE SENSING SOFTWARE

Remote Sensing data is processed and analysed with computer software, known as a remote sensing application. A large number of proprietary and open source applications exist to process remote sensing data. An NOAA Sponsored Research by Global Marketing Insights, Inc. the most used applications among Asian academic groups involved in remote sensing are as follows: ERDAS 36%; ESRI 30%; ITT Visual Information Solutions ENVI 17%; MapInfo 17%. Among Western Academic respondents as follows: ESRI 39%, ERDAS IMAGINE 27%, MapInfo 9%, AutoDesk 7%, ITT Visual Information Solutions ENVI 17%. Other important Remote Sensing Software packages include: TNTmips from MicroImages, PCI Geomatica made by PCI Geomatics, the leading remote sensing software package in Canada, IDRISI from Clark Labs, Image Analyst from Intergraph, RemoteView made by Overwatch Textron Systems, and the original object based image analysis software eCognition from Definiens. Dragon/ips is one of the oldest remote sensing packages still available, and is in some cases free. Open source remote sensing software includes GRASS GIS, ILWIS, QGIS, OSSIM, Opticks (software) and Orfeo toolbox.

ENERGY OF REMOTE SENSING

The first step in remote sensing is to have a source of energy that will be beamed towards the target. The energy comes in the form of light waves of different sizes. Like the waves in an ocean, energy waves can range from waves whose top point (crest) to lowest point (trough) are very tiny to those that are hundreds of feet (meters) long. The distance of one full wave, from crest to crest or trough to trough, is known as the wavelength. The range of waves is known as the electromagnetic spectrum. At one end of the electromagnetic spectrum lie the tiny waves such as gamma rays and X rays. These waves tend to carry large amounts of energy and can penetrate into solid or liquid material more so than other waves. That is why X rays can pass right through skin to reveal images of the bones and teeth underneath.

At the other end of the spectrum lie waves such as the microwaves that can penetrate a short distance to heat up foods, and radio waves that beam music through a radio speaker. Radio waves are not efficient for remote sensing operations. Microwaves are the longest waves with enough energy to be used for remote sensing. The regions of the electromagnetic spectrum that is useful for remote sensing contain the waves known as ultraviolet rays (the same rays that give a suntan or sunburn). The term ultraviolet means that the waves are just beyond the portion of the spectrum that contains the waves that are visible, in particular the region of the spectrum that contain violet–coloured waves.

Indeed, for the visible portion of the electromagnetic spectrum, our eyes are the remote sensors! Shorter, higher energy wavelengths are preferred for remote sensing because the waves have to move through air or water on their way to the target. Passing through air and water causes some of the waves to be absorbed or deflected (bounced) off the target. (The deflection of different wavelengths of light as they pass through Earth's atmosphere, the mass of air surrounding Earth, is the reason why the sky appears blue. Colours with relatively long wavelengths pass straight through the atmosphere. Blue light has a shorter wavelength and the atmosphere scatters it.) A higher energy wave will be better able to blast through any interference to the target, and to bounce back from the target.

The absorption of waves can be useful when trying to figure out the nature of the target. For example, microwaves tend to be absorbed by the gas form of water known as water vapour. The pattern of absorption detected by scientists on their instruments can provide important clues about the amount of water contained in the air above the ground or water.

HOW REMOTE SENSING WORKS

In order to illustrate how remote sensing works, imagine a bathtub full of water. If a bar of soap is dropped into the water, waves will move outward over the surface of the water. As the waves contact the sides of the tub, some the energy will rebound back into the tub. So it is with the energy that is beamed from a satellite, ship or plane. The returning energy is captured by a detector (also known as a sensor). Instruments and computers that are connected to the sensor can analyse the pattern of the returning waves to help scientists understand the distance and shape of the object on the ground or the ocean floor that deflected the waves.

SENDING ENERGY UNDERWATER

To chart the depth of a lake or ocean bottom, a transmitter on a boat will beam energy for a short time (a pulse transmission) straight down into the water. A sensor on the boat detects the returning signal. Using a mathematical formula to account for the presence of water, scientists can then determine the one–way distance of the signal. Other uses of vertical (up and down) sonar include detecting other ships and as an aid in navigating. The energy pulse can also be sent out horizontally through the water, rather than straight down. This is called sidescan sonar, and is useful in determining what lies around a ship. Some systems are so sensitive that they can detect an object in the water that is less than 0.4 inches (1 centimeter) in size. Sidescan sonar is also useful in investigating underwater archaeological sites.

SATELLITES AND REMOTE SENSING

The oceans are monitored by satellites (in orbit around Earth) by means of their remote sensors—instruments that gather information on the features of Earth without being in physical contact with it. These satellites send back information on ice cover, ocean depth, water temperature, and weather conditions. The *TOPEX/Poseidon* satellite, which was launched in 1992, monitors global ocean circulation, sea surface temperatures, and sea surface height.

Sea surface temperatures are shown in computer-generated colours; red and yellow represent warm water, while blue and green represent cold water. Computer-generated colours also reveal the presence of plankton, an indicator of ocean productivity.

Fig. Satellites such as the TOPEX/Poseidon are Used to Study Earth's Oceans.

In 1997, NASA launched the *SEASTAR*, which carries a colour scanner that measures the ocean's entire oxygen-producing plankton population on a weekly basis. NASA's latest Earth-observing satellite, *AQUA*, launched in 2002, gathers information on precipitation and evaporation, in order to determine if Earth's water cycle is being affected by climate change. Remote-sensing satellites are also used in search-and-rescue operations. The National Oceanographic and Atmospheric Administration (NOAA), a federal agency that monitors the ocean, operates SARSAT (*S*earch-*a*nd-*R*escue *S*atellite-*A*ided *T*racking), which locates people in distress on land and on sea. As of March 2002, nearly 13,000 people had been rescued by SARSAT worldwide.

PROBING THE DEPTHS

Geologists are scientists who study the physical structure of Earth. Marine geologists study the characteristics of, as well as any changes that occur in, the seafloor.

They drill into the seafloor and remove rock samples. The first drilling ship, the *Glomar Challenger,* began taking core rock samples in 1968. This drilling ship has been replaced by the modern vessel *Resolution,* which can drill into the seafloor at water depths of 6500 meters. Remote sensors also gather data on the deep ocean floor. These sensors replace submersibles operated by people, which are much more expensive to use.

The Remote Underwater Manipulator (RUM III) can pick up objects from the ocean floor, take video pictures, and use sonar to determine water depths. The

RUM III stays tethered to a research vessel on the surface. Smaller remote-controlled robots have been used to examine underwater shipwrecks such as the *Titanic*.

CENTERS OF LEARNING

As they investigate the marine world, scientists are always learning. New discoveries are reported in the scientific literature. The *Journal of Marine Biology* and the *Journal of Marine Science* are two of many scientific journals that publish original research and make it available to the scientific community as well as to the public. Much research is done in college and university laboratories around the country.

Undergraduate and graduate programmes in marine science are offered at the University of Miami, the University of Texas, the University of Hawaii, Oregon State University, and the University of Rhode Island, to name a few. Research institutes are important in the field of marine science.

Three of the most famous centers of learning devoted to marine research are the Scripps Institution of Oceanography in La Jolla, California, the Woods Hole Oceanographic Institution in Woods Hole (Cape Cod), Massachusetts, and the Lamont-Doherty Geological Observatory (of Columbia University) in New York. The submersible *Alvin*, which has been used to collect volumes of information about the undersea world, is maintained and operated by Woods Hole. Since 1963, Scripps, Woods Hole, and Columbia University have cooperated in a joint venture in underwater research by using the Floating Instrument Platform (FLIP), shown in Figure.

Fig. The FLIP Vessel is Used for Oceanographic Research.

FLIP is a 100-meter hollow tube with a research station at one end. After FLIP is towed out to a research site, its ballast tanks are filled with water. The floating instrument platform then "flips" over on itself. When FLIP is in a stable vertical position, scientists can conduct a variety of underwater tests.

MONITORING OUR MARINE RESOURCES

Environmental concerns are a very important part of marine science. How can we protect the marine environment and conserve its resources? The National Oceanographic and Atmospheric Administration (part of the U.S. Department of

Commerce) conducts research, monitors the atmosphere and the oceans, and helps conserve, manage, and protect our marine resources.

POPULAR MARINE SCIENCE

The marine world has become an important part of popular culture. You can snorkel or scuba dive around shipwrecks and coral reefs, swim with captive dolphins in Florida, or view a fantastic assortment of marine creatures in public aquariums. Whale-watching trips in the Atlantic and the Pacific have become increasingly popular as a means of getting close to marine life in its natural setting.

Fig. Passengers on a Whale-watch Boat View a Humpback Whale as it Dives Below the Ocean's Surface.

Recently, small passenger submersibles have been developed. These vessels can take you below the surface to view sea life, while you remain in dry comfort. Aquariums have been built in many cities to satisfy people's interest in the marine world. Public aquariums not only display a great variety of marine organisms, they also offer educational programmes to students and interested adults. Some public aquariums also conduct scientific research. The fields of marine biology and oceanography have been made popular in recent years due to the achievements of modern pioneers such as Captain Jacques-Yves Cousteau, Dr. Robert D. Ballard, and Dr. Sylvia Earle.

Like the astronauts who explore outer space, these aquanauts who explore the inner space of the ocean have helped to uncover and explain many of its mysteries. Who are these ocean pioneers and what have they achieved? See the chart below for a list of some of their important accomplishments.

THE SCIENTIFIC METHOD

In many cases, we learn best by doing. Marine biologists learn about life in the sea by performing experiments. An experiment is an investigation that is conducted according to an organized step-bystep problem-solving approach called the scientific method. The steps in the scientific method are: state the problem, collect relevant information, form your hypothesis, test your hypothesis using selected materials and methods, record observations, tabulate results, and draw a conclusion. The first step in the scientific method is actually to observe a selected part of nature and then state the problem you wish to solve in the form of a testable hypothesis. You select a problem that requires you to make observations, which are descriptions of events in the environment that are based on the use of your senses. The experiment that tests the hypothesis generates data that can be analysed.

STATING THE PROBLEM

Let's use an example to help you understand how the scientific method works. Students in one class observed the behaviour of marine snails in an aquarium. The tank contained several snails at one end and food (an opened mussel) at the other end. The students recorded their observations in a logbook, a notebook in which such data are kept. Read the following entries made by one student.

- *Nov. 1*: Observed five snails climbing up the side of the aquarium tank.
- *Nov. 2*: Opened a mussel and put it into the aquarium. In a few minutes, snails swarmed all over the food.

Based on these observations, can you suggest a problem to solve? Remember, it must be testable and in the form of a question. One student wondered whether snails would respond in some measurable way to the presence of food. She came up with the following problem to study: Does the presence of food affect the rate of movement in snails?

FORMING AND TESTING THE HYPOTHESIS

Identifying a problem to study is the first important step in conducting an experiment. When a problem has been identified, the scientist should do library research to learn what, if anything, has already been discovered about that topic. The next step in the scientific method is to offer a possible solution to the problem. A possible solution to a problem is called a hypothesis. The student suggested the following hypothesis:

- *Hypothesis:* If food is present in the aquarium, then snails will move with greater speed (towards the food).

A hypothesis is often called an "educated guess" because previous knowledge is used to formulate it. (Scientific advances always build on a foundation of previously learned knowledge.) The hypothesis is often written with the words *if* and *then*. When the hypothesis is stated in this form, it is easier to devise an experiment to test it.

Testing the hypothesis is the third step in the scientific method. To find out if a hypothesis is correct, an experiment is designed and carried out. However, before you can do the experiment, you must first select the appropriate materials.

SELECTING THE MATERIALS

Materials are the items or pieces of equipment needed to carry out an experiment. In our example, one student chose the following materials: aquarium, seawater, food, graduated cylinders, snails, and a metric ruler. Then the equipment must be assembled in the proper manner so that the hypothesis can be tested. The part of the scientific method in which materials are set up is called the method, or procedure. In this part of the project, you can use your creativity and originality in developing a good experimental design.

METHOD (PROCEDURE)

Look at Figure to see how the student scientist designed her experiment. She wanted to see if food affected snail locomotion. In this investigation, food is called the variable. A variable is any factor that could affect the outcome of an experiment. In this investigation, the seawater could also be a variable. The salt content and temperature of the water might affect the movement of a snail. The size and shape of the container could also be a variable. There are many possible variables in an experiment. However, you can test only one variable at a time, and all other variables must be kept the same (constant). Otherwise you could not be certain which variable, or combination of variables, produced the effects you observed.

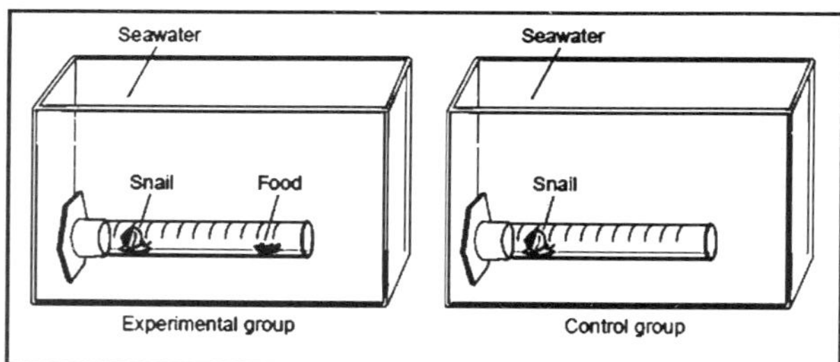

Fig. An Experimental Set-up in which Food is the Variable.

To determine if a particular variable affects the outcome of an investigation, a scientist carries out a controlledexperiment in which two groups are tested: one that is exposed to the variable and one that is not. The group that is exposed to the variable is called the experimental group. The experimental group in Figure is the group of snails in the container with the food. The scientist must also set up a control group.

This would be the group of snails in the container without the food. The control group contains exactly the same conditions as the experimental group, except for the variable being tested—in this case, the food. Since all other conditions are kept the same in both the experimental and the control groups, any difference in snail locomotion would have to be caused by the presence of food. To understand how the materials were set up to measure snail locomotion, read the instructions listed in the procedure below, which were developed by the student who carried out this experiment.

- Using a metric ruler, mark off the length of two graduated cylinders in centimeters.
- Fill the cylinders with seawater. Label one cylinder "experimental" and the other "control."
- Place a snail at the bottom of each cylinder. Submerge the graduated cylinders in the seawater in each aquarium. Be sure to place the cylinders flat on their sides on the bottom of each aquarium, and be sure the centimeter measurements are visible.
- Put a piece of food at the entrance of the cylinder labeled "experimental." Remember that in this experiment the control cylinder receives no food.
- Start timing the snails as soon as they begin to move. In each case, record the time it takes for the snail to move the length of the cylinder. Enter the time and distance traveled in Table. Perform as many trials as you can in the allotted time. Try to use different snails each time, so that you have a control group and an experimental group.

Table. Snail Locomotion in the Presence and Absence of Food.

Trial	Experimental Group (with Food)			Control Group (without Food)		
	Time (min)	Distance (cm)	Speed (cm/min)	Time (min)	Distance (cm)	Speed (cm/min)
1	3.5	20	5.7	7.0	20	2.9
2	4.0	20	5.0	9.5	20	2.2
3	6.2	20	3.2	8.0	20	2.5
4	2.5	20	8.0	11.0	20	1.8
5	5.0	20	4.0	8.5	20	2.4
6	3.0	20	6.7	7.5	20	2.7
Total	24.2	120	32.6	51.5	120	14.5
Av. Speed			5.4			2.4

OBSERVATIONS AND RESULTS

Accuracy is most important in making observations on snail locomotion. Saying that one snail moves faster or slower than another snail is not very precise. How fast is fast? (In the case of a snail, not very fast at all!) When you use exact measurements, you are being more precise. Measurements that are recorded during an experiment are called *data.* The data collected make up the *results* of your experiment, which can be analysed. The results represent the next step in the scientific method. Numerical data from an experiment are best recorded in a table, which is organized and easy to read. Look at the data collected in the student's experiment, shown in T able. For the sake of accuracy, experiments are usually repeated. Each time that an experiment is carried out, it is referred to as a trial. (The average of several trials will give the most accurate results.) In this case, there were six trials conducted. The results of an experiment also include calculations.

In this experiment, the speed of the snail is computed by using the following formula:

- Speed = Distance/Time

Scientists use mathematics in solving problems. Compute the speed of the snail in Trial 1 of the experimental group.

Substituting data from the table into the formula, you get speed in centimeters per minute (cm/min):

- Speed = Distance/Time
- Speed = 20 cm/3.5 min
- Speed = 5.7 cm/min

The average speed for the six trials was computed for the experimental and control groups. To compute the average speed, find the sum of the speeds in all the trials and then divide by the number of trials. In this case, the average speed for the experimental group was 5.4 cm/min. The control group had an average speed of 2.4 cm/min. The data from an experiment can also be displayed in the form of a graph. A graph is a pictorial representation of data that shows relationships at a glance. Suppose you want to compare the average speed of snails in the presence and absence of food. You can construct a bar graph. Look at the bar graph shown in Figure.

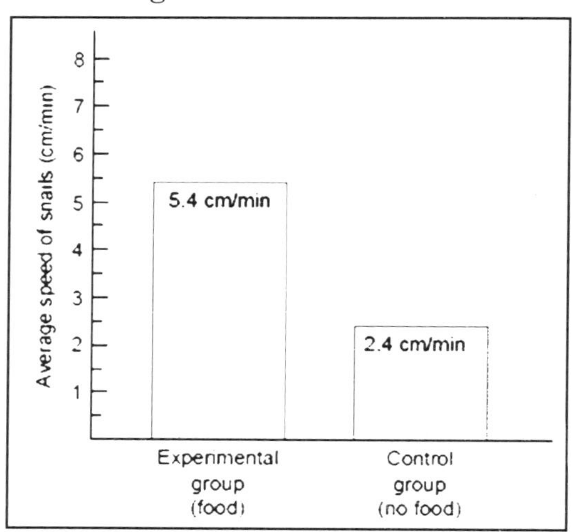

Fig. A Bar Graph is Used to Compare Data and Show Relationships at a Glance.

The scale on the vertical axis is numbered from 1 to 8. These numbers represent the range of snail speeds, in cm/min. According to the bar graph, which group of snails has the higher average speed? The bar graph clearly shows that the experimental group with food moves faster than the control group without food. In another experiment, a student wanted to know if water temperature could affect the speed of snail movement.

The student formulated the following hypothesis:

- *Hypothesis:* If the temperature is increased, then the snails will move more slowly.

The average speed of snails at varying temperatures was collected, as shown in Table.

Table. Average Speed of Snails at Different Temperatures (CM/MIN).

Trial	13°C	18°C	24°C	30°C
1	.75	1.0	2.0	2.5
2	.90	1.1	1.8	2.8
3	.50	1.2	1.9	2.4
4	.25	1.3	2.0	2.3
5	.80	1.1	2.1	2.2
6	.65	1.3	2.2	2.0
Total	3.85	7.0	12.0	14.2
Av. Speed	.64	1.2	2.0	2.4

The data from the table can be plotted on a line graph. Look at the line graph shown in Figure.

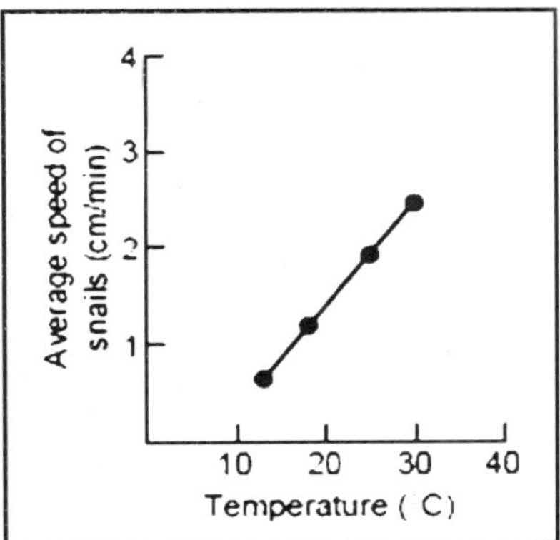

Fig. A Line Graph Shows the Relationship between Variables.

The line graph contains two lines, or axes. Temperature data are placed on the horizontal axis, and snail speed data are placed on the vertical axis. The temperature is the "independent variable," while the snail's speed, which is affected by it, is the "dependent variable." Notice that the units on each axis are marked off in equal intervals from the lowest to the highest value. The average speed for each temperature is given at the bottom. Plot the first point by locating 13 on the horizontal axis. Then find (or approximate) 0.64 on the vertical axis. Make a dot on the graph where the two points intersect. Do the same for the other readings. Connect the dots by drawing a straight line between them with a ruler.

DRAWING CONCLUSIONS

Now that you have recorded data in tables and displayed data on graphs, you are ready to analyse and interpret this information. The part of the scientific method in which data are analysed and interpreted is called the conclusion. What do you conclude from your results? First, you must determine if the results in the experiment support your hypothesis. In the first experiment, it was hypothesized that if food is present, the snails will move faster. According to the results, the average speed of the experimental group (with food) was 5.4 cm/min. The average speed of the control group was 2.4 cm/min. What do you conclude from these results? The results show that snails move faster in the presence of food. The student would therefore conclude that her hypothesis was supported by the data. A hypothesis that is supported by the data is called a *valid hypothesis.* To check the validity of a hypothesis, scientists will repeat the same controlled experiments several times to see if they obtain similar results. An important tool used in the process of drawing conclusions is statistical analysis. To determine whether certain results are valid, statistics may be used to calculate the significance of the various numbers obtained. Figures that appear insignificant at first may prove to show important differences when analysed with statistics. In the second investigation, it was hypothesized that if the temperature of the water is increased, then the snails will move more slowly. Do the data support this hypothesis? You can see in the table and in the graph that as temperature increases, the average speed of the snails also increases. Therefore, the hypothesis would have to be rejected because it is not supported by the data.

9

Conservation and Management of Biodiversity

CONSERVATION OF BIODIVERSITY

Conservation and sustainable use of biodiversity have been an integral part of Indian ethos. The varied eco-climatic conditions coupled with unique geological and cultural features have contributed to an astounding diversity of habitats, which harbour and sustain immense biological diversity at all levels. With only 2.4% of world's land area, India accounts for 7-8% of recorded species of the world.

In terms of species richness, India ranks seventh in mammals, ninth in birds and fifth in reptiles. In terms of endemism of vertebrate groups, India's position is tenth in birds with 69 species, fifth in reptiles with 156 species and seventh in amphibians with 110 species.

India's share of crops is 44% as compared to the world average of 11%. India also has 23.39% of its geographical area under forest and tree cover. Of the 34 globally identified biodiversity hotspots, India harbour 3 hotspots, i.e., Himalaya, Indo-Burma, Western Ghats and Sri Lanka. Western Ghats are recently included in World Heritage list. It is very rich in flora and fauna and serves as cradle of biodiversity.

One of the most pressing environmental issues today is the conservation of biodiversity. Many factors threaten the world's biological heritage. The challenge is for nations, government agencies, organisations and individuals to protect and enhance biological diversity, while continuing to meet people's needs for natural resources. Efforts have been initiated to save biodiversity both by ex-situ and in-situ conservation. International Biodiversity day is celebrated across the globe on 22nd May every year.

BIODIVERSITY ACT 2002

The Biological Diversity Act, 2002 is a federal legislation enacted by the Parliament of India for preservation of biological diversity in India, and provides mechanism for equitable sharing of benefits arising out of use of traditional biological resources and knowledge. The Act was enacted to meet the obligations under Convention on Biological Diversity (CBD), to which India is a party.

The National Biodiversity Authority (NBA) was established in 2003 to implement India's Biological Diversity Act (2002). The NBA is a Statutory, Autonomous Body and it performs facilitative, regulatory and advisory function for the Government of India on issues of conservation, sustainable use of biological resources and fair and equitable sharing of benefits arising out of the use of biological resources.

LEVELS OF BIODIVERSITY

- *Marine Biodiversity* refers to 'Life in the Seas and Oceans. The marine environment has a very high biodiversity because 32 out of the 33 described animal phyla are represented in there. Marine organisms contribute to many critical processes that have direct and indirect effects on the health of the oceans and humans.
- *Forest biological diversity* is a broad term that refers to all life forms found within forested areas and the ecological roles they perform. As such, forest biological diversity encompasses not just trees, but the multitude of plants, animals and micro-organisms that inhabit forest areas and their associated genetic diversity.
- *Genetic diversity* refers to the total number of genetic characteristics in the genetic makeup of a species. Genetic diversity serves as a way for populations to adapt to changing environments. With more variation, it is more likely that some individuals in a population will possess variations of alleles that are suited for the environment. The population will continue for more generations because of the success of these individuals.
- *Species Diversity* is the effective number of different species that are represented in a collection of individuals (a dataset). Species diversity consists of two components: species richness and species evenness.
- *Ecosystem Diversity* refers to the combination of communities of living things with the physical environment in which they live. There are many different kinds of ecosystems like deserts, mountain slopes, the ocean floor, Antarctic etc. Each ecosystem provides many different kinds of habitats or living places.
- *Agriculture Biodiversity* includes all forms of life directly relevant to agriculture: rare seed varieties and animal breeds (farm biodiversity), but also many other organisms such as soil fauna, weeds, pests, predators, and all of the native plants and animals (wild biodiversity) existing on and flowing through the farm.

BIOSPHERES AND BIODIVERSITY RESERVES

The Indian government has established 18 Biosphere Reserves in India, which protect larger areas of natural habitat and often include one or more National Parks and Reserves, along buffer zones that are open to some economic uses. Protection is granted not only to the flora and fauna of the protected region, but also to the human communities who inhabit these regions, and their ways of life. Animals are protected and saved here.

HOTSPOTS

A biodiversity hotspot is a biogeographic region with a significant reservoir of biodiversity that is under threat from humans. Around the world, 25 areas qualify under definition of hotspots. These sites support nearly 60% of the world's plant, bird, mammal, reptile, and amphibian species, with a very high share of endemic species.

The biodiversity hotspots hold especially high numbers of endemic species, yet their combined area of remaining habitat covers only 2.3 percent of the Earth's land surface. Each hotspot faces extreme threats and has already lost at least 70 percent of its original natural vegetation. Over 50 percent of the world's plant species and 42 percent of all terrestrial vertebrate species are endemic to the 34 biodiversity hotspots.

UNO EFFORTS FOR CONSERVING BIODIVERSITY

- *Convention on International Trade in Endangered Species of Wild Fauna and Flora (CITES)* was signed in Washington, DC, on 3 March 1973. In August 2000, 152 States were parties to this Convention. The aim of CITES is to put a ban on international trade in wildlife.
- *The World Conservation Union IUCN* brings together States, government agencies and a diverse range of non-governmental organizations in a unique world partnership. IUCN seeks to influence, encourage and assist societies throughout the world to conserve the integrity and diversity of nature and sustainable use of natural resources.
- *International Treaty on Plant Genetic Resources for Food and Agriculture* was adopted in Rome in November 2001 to create a legally binding framework for the protection and sustainable use of all plant genetic resources for food and agriculture.
- *The United Nations Convention on Biological Diversity (CBD), 1992* known informally as the Biodiversity Convention, is a multilateral treaty. The Convention has three main goals like conservation of biological diversity (or biodiversity); sustainable use of its components; and fair and equitable sharing of benefits arising from genetic resources.

- The most significant feature of 1972 *World heritage Convention* is that it links together in a single document the concepts of nature conservation and preservation of cultural properties. The Convention recognises the way in which people interact with nature and fundamental need to preserve the balance between the two. The law of sea 1982, envisaged by UNO aims at protecting marine biodiversity and to control marine pollution.

DESERT NATIONAL PARK

Desert National Park is a unique biosphere reserve for conservation and development of biodiversity in India. It is situated in the West Indian state of Rajasthan near the town of Jaisalmer. This is one of the largest national parks, covering an area of 3162 km^2. The Desert National Park is an excellent example of the ecosystem of the Thar Desert. Sand dunes form around 20% of the Park.

ROLE OF WILDLIFE CORRIDORS IN BIODIVERSITY CONSERVATION

A habitat corridor, wildlife corridor or green corridor is an area of habitat connecting wildlife populations separated by human activities such as roads, development, or logging. This allows an exchange of individuals between populations, which may help prevent the negative effects of inbreeding and reduced genetic diversity that often occur within isolated populations.

WETLANDS REPOSITORIES OF BIODIVERSITY

Wetlands are complex ecosystems and encompass a wide range of inland, coastal and marine habitats. They include flood plains, swamps, marshes, fishponds, tidal marshes natural and man-made wetlands. The Convention on Wetlands, signed in Ramsar, Iran, in 1971, is an intergovernmental treaty which provides the framework for national action and international cooperation for the conservation and wise use of wetlands and their resources.

BENEFITS OF BIODIVERSITY

Biodiversity provides *food from crops, livestock, forestry and fish*. Biodiversity is of use to modern agriculture as a source of new crops, as a source material for breeding improved varieties and as a source of new biodegradable pesticides. Biodiversity is a rich source of substances with *therapeutic properties*. Several important pharmaceuticals have originated as plant-based substances, which are of incalculable value to human health. The industrial products like timber, oils, lubricants, food flavours, *industrial enzymes*, cosmetics, perfumes, fragrances, dyes, paper, waxes, rubber, latexes, resins, poisons and cork can all be derived from various plant species. Biodiversity is a source of economical wealth for many areas, such as many parks and forests, where wild nature and animals are a source of beauty and joy, attract many visitors. *Ecotourism* in particular, is a growing outdoor recreational activity. Biodiversity has also great *aesthetic value*. Examples of aesthetic rewards include ecotourism, bird watching, wildlife, pet keeping, gardening, etc. Biodiversity is also essential for the *maintenance and sustainable utilization of goods and services from ecological systems* as well as from the individual species. These services include maintenance of gaseous composition of the atmosphere, climate control by forests and oceanic systems, natural pest control, pollination of plants by insects and birds, formation and protection of soil.

THREATS TO BIODIVERSITY

The destruction of habitats is the primary reason for the loss of biodiversity in terrestrial and coastal ecosystems. Habitat loss could be attributed to conversion, habitat degradation and fragmentation. When people cut down trees, fill a wetland, plough grassland or burn a forest, the natural habitat of a species is changed or destroyed. Introduction of invasive species may cause disappearance of native species through biotic interactions. Invasive species are considered second only to habitat destruction as a major cause of extinction of species. Communities are affected by natural disturbances, such as fire, tree fall, and defoliation by insects. Man-made disturbances differ from natural disturbances in intensity, rate and spatial extent. For example, man by using fire more frequently may change species richness of a community. Exploitation, including hunting, collecting, fisheries and fisheries by-catch, and the impacts of trade in species and species' parts, constitute a major threat for globally threatened birds (30% of all), mammals (33% of all), amphibians (6% of those assessed), reptiles and marine fishes (Baillie et al. 2004). Trade affects 13% of both threatened birds and mammals.

Extinction is a natural process. Species have disappeared and new ones have evolved to take their place over the long geological history of the earth. It is useful to distinguish three types of extinction processes. Over-fishing, habitat destruction, widespread marine pollution and human induced climate change threaten the survival of marine biodiversity. Pollution, oil and gas drilling and oil spills may increase the risks of extinction by increasing mortality of marine organisms. The Silent Valley Project in Kerala was abandoned because it was considered as a threat to biodiversity in the region.

BIODIVERSITY AND FOOD SECURITY

In a recent estimate it was speculated that over 25 per cent of the world's plant species might be lost by the year 2025 AD, if the current rate of plant genetic erosion continues. Preserving this germ pool is an integral part of food security. It is evident that preservation of wide range of germ pool is an integral part of breeding programme.

If we are unable to combat the problems of genetic erosion, it may lead to losing sources of resistance to pests, diseases and climatic stress and, finally, leading to crop failure in future. It is well-known that out of over 20,000 edible species only a few dozen of plants are domesticated and now feed most of the people.

All types of protected area constitute over 12% of the total forest area of the country. This network of protected areas covers most of the representative habitat types in the country and affords protection both to the wild flora and fauna.

KNOWLEDGE ON BIODIVERSITY CONSERVATION

In order to be effective, efforts on biodiversity conservation can learn from the context-specific local knowledge and institutional mechanisms such as cooperation and collective action; intergenerational transmission of knowledge, skills and strategies; concern for well-being of future generations; reliance on local resources; restraint in resource exploitation; an attitude of gratitude and respect for nature; management, conservation and sustainable use of biodiversity outside formal protected areas; and, transfer of useful species among the households, villages and larger landscape. These are some of the useful attribute of local knowledge systems. Traditional knowledge on biodiversity conservation in India is as diverse as 2753 communities and their geographical distribution, farming strategies, food habits, subsistence strategies, and cultural traditions.

Local Vegetation Management

Over thousands of years local people have developed a variety of vegetation management practices that continue to exist in tropical Asia, South America, and other parts of the world.

People also follow ethics that often help them regulate interactions with their natural environment. Such systems are often integrated with traditional rainwater harvesting that promotes landscape heterogeneity through augmented growth of trees and other vegetation, which in turn support a variety of fauna.

In India these systems can be classified in several ways:

- Religious traditions: temple forests, monastery forests, sanctified and deified trees.
- Traditional tribal traditions: sacred forests, sacred groves and sacred trees.
- Royal traditions: royal hunting preserves, elephant forests, royal gardens etc.
- Livelihood traditions: forests and groves serving as cultural and social space and source of livelihood products and services.

The traditions are also reflected in a variety of practices regarding the use and management of trees, forests and water. These include:

- Collection and management of wood and non-wood forest products.
- Traditional ethics, norms and practices for restraint use of forests, water and other natural resources.
- Traditional practices on protection, production and regeneration of forests.
- Cultivation of useful trees in cultural landscapes and agroforestry systems.
- Creation and maintenance of traditional water harvesting systems such as tanks along with plantation of the tree groves in the proximity.

These systems support biodiversity, which is although less than natural ecosystems but it helps reduce the harvest pressure. For instance, there are 15 types of resource management practices that result in biodiversity conservation and contribute to landscape heterogeneity in arid ecosystems of Rajasthan. Environmental ethics of Bisnoi community suggest compassion to wildlife, and forbid felling of Prosopis cineraria trees found in the region. Bisnoi teachings proclaim: "If one has to lose head (life) for saving a tree, know that the bargain is inexpensive". In India, local practices of vegetation management perhaps emanate from the basic ecological concepts of local communities reflected in "ecosystem-like concepts in traditional societies". Two key characteristics of these systems are that the unit of nature is often defined in terms of a geographical boundary; and abiotic components, plants, animals, and humans within this unit are considered to be interlinked. Many local knowledge systems are similar in temperament to the emerging scientific view of ecosystems as unpredictable and uncontrollable, and of ecosystem processes as nonlinear, multiequilibrium, and full of surprises.

Biodiversity in Sacred Cliffs

Cliffs are completely forgotten cultural landscape elements that support a variety of species of plants and animals in India. As humans have special fascinations to such areas often cliffs across the country are considered sacred. Cliffs elsewhere have been found to support undisturbed ancient woodland, dominated by tiny, slow-growing and widely spaced trees. Vertical cliffs often support populations of widely spaced trees that are exceptionally old, deformed and slow growing. Some of the most ancient and least-disturbed wooded habitats on Earth are found on cliffs, even if such sites are close to intensive agricultural and industrial development. The age of the trees on cliffs may indicate the age and growth rates of the entire plant communities on the cliffs. Cliffs across the world may support ancient, slow-growing, open woodland communities that have escaped major human disturbance, even when they are situated close to agricultural and industrial activity, which has destroyed or altered most other natural habitats. Examples of such habitat in India abound. Cliffs in Udaipur and Kota districts of Rajasthan were surveyed (7 cliff with ancient vegetation).

Cliffs were found to have more than 25 species of trees, several species of shrubs and herbs. Areas close to Bhopal have more than 50 cliffs in central India in a radius of about 100 kms. All the 7 cliffs surveyed in Rajasthan are sacred. They are often part of the sacred corridors along the riverbank escarpment with several meters of precipitous fall. Attempts have been made to regenerate the Gaipernath Cliff with the traditional species occurring in the area. The result was very poor initially. But local ethnoforestry techniques of tucking the branch cuttings of coppicing species in whatever little crevices area may have were successful. Also, depositing the seeds (same species that occur) in crevices with the ball of moist earth has been found promising.

Farm Biodiversity

Throughout the Indian farms and field one finds strips of vegetation containing several species of plants and small animals. These strips are beneficial in several ways. Such strips on tropical lands have been found to accelerate natural successional processes by attracting seed-dispersing animals and increasing the seed rain of forest plants. Effects of these strips resemble the windbreaks on seed deposition patterns.

Isolated trees provide seed in the area for natural regeneration. The strips enhance seed rain, and connectivity. Because such strips trap large number of seeds of several species they help in further tree growth. Compared to open fields, farm boundaries with vegetation receive seed in greater densities and species-richness than open farms and pastures.

All forms of seed dispersal help in the process but animal-dispersed (birds, bats, mammals etc.) seeds often occur in greater densities and species numbers. Presence of isolated trees and shrubs or remnant trees helps. Farm boundaries maintained throughout the country are often self regenerating and require only management as these barriers considerably increase the deposition of tree and shrub seeds within the cultural landscape. Indeed considerable biodiversity is found within these strips. This is a practice that needs to be maintained as it has several socio-economic benefits as well.

Value of traditional agroecosystems in supporting the plant and animal diversity is immense. Tree diversity in farms and agroecosystems is often the product of interaction of local and formal knowledge. A recent study by Shastri et al. provides interesting insights on the tree-growing practices and associated biodiversity in Karnataka. Shastri et al. found trees belonging to 93 species in a sampled area of 1.7 ha of Sirsimakki agro-ecosystem. Additional 44 species were noted on non-agricultural lands in the village ecosystem, which included soppina betta, minor forest and reserve forest. The overall agroecosystem had 556 trees/ha, while the non-agroecosystem had only 354 trees/ha. The overall, tree density of 418.8 per ha was present in the village. There were 144 species in the village ecosystem with 2238 individuals in the sampled area of 5.34 ha. The total number of species in non-agro ecosystem was 104 with 1286 individuals. Home-gardens are notable with 93 tree species in just about 1.7 ha. The number of tree species varies between 20 and 40 in home-gardens, indicating that home-gardens in Karnataka villages are highly biodiverse in comparison to those in Mexico and Brazil.

Farms themselves have domesticated biodiversity essential for survival and subsistence. One such example is by Kimata et al. form South India on the cultivation and process of domestication of Brachiaria ramosa cultivated in pure stands. Its grains are used in nine traditional food preparations in South India. Another crop Setaria glauca is cultivated in mixed stands along with little millet (Panicum sumatrense).

In Orissa state and in Southern India the grains are used to make at least six traditional supplementary foods. The weedy forms of these species were found by the researchers growing with upland rice and some millets in diverse agro-ecological niches. The domestication process is supposed to have gone through three phases: first growing in association with weed and with upland rice and other millets; a secondary crop mixed with kodo millet; and finally as an independent crop.

Cultivation of Medicinal Plants

There are numerous examples of medicinal plant cultivation by local people in India. Socio-culturally valued species find place in home gardens and courtyards. For example, Around the Nanda Devi Biosphere Reserve in the western Himalaya, the Bhotiya community, whose livelihood is depends on local natural resources, practices seasonal and altitudinal migration and stay inside the buffer zone for only 6 months (May-October). A survey in 5 villages in Pithoragarh District, found that Bhotiya people cultivate medicinal plants on their agriculture fields.

Of a total of 71 families, 90% cultivated medicinal plants on 78% of the total reported cultivated area (15.29 ha). Around 12 species of medicinal plants were under cultivation. Survey also found that a family earned about Rs. 2423 +/- 376.95 per season from the sale of medicinal plants in 1996 (Rs.38 = US$1 in 1996).

Thus, supporting medicinal plant cultivation at high altitudes in the Himalayas may help to generate additional support to people as well as conserve the species in the wild. Another study on traditional knowledge of Kumaun Higher Himalaya found that Bhotia tribes use 34 species of medicinal plants native to the region. Among these, *Angelica glauca and Allium stracheyi* are narrow range endemic and *Allium stracheyi, Picrorhiza kurrooa* and *Nardostachys grandiflora* have been recorded in the Red Data Book of Indian Plants.

Interestingly, the annual production of medicinal plants has been found to be comparable with the annual production of traditional crops. Thus, cultivation, and harvesting can help in livelihood security and in situconservation of these species.

Similarly, juang and Munda tribes of the Keonjhar district of eastern India use 215 plants, belonging to 150 genera and 82 families. This suggests a wealth of traditional knowledge on biodiversity and herbal health care in tribes of eastern India. Tribes in the region are dependent on forests for other species as species of mushrooms, wild berries, tubers, and flowers that are included in their diet including cooking oil. Understanding of traditional knowledge on biodiversity of the region will be most helpful in planning for sustainable forest management.

Traditional Ethos

Similarly, in spite of the modernization, traditional ecological ethos continue to survive in many other local societies, although often in reduced forms. Investigations into the traditional resource use norms and associated cultural institutions prevailing in rural Bengal societies demonstrate that a large number of elements of local biodiversity, regardless of their use value, are protected by the local cultural practices.

Some of these may not have known conservation effect, yet may symbolically reflect, a collective appreciation of the intrinsic or existence value of life forms, and the love and respect for nature. Traditional conservation ethics are still capable of protecting much of the country's decimating biodiversity, as long as the local communities have even a stake in the management of natural resources. Traditional ethos is reflected in a variety of practices including sacred groves and sacred landscapes. They are fairly well described.

One example from northeast India is particularly notable. The tribal communities of Meghalaya – Khasis, Garos, and Jaintias – have a tradition of environmental conservation based on various religious beliefs. As elsewhere in India, particular patches of forests are designated as sacred groves under customary law and are protected from any product extraction by the community. Such forests are very rich in biological diversity and harbour many endangered plant species including rare herbs and medicinal plants. Tiwari et al. identified 79 sacred groves and their floristic survey revealed that these sacred groves are home to at least 514 species representing 340 genera and 131 families. The status of sacred groves was ascertained through canopy cover estimate. About 1.3% of total sacred grove area was undisturbed, 42.1% had relatively dense forest, 26.3% had sparse canopy cover, and 30.3% had open forest. Notably, the species diversity indices were higher for the sacred grove than for the disturbed forest. Another notable example is from peninsular India. Study on two sacred groves, Oorani and Olagapuram, situated on the north-west of

Pondicherry found a total of 169 angiosperms from both sites. The Oorani grove (3.2 ha) had 74 flowering plant species distributed in 71 genera and 41 families; 30 of them are woody species, 8 are lianas and 4 are parasites. The Olagapuram grove (2.8 ha) was more species-rich with 136 species in 121 genera of 58 families; woody species were fewer while 9 lianas and 3 parasites occurred. Associated local knowledge, cultural and religious rituals of local people sustain such diversity.

Another tradition worth mention is use of plants in mural painting. Such paintings are found, for example, in the Ajantan mural art. The practice spanned a whole millennium from the second century B.C. to the eighth century A.D. The tradition continued up to the nineteenth century under the support of different dynasties in India, but declined by the end of that century. Nayar et al. note that the art is kept alive by a few artists in Kerala who practice even today the methods and techniques of mural paintings similar to those practised by the Ajantan mural painters. Various plant species provided materials for mural painting. Such knowledge can be very helpful in providing livelihood security to practitioners.

Traditional water harvesting structures too are also habitat for a variety of species. Even if pond size is small, as is the case in about 60% (out of 1.5 million total tanks) in India it may still be useful habitat for many species in rural ecosystems. Indeed, the island biogeography theory – valid in numerous cases – suggesting that larger areas support more species did not stand in case of 80 ponds in Switzerland. Theoretical predictions and empirical support suggests that although intentional conservation may be rare among small-scale societies as Smith and Wishnie have pointed out, but practices that actually result in what we today call 'sustainable use and management' of resources and habitats by local people is widespread globally that contribute to in biodiversity conservation and enhancement through creation of habitat mosaics.

Formal conservation efforts in India have relied heavily on the recently declared official protected areas in various categories for biodiversity conservation. However, ancient and widespread human practice to set aside areas for the preservation of natural values in India can be seen in several examples of sacred groves, royal hunting forests, and sacred gardens. Several of these areas became national parks and wildlife sanctuaries in India and elsewhere.

It must be noted here that much of the India's biodiversity lies outside the officially declared protected areas. Indeed, biodiversity occurs in landscape continuum. Other areas protect ecosystem services such as the delivery of clean water or the supply of timber, or mitigate the expected adverse effects of over-clearing. Others protect recreational and scenic values and some have been planned to foster international cooperation. Many of these areas meet the World Conservation Union's definition of a strictly protected area.

In view of accelerating biological and cultural landscape degradation, a better understanding of interactions between landscapes and the cultural forces driving them is essential for their sustainable management. We need environmental and cultural revolution, aiming at the reconciliation of human society with nature.

TRADITIONAL KNOWLEDGE, WATER, AND BIODIVERSITY

Simple local technology and an ethic that exhorts "capture rain where it rains" have given rise to 1.5 million traditional village tanks, ponds and earthen embankments that harvest substantial rainwater in 660,000 villages in India, and encourage growth of vegetation in commons and agroecosystems. If India were to simply build these tanks today it would take at least US $ 125 billion.

Humans have virtually appropriated fresh water. Humanity now uses 26 percent of total terrestrial evapotranspiration and 54 percent of runoff that is geographically and temporally accessible. New dam construction could increase accessible runoff by about 10 percent over the next 30 years, whereas population is projected to increase by more than 45 percent during that period.

Over thousands of years societies have developed a diversity of local water harvesting and management regimes that still continue to survive, for example, in South Asia, Africa, and other parts of the world. Such systems are often integrated with agroforestry and ethnoforestry practices. Recently it has been suggested that market mechanisms for sustainable water management such as taxing users to pay commensurate costs of supply and distribution and of integrated watershed management and charging polluters for effluent treatment can solve the problem. Such measures are essential although, but they are insufficient and would need to draw on the local knowledge on rainwater harvesting across different cultures.

Rainwater harvesting in South Asia is different from other parts of the world in that it has a continued history of practice for at least over 5000 years. Similarly, Balinese water temple networks as complex adaptive systems are also very useful systems. Although hydraulic earthworks are known to have occurred in ancient landscapes in many regions, they are no longer an operational systems among the masses in the same proportion as in South Asia. For instance, remains of earthworks and water storage adaptations are found in Mayan lowlands in South America. Such systems had been used for prehistoric agriculture in Mayan lowlands, and for fish culture in Bolivian Amazon.

Rainwater harvesting have been found to be scientific and useful for rainfed areas. For instance, a validation comes from the Negev. Ancient stone mounds and water conduits are found on hillslopes over large areas of the Negev desert. Field and laboratory studies suggest that ancient farmers were very efficient in harvesting water. A comparison of the volume of stones in the mounds to the volume of surface stones from the surrounding areas indicates that the ancient farmers removed only stones that had rested on the soil surface and left the embedded stones untouched. According to results of simulated rainfall experiments, this selective removal increased the volume of runoff generated over one square meter by almost 250% for small rainfall events compared to natural untreated soil surfaces.

One of the principle tree genus growing in association with tanks and ponds in India is Ficus which is culturally valued throughout the country. It is a keystone genus and supports a variety of other species. Records of frugivory from over 75 countries for 260 Ficus species (approximately 30% of described species) suggest that in addition to a small number of reptiles and fishes, 1274 bird and mammal species in 523 genera and 92 families are known to eat figs.

CONSERVATION PRINCIPLES IN ANCIENT TEXTS

Natural Resource Management has been in the traditions of the Indian society, expressing itself variously in the management and utilization practices. This evolved through the continued historical interaction of communities and their environment, giving rise to practices and cultural landscapes such as sacred forests and groves, sacred corridors and a variety of ethnoforestry practices. This has also resulted in conservation practices that combined water, soil and trees. Nature-society interaction also brought about the socio-cultural beliefs as an institutional framework to manage the resultant practices arising out of application of traditional knowledge. The attitude of respect towards earth as mother is widespread among the Indian society.

Local knowledge has proved useful for forest restoration and protected area management in Rajasthan – one of the driest regions of India with scanty rainfall. Cultural landscapes in rural and urban areas and agroecosystems, created by the application of scientific and local knowledge, also support a variety trees, birds and other species, and provide opportunity of integration of nature and society.

Ancient texts make explicit references as to how forests and other natural resources are to be treated. Sustainability in different forms has been an issue of development of thought since ancient times. For example, robust principles were designed in order to comprehend whether or not the intricate web of nature is sustaining itself. These principles roughly correspond with modern understanding of conservation, utilization, and regeneration.

Conservation Principles

Atharva Veda (12.1.11) hymn, believed to have been composed sometime at around 800 BC, somewhere amidst deep forests reads: "O Earth! Pleasant be thy hills, snow-clad mountains and forests; O numerous coloured, firm and protected Earth! On this earth I stand, undefeated, unslain, unhurt."

Implicit here are the following principles:

- It must be ensured that earth remains forested.
- It must be understood that humans can sustain only if the earth is protected.
- To ensure that humans remain 'unslain' and 'unhurt', the ecosystem integrity must be maintained.
- Even if vaguely, it also makes reference to ecology, economy and society concurrently.

Utilization and Regeneration Principles

Another hymn from Atharva Veda (12.1.35) reads: "Whatever I dig out from you, O Earth! May that have quick regeneration again; may we not damage thy vital habitat and heart".

Implicit here are the following principles:

- Human beings can use the resources from the earth for their sustenance,
- Resource use pattern must also help in resource regeneration,
- In the process of harvest no damage should be done to the earth,
- Humans are forewarned not against the use of nature for survival, but against the overuse and abuse.

Although not in modern terminology, the three segment of sustainability – ecology, economy and society seem to get addressed simultaneously.

Similarly, water management and associated tree growing has been the subject of ancient text. Tanks have been the most important source of irrigation in India. Some tanks may date as far back as the Rig Vedic period, around 1500 BC. The Rig Veda refers to lotus ponds (5.78.7), ponds that give life to frogs (7.103.2) and ponds of varying depths for bathing (10.71.7). Reference to the tanks is also found in the Arthashastra of Kautilya written around 300 BC. The Arthashastra refers to the ownership and management of the village tanks in the following verses:

- Waterworks such as reservoirs, embankments and tanks can be privately owned and the owner shall be free to sell or mortgage them (3.9.33).
- The ownership of the tanks shall lapse, if they had not been in use for a period of five years, excepting in case of distress (3.9.32).
- Anyone leasing, hiring, sharing or accepting a waterworks as a pledge, with a right to use them, shall keep them in good condition (3.9.36).
- Owners may give water to others in return for a share of the produce grown in the fields, parks or gardens (3.9.35).
- In the absence of owners, either charitable individuals or the people in village acting together shall maintain waterworks (3.10.3).
- No one will sell or mortgage, directly or indirectly, a bund or

embankment built and long used as a charitable public undertaking except when it is in ruins or has been abandoned (3.10.1,2).

- The earliest scholar to have commented on the relationship of tanks and trees is Varahamihira who described the detailed technical instructions for the tank constructions in his famous work Brahatsamhita (550 AD):
- Without the shade of the trees on their sides, water reservoirs do not look charming; therefore, one ought to plant the gardens on the banks of the water (55.1)7

Commenting on the species to be planted on the embankments of the tank, after its construction, Varahamihira writes:

- The shoreline (banks) of the tanks should be shaded (planted) with the mixed stands of Arjun (Terminalia arjuna), Vata (Ficus benghalensis), Aam (Mangifera indica), Pipal (Ficus religiosa), Nichul (Nauclea orientalis), Jambu (Syzygium cuminii), Vet (Calamus), Neep (Mitragyna parvifolia), Kurvak, Tal (Borassus flabellifer), Ashok (Saraca asoka), Madhuk (Madhuca indica), and Bakul (Mimusops elengi) (54.119).

For example, there is a considerable overlap in the formal and scientific forestry policy and practice, which provides hope that traditional knowledge systems can contribute to the management of natural resources. It would be pertinent to quote Gadgil and Guha (1992: 51) in this context:

- "Indeed one could argue that scientific prescriptions in industrial societies show little evidence of progress over the simple rule-of-thumb prescriptions for sustainable resource use and the conservation of diversity which characterized gatherer and peasant societies. Equally, the legal and codified procedures which are supposed to ensure the enforcement of scientific prescriptions work little better than earlier procedures based on religion or social convention".

INTEGRATION OF TRADITIONAL AND FORMAL SCIENCE

Are there any possibilities of integration of science and ethnoscience? Empirical evidence suggests in affirmative. Traditional knowledge may indeed complement scientific knowledge by providing practical experience in living within ecosystems and responding to ecosystem change. But, as Berkes et al. note the "language" of traditional ecology is different from the scientific and generally includes "metaphorical imagery and spiritual expression, signifying differences in context, motive, and conceptual underpinnings".

Indic traditions and local knowledge have often paved the way for many discoveries in science. For example, progress of science in India has built on the foundations of knowledge and wisdom that was created in ancient times on a variety of disciplines including metallurgy, mathematics, medicine, surgery and natural resource management. Traditional skills, local techniques and rural craft provide a wide spectrum of knowledge in India, and since "knowledge cannot be fragmented" we have to take the validated local knowledge into account together with science for evolving a robust sustainability science.

Sharp boundaries between formal and local systems of knowledge, and natural sciences and social sciences may indeed be imaginary. Perceived confines may just be the unexplored domain that defies cognition for want of interdisciplinary explorations. This is however changing, as Wilson notes, disciplines are being rendered "consilient". Scientific community is increasingly realizing that "there is a continuum between artificially dichotomized aspects of science: objective versus subjective, value free versus value laden, neutral versus advocacy". This disciplinary mosaic will have profound impact on science and policy development.

Since local knowledge systems in India are still being practised among the masses, they can contribute to address the challenges of forest management, sustainable water management, biodiversity conservation, and mitigation of global climate change. Ecological consequences of climate change require that we access all stocks of knowledge for mitigation strategies.

BIODIVERSITY CONSERVATION

Strategies employed for conservation and management of natural resources prominently rely on nature reserves, national parks, wildlife sanctuaries and other such categories of protected areas. Protected-area-alone approach for nature conservation, however, has serious flaw as it has further exacerbated the problem of human-animal conflicts, and a majority of reserves have failed to achieve the conservation goals in marine as well as terrestrial ecosystems. Such an approach has also "led to conflicts between the local communities and the management authorities".

Further, application of island biogeography theory to conservation practice has been contended since long. As Simberloff and Abele note "theoretically and empirically, a major conclusion of such applications – that refuges should always consist of the largest possible single area – can be incorrect under a variety of biologically feasible conditions. The cost and irreversibility of large-scale conservation programs demand a prudent approach to the application of an insufficiently validated theory." Protecting biodiversity in protected areas indeed has remained a challenge across nations.

On the other hand there are detailed accounts of a variety of mechanisms and contexts through which local people conserve and maintain biodiversity across landscape continuum. Practice to set aside areas for the preservation of natural values such has sacred groves of Asia and Africa and royal hunting forests in India are some historical examples of nature conservation. Several of these areas became national parks and wildlife sanctuaries in India and elsewhere.

Consensus that seems emerging is that we might need multiple conservation and sustainable management approaches Under these circumstances, instead of an exclusive approach, both protected areas and community areas seem complementary strategies. As the human and livestock population grows and natural resources decline command-and-control management of natural resources tends to become the norm. Stricter enforcement of protected areas again is gaining currency as a management proposal due to perceived failure of people-oriented approaches to safeguard biodiversity. Unfortunately, such an approach usually results in adverse consequences for natural ecosystems and human welfare in the form of collapsing resources, social and economic conflict, and loss of biological diversity.

Additionally, this resurgent focus on authoritarian protection practices largely overlooks key aspects of social and political process including clarification of moral standpoint, legitimacy, governance, accountability, learning, and external forces. A single stock of knowledge is inadequate to address the challenges that sustainability science faces today.

WATER HARVESTING AND BIODIVERSITY CONSERVATION

Revival of local rainwater harvesting globally could provide substantial amounts of water for nature and society. For example, a hectare of land in Jaisalmer, one of India's driest places with 100 millimetres of rainfall per year, could yield 1 million litres of water from harvesting rainwater. Even with the simple technology such as ponds and earthen embankments called tanks, at least half a million litres a year can be harvested from rain falling over one hectare of land, as is being done in the Thar desert, making it the most densely populated desert in the world. Indeed, there are 1.5 million village tanks in use and sustaining everyday life in the 660,000 villages in India.

In the Negev Desert, decentralized harvesting through the collection of water in microcatchments from rain falling over a 1-hectare watershed yielded 95 cubic meters of water per hectare per year, whereas collection efforts from a single large unit-rather than small microcatchments – 345-hectare watershed yielded only 24 cubic meters per hectare per year. Thus, 75% of the collectible water was lost as a result of the longer distance of runoff in larger watershed. Indeed, this is consistent with local knowledge distilled in Indian proverbs: "capture rain where it rains". This is also inconsonance with Water and civilizations with a promise of using history to reframe water policy debates and to build a new ecological realism.

There is an urgent need to policy innovations on rainwater harvesting that has been found useful by many studies. In the cities, rainwater could be harvested from building rooftops for residential use, and any surplus could be channelled through bore wells to replenish the groundwater, avoiding loss to runoff. However, if rainwater harvesting is to be used to their full potential, policy innovations must include institutional changes so that such resources are effectively managed. In Rajasthan, tanks and ponds have been a mainstay of rural communities for centuries. Strategies for tank rehabilitation (such as proposed for 1200 large tanks in Rajasthan) must not treat tanks only as flow irrigation systems; such an approach is very likely to result in a flawed strategy. A strategy that considers tanks as multiple-use socio-ecological entities, and which recognizes multiple stakeholder groups is more likely to enhance the social value of tanks. In order to fully reward the context specific cultural resources, such as local knowledge, government subsidies need to be removed to allow market mechanisms to run their course and surplus revenue generated can be given to the communities who own the systems such as tanks.

LOW INTENSITY-AGRICULTURE

Since low-intensity agriculture promotes biodiverse farms across landscape, such systems need to be supported and promoted. Agricultural intensification has been found to impact biodiversity in farms badly. Crop-animal systems in Asia, where 95% of ruminants are found in the mixed farming systems is famous for diversity. Crop-animal systems are projected to see growth and remain the dominant system in Asia.

Biodiversity in such mixed farming systems are vital for food production. Crop-animal systems, in which livestock play a multi-purpose role, are the backbone of Asian agriculture. Increased productivity from livestock will be necessary in these systems to meet the increased demand for animal products, to alleviate poverty and to improve the livelihoods of resource-poor farmers. In the face of land degradation native farm vegetation will play a major role in the sustainability of the farming systems.

INCORPORATING TRADITIONAL KNOWLEDGE IN PRACTICE

Any attempt, endeavouring to integrate traditional knowledge for biodiversity conservation and sustainability of natural resources should be based on the principle that traditional knowledge often cannot be dissociated from its cultural and institutional setting. Regarding the cultural and institutional the following suggestions may be useful:

1. Each programme aiming at the promotion of traditional knowledge should be based on the recognition that natural resource rights and tenurial security of local communities forms the fundamental basis of respecting traditional knowledge.
2. More attention is needed on protection of intellectual property rights of traditional people.
3. Innovative projects may need to be developed that aim at the enhancement of the capacity of local communities to use, express

and develop their traditional knowledge on the basis of their own cultural and institutional norms.

There is an urgent need for the integration of Traditional and formal sciences. Following considerations may be useful in this regard:

1. Development of methods for mutual learning between local people and the formal scientists.
2. State forest policies and sustainable forest management processes need to give full attention to ethforestry and local institutional arrangements to incorporate traditional knowledge in forest management and development projects.
3. Traditional knowledge and traditions can contribute to the preparation of village microplans, which are prepared for eco-development, joint forest management and rural development. The plans should be based on both geographic and traditional community boundaries rather than only on administrative boundaries.
4. Revival of the traditional water management systems that have served the society for hundreds of years but are currently threatened
5. There is a clear need to integrate traditional and formal sciences for participatory monitoring, and taking feedback to achieve adaptive strategies for management of natural resources.

In spite of the value of traditional knowledge for biodiversity conservation and natural resource management there still is a need to further the cause. The following consideration may be useful in this respect:

1. Encouraging the documentation of indigenous knowledge and its use in natural resource management. Such documentation should be carried out in participation with the communities that hold the knowledge. Due attention should be given to document the emic perspectives regarding IK rather than only the perspectives of professional outsiders. The documentation should not only consist of descriptions of knowledge systems and its use, but also information on the threats to its survival. People's biodiversity registers are a case in point. The program of People's Biodiversity Registers promotes folk ecological knowledge and wisdom by devising a formal means for their maintenance, and by creating new contexts for their continued practice. PBRs document traditional ecological knowledge and practices on use of natural resources, with the help of local educational institutions, teachers, students and NGOs working in collaboration with local, institutions. Such a process and the resulting documents, could serve a significant role in "promoting more sustainable, flexible, participatory systems of management and in ensuring a better flow of benefits from economic use of the living resources to the local communities".

2. Facilitating the translation of available and new documents describing Indic traditions such as ancient texts on medicinal plants, into local languages and dissemination of these documents amongst local people. Such a translation is indeed required because texts are often available in languages (e.g. Sanskrit) not understood by many in contemporary India. On the other hand, translation of local knowledge into formal scientific terminology will provide space to external researchers, policy makers, and practitioners to comprehend and support people's knowledge systems and initiatives.
3. Facilitating the exchange of information amongst practitioners of local knowledge.
4. Developing clear and concise educational material on traditional knowledge systems to be used in communication programmes to impart information regarding the merits and threats to indigenous knowledge systems to both policy makers and the general public.

Scientific institutions have an important role to play in supporting the knowledge systems. As has been pointed out earlier, it is now recognised that a dichotomy between local and formal systems of knowledge is not real, and that any knowledge is based on a set of basic values and beliefs and paradigms. Therefore, there is a definite need to further develop systematic insight into the nature and scope of traditional knowledge. The following activities may be useful in this regard:

1. Developing curricula and methods for providing formal training and education in traditional knowledge systems to agencies, researchers and practitioners who work in collaboration with communities. In this context, the Indian Himalayan Region, which represents a unique biogeographic entity, new initiatives by G.B. Pant Institute of Himalayan Environment and Development have yielded positive results.
2. Developing research projects aimed at assessing the possibilities and constraints of using traditional knowledge under specific conditions. Such research projects should move beyond the first generation research projects, which aimed at demonstrating the value of local knowledge systems by focusing on successful cases of application. Second generation research projects shall focus on comparing application of knowledge systems across a range of circumstances and across disciplines to craft the traditional sustainability science.
3. Developing new methods for incorporating local knowledge systems in natural resource management regimes through action research

CONSERVATION OF BIODIVERSITY IN THE INDIAN CONTEXT

Biological diversity – or biodiversity – is a term we use to describe the variety of life on Earth. It refers to the wide variety of ecosystems and living organisms: animals, plants, their habitats and their genes, it is degree of variation within a given ecosystem, biome or entire planet. Biodiversity is the foundation of life on Earth. It is crucial for the functioning of ecosystems which provide us with products and services without which we couldn't live. Oxygen, food, fresh water, fertile soil, medicines, shelter, protection from storms and floods, stable climate and recreation – all have their source in nature and healthy ecosystems. But biodiversity gives us much more than this. We depend on it for our security and health; it strongly affects our social relations and gives us freedom and choice.

Biodiversity which is the 'the variability among living organisms from all sources, including terrestrial, marine and other aquatic ecosystems and the ecological complexes of which they are a part and also includes diversity within species, between species and of ecosystems'. Conservation and sustainable use of biodiversity is fundamental to ecologically sustainable development. At the same time, no other feature of the Earth has been so dramatically influenced by man's activities. By changing biodiversity, we strongly affect human well-being and the well-being of every other living creature.

Biodiversity is extremely complex, dynamic and varied like no other feature of the Earth. Its innumerable plants, animals and microbes physically and chemically unite the atmosphere (the mixture of gases around the Earth), geo-sphere (the solid part of the Earth), and hydrosphere (the Earth's water, ice and water vapour) into one environmental system which makes it possible for millions of species, including people, to exist.

Biodiversity supports many ecosystem services that are often not readily visible. It plays a part in regulating the chemistry of our atmosphere and water supply. Biodiversity is directly involved in water purification, recycling nutrients and providing fertile soils. Experiments with controlled environments have shown that humans cannot easily build ecosystems to support human needs; for example insect pollination cannot be mimicked, and that activity alone represents tens of billions of dollars in ecosystem services per year to humankind.

The diversity of genes, species and ecosystem is a valuable resource that can be tapped as human needs and demands change, the still more basic reasons for conservation are the moral, cultural and religious values. The importance of biodiversity can be understood, it is not easy to define the value of biodiversity, and very often difficult to estimate it. The value of biodiversity is classified into direct and indirect values.

Biodiversity is part of our daily lives and livelihood, and constitutes resources upon which families, communities, nations and future generations depend. Every country has the responsibility to conserve, restore and sustainably use the biological diversity within its jurisdiction. Biological diversity is fundamental to the fulfilment of human needs. An environment rich in biological diversity offers the broadest An environment rich in biological diversity offers the broadest array of options for sustainable economic activity, for sustaining human welfare and for adapting to change.

INDIA AND BIODIVERSITY

India has 47 000 species of flowering and non-flowering plants representing about 12% of the recorded world's flora. Out of 47 000 species of plants, 5 150 are endemic and 2 532 species are found in the Himalayas and adjoining regions and 1 782 in the peninsular India. India is also rich in the number of endemic faunal species it possesses, while its record in agro-biodiversity is very impressive as well. There are 166 crop species and 320 wild relatives along with numerous wild relatives of domesticated animals. Overall India ranks seventh in terms of contribution to world agriculture

India is one of the 17 "mega diverse" countries and is composed of a diversity of ecological habitats like forests, grasslands, wetlands, coastal and marine ecosystems, and desert ecosystems. Almost 70% of the country has been surveyed and around 45,000 plant species (including fungi and lower plants) and 89,492 animal species have been described, including 59,353 insect species, 2,546 fish species, 240 amphibian species, 460 reptile species, 1,232 bird species and 397 mammal species. Endemism of Indian biodiversity is significant with 4950 species of flowering plants, 16,214 insects, 110 amphibians, 214 reptiles, 69 birds and 38 mammals endemic to the country.

India's contribution to agro-biodiversity has been impressive. India stands seventh in the world as far as the number of species contributed to agriculture and animal husbandry is concerned. In qualitative terms too, the contribution has been significant, as it has contributed such useful animal species as water buffalo and camel and plant species such as rice and sugarcane. India has also been a secondary centre of domestication for animal species such as horse and goat, and such plant species as potato and maize.

Biodiversity Contribution to Indian Economy

Biodiversity products have obtained a commercial value and have been increasingly exchanged in the markets having a monetary value, from which their share in the national economy can be judged. In the Indian context it is difficult to put a value on diversity as such because the marketable products are of various kinds both legal and illegal e.g. wood and non-wood products from forests where wood comprises the major commercial produce is both legally exported as well as presumed to be illegally smuggled out of the country.

The contribution of natural and agricultural biodiversity in terms of crops, live stock, fisheries etc. is very substantial in terms of commercial value. Such biodiversity has a major contribution to make to the Indian GDP (gross domestic product). The large economic implications of biodiversity in its wild and domesticated forms is the rice improvement programme. Rice accounts for 22% of the total cropped area and 39% of the total area under cereals, which reflects its importance in the country's struggle to attain self-sufficiency in food.

LOSS OF BIODIVERSITY IN INDIA

Loss of biodiversity – the variety of animals, plants, their habitats and their genes – on which so much of human life depends, is one of the world's most pressing crises. It is estimated that the current species extinction rate is between 1,000 and 10,000 times higher than it would naturally be. The main drivers of this loss are converting natural areas to farming and urban development, introducing invasive alien species, polluting or over-exploiting resources including water and soils and harvesting wild plants and animals at unsustainable levels.

Rapid environmental changes typically cause extinctions. One estimate is that less than 1% of the species that have existed on Earth are extinct. Ecosystem stability is also positively related to biodiversity, protecting against disruption by extreme weather or human exploitation. One of the major causes for the loss of biodiversity in India is the expansion of agriculture in previously wild areas. Other impacts include: unplanned development, opening of roads, overgrazing, fire, pollution, introduction and spread of exotics, excessive siltation, dredging and reclamation of water bodies, mining and industrialization.

In this century, the Indian cheetah, Lesser Indian rhino, Pink-headed duck, Forest owlet and the Himalayan mountain quail are reported to have become extinct and several other species (39 mammals, 72 birds and 1,336 plants) are identified vulnerable or endangered. Pressure Habitat destruction, overexploitation, pollution, and species introduction are the major causes of biodiversity loss in India. Other factors included fires, which adversely affect regeneration in some cases, and such natural calamities as droughts, diseases, cyclones, and floods, tsunamis etc.

Habitat destruction, decimation of species, and the fragmentation of large contiguous populations into isolated, small, and scattered ones has rendered them increasingly vulnerable to inbreeding depression, high infant mortality, and susceptibility to environmental degradation of a chaste city and, in the long run, possibly to extinction. Besides these, the failure to stem this tide of destruction results from an amalgamation of lacunae in economic, policy, institutional, and governance systems.

Among others, these include:

- Management with limited local community participation and involvement and inadequate implementation of eco development programmes;
- Poor implementation of various environmental legislations including Wildlife (Protection) Act of 1972 as amended in 1991 has lead to a major loss of biological diversity.

- Unani, Ayurveda and Sidha. Many indigenous medicines also utilize animals and their parts or extracts as remedies for various diseases. Diverse habitats and species also have non-consumptive use-value.
- Tourism, recreation and scientific research are the major examples. The indirect use-value of biodiversity includes ecosystem process of biological diversity, which provides valuable ecological services to the biosphere; some Poor conviction rates of wildlife cases due to inadequate legal competence in the forest department, and the lackadaisical or apathetic approach of courts with cases pending for years.
  - Loss of biodiversity has serious economic and social costs for a country. The experience of the past few decades has shown that as industrialization and economic development in the classical sense takes place, patterns of consumption, production and needs, change, straining, altering and even destroying ecosystems.
  - India, a mega biodiversity country, while following the path of development, has been sensitive to needs of conservation and hence is still rich in biological resources. Ethos of conservation and harmonious living with nature is very much ingrained in the lifestyles of India's people. If followed diligently can still help in the conservation of biodiversity.
  - There are innumerable species, the potential of which is not as yet known. It would therefore be prudent to not only conserve the species we already have information about, but also species we have not yet identified and described from economic point of view. Taxus baccata, a tree found in the Sub-Himalayan regions, once believe to be of no value is now considered to be effective in the treatment of certain types of cancer.
  - Biodiversity is part of our daily lives and livelihood, and constitutes resources upon which families, communities, nations and future generations depend. Every country has the responsibility to conserve, restore and sustainably use the biological diversity within its jurisdiction. Biological diversity is fundamental to the fulfilment of human needs. An environment rich in biological diversity offers the broadest array of options for sustainable economic activity, for sustaining human welfare and for adapting to change.
  - The long-term fluctuations in temperature, precipitation, wind, and all other aspects of the Earth's climate. It is also defined by the United Nations Convention on Climate Change as "change of climate which is attributed directly or indirectly to human

activity that alters the composition of the global atmosphere and which is in addition to natural climate variability observed over comparable time periods"

Major problems with biodiversity conservation are:

- Low priority for conservation of living natural resources.
- Exploitation of living natural resources for monetary gain.
- Values and knowledge about the species and ecosystem inadequately known.
- Unplanned urbanization and uncontrolled industrialization.

Major threats to biodiversity are:

- Habitat destruction.
- Extension of agriculture.
- Filling up of wetlands.
- Conversion of rich bio-diversity site for human settlement and industrial development.

Five main threats to biodiversity are commonly recognized in the programmes of work of the environmental conventions: invasive alien species, climate change, nutrient loading and pollution, habitat change, and overexploitation. Unless we successfully mitigate the impacts of these direct drivers of change on biodiversity, they will contribute to the loss of biodiversity components, negatively affect ecosystem integrity and hamper aspirations towards sustainable use.

STRATEGIES MANAGEMENT IN BIODIVERSITY CONSERVATION

The Biodiversity Conservation Strategy is the overarching strategy for the protection of biodiversity in the growth corridors and is a significant step towards finalising the planning for biodiversity required for the Melbourne Strategic Assessment.

The strategy:

- Addresses all relevant matters of state significance, as well as matters of national environmental significance protected under the Commonwealth Environment Protection and Biodiversity Conservation Act 1999
- Ensures the long term protection of biodiversity in the growth corridors, by setting up a network of conservation areas
- Sets out conservation measures to protect important biodiversity outside Melbourne to complement actions within the growth corridors.

The Biodiversity Conservation Strategy informed the Growth Corridor Plans prepared by the Growth Areas Authority. These plans set the strategic direction for future urban development of land in the growth areas over the next 20 to 30 years.

STRATEGIES FOR BIODIVERSITY CONSERVATION IN FORESTS

There are several approaches to nature conservation in forest landscapes. Areas have since long been set aside as national parks or nature reserves. During the last decades new steps have been taken by incorporation of conservation actions into forestry operations, mainly by retaining trees to promote biodiversity at logging. Thus, conservation today implies that trees are set aside to benefit the environment at widely different scale levels, from individual trees at logging to millions of trees in large protected areas. We analyse ecological aspects of different strategies, and also include studies on their cost-efficiency.

Retention Approaches in Forestry

The research on tree retention is mainly conducted within the large multidisciplinary research programme Smart Tree Retention with researchers on ecology, political science, forest planning, conservation planning, and landscape architecture, and with participants from SLU Uppsala, Alnarp, Umeå, Umeå university and Skogforsk. The research at our unit here at SLU Uppsala includes a global overview on practice and research on retention forestry, and structural changes in young forests (National Forest Inventory data).

Ecological studies embrace saproxylic beetles on clearcuts as well as lichen epiphytes on retention trees. We are also conducting a study on the survival of red-listed species on clearcuts, based on observations before and after logging, and with a special focus on the role of retained structures. Further, we are working with a systematic review embracing the large number of studies on the retention approach conducted during the last two decades, on the relation between retention amounts and biodiversity response. Apprehensions have been raised that biofuel harvest (slash and stumps) may damage and reduce living as well as dead retained trees. We are studying this in a nation-wide field study by comparing types and amount of retention in a number of stands with and without biofuel extraction.

IN-SITU CONSERVATION THROUGH PARTICIPATORY CONSERVATION

For much of the time man lived in a hunter-gather society and thus depended entirely on biodiversity for sustenance. But, with the increased dependence on agriculture and industrialisation, the emphasis on biodiversity has decreased. Indeed, the biodiversity, in wild and domesticated forms, is the sources for most of humanity food, medicine, clothing and housing, much of the cultural diversity and most of the intellectual and spiritual inspiration. It is, without doubt, the very basis of life. Further that, a quarter of the earth's total biological diversity amounting to a million species, which might be useful to mankind in one way or other, is in serious risk of extinction over the next 2-3 decades. On realisation that the erosion of biodiversity may threaten the very extinction of life has awaked man to take steps to conserve it. During the last decade, the IUCN have developed strategies for Conservation of Nature and Natural Resources with hold up from World Bank and other institutions. On the whole, the conservation plan has a holistic approach and encompasses whole spectrum of biota and activities ranging from ecosystems at the macro level (in-situ conservation) to DNA libraries at the molecular level (ex-situ conservation).

Definition

Ex-situ conservation literally means, "off-site conservation". It is the process of protecting an endangered species of plant or animal by removing part of the population from a threatened habitat and placing it in a new location, which may be a wild area or within the care of humans. While ex-situ conservation comprises some of the oldest and best known conservation methods, it also involves newer, sometimes controversial laboratory methods. Ex situ conservation, using sample populations, is done through establishment of gene banks, which include genetic resources centres, zoo's, botanical gardens, culture collections etc.

ADVANTAGES OF EX-SITU CONSERVATION

The conservation of biodiversity can be achieved through an integrated approach balancing in situ and ex situ conservation strategies. The preservation of species in situ offers all the advantages of allowing natural selection to act, which cannot be recreated ex situ.

The maintenance of viable and self-sustainable populations of wild species in their natural state represents the ultimate goal, but habitat destruction is inevitable and endangered species need to be preserved before they become extinct. Ex situ conservation provide the opportunity to study the biology of, and understand the threats to, endangered species in order to eventually consider successful species recovery programmes, which would include restoration and reintroduction. It also has the advantage of preserving plant material and making it available for research purposes, without damaging the natural populations. Their conservation ex situ is therefore complementary to in situ conservation and can act as an "insurance policy" when species are threatened in their natural habitats It is the process of protecting an endangered species of plant or animal by removing part of the population from a threatened habitat and placing it in a new location, which may be a wild area or within the care of humans.

Ex-situ conservation has several purposes:

- Rescue threatened germplasm.
- Produce material for conservation biology research.
- Bulk up germplasm for storage in various forms of ex situ facility.

- Supply material for various purposes to remove or reduce pressure from wild collecting.
- Grow those species with recalcitrant seeds that cannot be maintained in a seed store.
- Make available material for conservation education and display.
- Produce material for reintroduction, reinforcement, habitat restoration and management.

Objective

The main objective of this unit is to give you a picture of present scenario of *ex-situ* conservation of biological resources and recently development in this field. The major objectives of present study are:

- To understand the various components of *ex-situ* conservation of biological diversity;
- To study the mechanism of *ex-situ* conservation and strategies for conservation of plants, animals and micro-organisms;
- To study the conservation process of threatened species;
- To study the institutional mechanism of *ex-situ* conservation in India.

Principles and Practices

Ex-situ conservation is the chief mode for preservation of genetic resources, which may include both cultivated and wild material. Generally seeds or in-vitro maintained plant cells, tissue and organs are preserved under appropriate conditions for long term' storage as gene bank. This requires considerable knowledge of the genetic structure of population sampling techniques, methods of regeneration and maintenance of variety of gene pools, particularly in cross-pollinated plants. These ex-situ collections of living organisms (living collections, seed banks, pollen, vegetative propagules, tissue or cell cultures) need to be managed according to strict scientific and horticultural standards to maximise their value for conservation purposes. Thus they need to be correctly identified, documented and managed and an efficient information management system put in place. Integrated conservation management can also ensure that ex-situ collections can support in-situ conservation, through habitat restoration and species recovery.

There are two ways or modes of Ex-situ conservation.

1. Conventional methods
2. Biotechnological aspects.

CONVENTIONAL METHODS OF EX-SITU CONSERVATION

Gene Banks

Plant genetic resources gene banks store, maintain and reproduce living samples of the world's huge diversity of crop varieties and their wild relatives.

They ensure that the varieties and landraces of the crops and their wild relatives that underpin our food supply are both secure in the long term and available for use by farmers, plant breeders and researchers.

Gene banks conserve genetic resources. The most fundamental activity in a gene bank is to treat a new sample in a way that will prolong its viability as long as possible while ensuring its quality. The samples (or accessions as they are called) are monitored to ensure that they are not losing viability. A cornerstone of gene bank operations is the reproduction-called regeneration-of its plant material. Plant samples must periodically be grown out, regenerated, and new seed harvested because, even under the best of conservation conditions, samples will eventually die.

To conserve and regenerate genetic resources, gene banks first must collect genetic resources. But gene banks aren't built just to conserve genetic resources; they are intended to ensure that these resources are used, whether it is in farmers' fields, breeding programmes or in research institutions. This means making sure the collections are properly characterised and documented; and that the documentation is available to those who need it. The information systems used by gene banks are becoming increasingly important tools for researchers and breeders seeking data on the distribution of crops and their wild relatives.

Community Seed Banks

Seed banks don't have to be high-tech and managed by governments or businesses. Germplasm conservation in the form of seed inmost convenient since seeds occupy a relatively small space. Their transport to various introduction centres and gene banks is also economical but the draw backs in conservation by seed are:

1. Loss of viability over passage of time and susceptibility to insect or pathogen attack.
2. Inability to maintain distinct clones except for inbreed and apomicts species.
3. Non-applicability to vegetatively propagated crop *e.g.* Dioscorea, Ipomoea, potato etc.

In many developing countries, farmers rely on informal seed systems based on local growers retention of seed from previous harvests, storage, treatment and exchange of this seed within and between communities. The informal seed sector is typically based on indigenous structures for information flow and exchange of seed. Seed banks managed within this local seed system operate on a small scale at the community level with few resources. These community seed banks and the institutions that support them are extremely important in the preservation of local varieties and for agricultural production. Much could be gained from learning more about these seed banks and working with communities to improve them. In spite of this, informal seed banks have until now received little attention or support from the scientific community or the state.

Seed Banks

Undeniably, the most cost-effective method of providing plant genetic resources for long-term ex situ conservation is through the storage of seeds under very specific conditions, following techniques well developed for crop plants by organisations such as the International Plant Genetic Resources Institute (IPGRI), previously the International Board of Plant Genetic Resources (IBPGR) and the Food and Agricultural Organisation of the United Nations (FAO). The main advantage of seed banking is that it allows large populations to be preserved and genetic erosion to be minimised by providing optimum conditions and reducing the need for regeneration (Given, 1987). Endangered plants may also be preserved in part through seedbanks or germplasm banks. The term seedbank sometimes refers to a cryogenic laboratory facility in which the seeds of certain species can be preserved for up to a century or more without losing their fertility. It can also be used to refer to a special type of arboretum where seeds are harvested and the crop is rotated. For plants that cannot be preserved in seedbanks, the only other option for preserving germplasm is in-vitro storage, where cuttings of plants are kept under strict conditions in glass tubes and vessels.

However, when a natural population still exists, it may be advisable to re-collect rather than regenerate a new supply from the previous collection as damage can occur such as mutations associated with the loss of viability during storage. The success of long-term conservation of seeds is dependant on continuous viability monitoring and regeneration or re-collection when the viability of the sample drops below a minimum level (Eberhart, Roos and Towill, 1991). It is important to realise that however much care is taken during seed collection, regeneration and storage, natural selection cannot be simulated and some artificial selection will be unavoidable, which inevitably leads to unpredictable genetic changes (Ashton, 1988).

The recommended preferred standards for long-term seed storage of orthodox species recommended by IPGRI is to dry the seeds to a moisture content of below 7 per cent and seal the dried seeds in a moisture-proof container such as laminated foil bags, aluminium cans or glass jars for storage at a low temperature of -18°C. Clearly this is only applied to true orthodox species of crops and their relatives. However because less is known about wild species, a temperature of -4°C and moisture content of 7-8 per cent is advisable to begin with. The activities in seed banks should take the following sequence: collection, seed preparation, seed drying, packaging, storage, periodic germination tests, seed regeneration, re-storage and documentation at each stage of activity. When plant species are recalcitrant or long-term conservation cannot be achieved through seed banking, different methods have been developed with their respective merits, such as field gene banks, *in vitro* germplasm collections, pollen and DNA banks. This is in many cases the main problem faced by botanic gardens dealing with many different species of which, a great proportion are be recalcitrant. They need to develop complementary conservation techniques and adopt different methods.

Ex-situ conservation of plant genetic resources can be achieved through different methods such as seed banks, field gene banks, *in vitro* storage methods, pollen banks and DNA banks. The major consideration for long-term conservation of germplasm collections is the determination of the seed behaviour of each individual species to be preserved during storage under dry conditions and cold temperatures. If the seeds can be dried to a low percentage humidity such as below 8 per cent, in the majority of cases, the seeds will then withstand very cold temperatures of below 20°C.

When the seeds tolerate these conditions and remain viable after many years of storage, they are classified as orthodox (desiccation tolerant) as opposed to recalcitrant (desiccation intolerant) when they do not. However, there might be many other reasons why a particular species cannot be preserved in seed banks, such as a very low production of seed or long-life cycle species, such as trees and perennials. Their conservation as seeds would make the study of their biology difficult.

Botanical Gardens

Botanical gardens and zoos are the most conventional methods of ex-situ conservation, all of which house whole, protected specimens for breeding and reintroduction into the wild when necessary and possible. These facilities provide not only housing and care for specimens of endangered species, but also have an educational value. They inform the public of the threatened status of endangered species and of those factors which cause the threat, with the hope of creating public interest in stopping and reversing those factors which jeopardise a species' survival in the first place. They are the most publicly visited ex-situ conservation sites.

The history of botanic gardens can be traced as far back as the Hanging Gardens of Babylon, built by Nebuchadnezzar in 570 BC as a gift to his wife. Early botanic gardens were designed mainly for the purpose of recreation. By the 16th Century, however, they had also become important centres for research. They promoted the study of taxonomy and became a focal point for the study of aromatic and medicinal plants. More recently, they have taken on significant conservation responsibilities and they often have conservation facilities, such as seed banks and tissue culture units.Botanical gardens hold living collections. Indeed botanical garden conservation could be considered as field gene bank or seed gene bank or both, depending on the conservation method being used. However, they tend to focus their conservation efforts on wild, ornamental, rare and endangered species. Most of the germplasm conserved in botanical gardens do not belong to the plant genetic resources for food and agriculture.

A botanic garden which wishes to start a small seed bank/gene bank would be advised to start with collecting germplasm that is very well documented from their living plant collection. This would allow them to experiment with a wide range of species and find suitable facilities and techniques for their particular needs. Once the set up is organised and functional, it would be advisable to collect accessions directly from the wild in order to distribute a wider genetic variability and to reduce the effect of domestication on the genetic make up of the accessions. Last two hundred years, efforts of botanic gardens in collecting plant material, and the great efforts on crop germplasm collection during the 1970s and the 1980s, there are a large number of gene banks and germplasm collections around the world. According to the FAO and World Information and Early Warning System (WIEWS) database, it is estimated that there are now more than 2000 botanic gardens known around the world in over 150 countries.

Together, they maintain more than 6 million accessions in their living collections and 142 million herbaria specimens in the botanic garden herbaria. 60 per cent of the total numbers of accessions are known to be stored in medium-term or long-term facilities, 8 per cent in short-term facilities and 10 per cent in field gene banks, *in vitro* and under cryopreservation. Clearly, seed storage is the predominant form of plant genetic resource conservation, accounting for about 90 per cent of the total accessions held ex situ.

It is very difficult to give an estimate of the type of collections stored around the world as such information is known for only a third of the accessions in the WIEWS database. However, it has been estimated that 48 per cent of all accessions are advanced cultivars or breeders' lines, while over a third are landraces or old cultivars and about 15 per cent are wild or weedy plants or crop relatives.

PROTECTED AREA NETWORK

Networks of protected areas are those that are linked by a common focus, often having similar values and management approaches. UNEP-WCMC works to identify, support and promote these networks.

About Protected Areas

Protected areas are internationally recognised as regions set aside primarily for nature and biodiversity conservation and are a major tool in managing species and ecosystems which provide a range of goods and services essential to sustainable use of natural resources. The term 'protected area' includes Marine Protected Areas (MPA) the boundaries of which will include some area of ocean.

Every nation in the world has designated protected areas, often with an extensive network of protected areas having been developed over many years, and the purpose of them is generally universal in the intention of limiting levels of human occupation and restricting the exploitation of natural resources. These systems vary considerably between countries, depending on national needs and priorities and on differences in legislative, institutional and financial support.

BIOSPHERE RESERVES

A biosphere reserve is an area proposed by its residents, ratified by a national committee, and designated by UNESCO's' Man and Biosphere' programme [MAB], which demonstrates innovative approaches to living and working in harmony with nature. One of the primary objectives of MAB is to achieve a sustainable balance between the goals of conserving biological diversity, promoting economic development, and maintaining associated cultural values.

The term 'biosphere' refers to all of the land, water and atmosphere that supply life on earth. The word 'reserve' means that it is a special area recognised for balancing conservation with sustainable use. The term 'reserve' does not mean that these places are set aside from human use and development. In fact, the study of human use is an important part of the biosphere reserve programme. Each biosphere reserve demonstrates practical approaches to balancing conservation and human use of an area. They are excellent examples of community-based initiatives that protect our natural environment while ensuring the continued healthy growth of the local economy. Biosphere reserves recognise that quality economies require quality environments, and that conservation is important for both.

The biosphere reserve programme is entirely voluntary. Authority over land and water use does not change when a biosphere reserve is designated in Canada. Government jurisdictions and private ownership rights remain as they were before designation.

DEFINITION

A Biosphere Reserve is a unique and representative ecosystem of terrestrial and coastal areas which are internationally recognised, within the framework of UNESCO's Man and Biosphere (MAB) programme. The biosphere reserve should fulfil the following three objectives:

- In-situ conservation of biodiversity of natural and semi-natural ecosystems and landscapes
- Contribution to sustainable economic development of the human population living within and around the Biosphere Reserve.

- Provide facilities for long term ecological studies, environmental education and training and research and monitoring.

In order to fulfil the above objectives, the Biosphere Reserves are classified into zones like the core area, buffer area. The system of functions is prescribed for each zone.

Zonation of the Biosphere Reserves

One or more core zones: Securely protected sites for conserving biological diversity, monitoring minimally distributed ecosystems and undertaking non-destructive research and other low-impact uses (such as eco-tourism and education) A well defined buffer zone(s): Usually surrounds or adjoins the core zones and is used for cooperative activities compatible with sound ecological practices, including environmental education, recreation and applied and basic research.

A flexible transition area or area of cooperation: May contain a variety of agricultural activities, settlements and other uses and in which local communities, management agencies, scientists, non-governmental organisations, cultural groups, economic interests and other stakeholders work together to manage and sustainably develop the area's resources. To fulfil the main objectives of a Biosphere Reserve, the local people's support is essential. In 1994, UNESCO recommended 10 important points for this purpose. They are:

Three Functions of a Biosphere Reserve

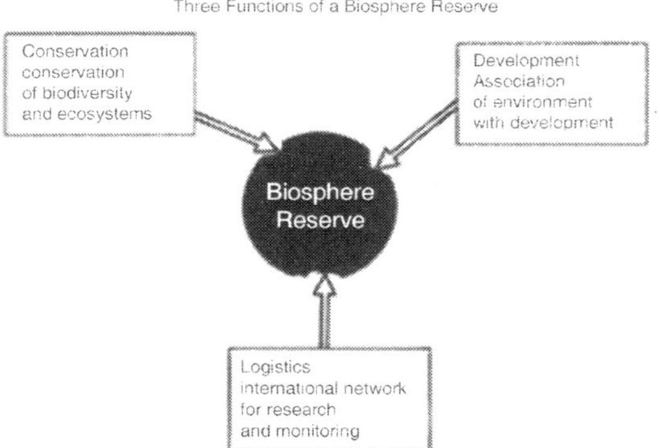

- Recognise that local support is fundamental to the long-term conservation and sustainable use of biodiversity
- Ensure that local populations participate as true partners in designing and managing conservation programmes.
- Allow local populations to identify their own socio-economic needs.
- Ensure that people who bear the costs of conservation projects (*e.g.* restrictions on fishing) also receive a huge proportion of benefits (*e.g.* tourist revenues).
- Initiate research activities that identify options for sustainable use of biodiversity

- Use indigenous knowledge to manage protected areas to the extent possible
- Ensure that local populations have maximum stewardship over local resources (rather than government agencies at the regional and national levels).
- Offer income-earning activities and/or services (*e.g.* improved access too markets, low interest credit, controlled access to resources) to local populations and others with a stake in conservation-development projects.
- Provide local population with the skills and resources needed to make life-style changes necessitated by conservation measures.
- Educate local populations about the rationale for conservation and the relationship between conservation actions and benefits.

BIOSPHERE RESERVE NETWORK

The first Biosphere Reserve Congress was held at Minsk (Belarus) in 1983 and reported 226 Biosphere reserves in 62 countries around the world. The second Congress was held at Seville in 1995 and reported 324 Biosphere Reserves in 82 countries. Presently there are 425 Biosphere Reserves are existing in 95 countries.

Biosphere Reserve Network in India

The Government of India constituted a committee of experts in 1979 to identify the potential areas for recognition as Biosphere Reserves as per the guidelines of UNESCO (MAB). The experts identified 14 sites to be declared as Biosphere Reserves. Out of 14 sites, 13 have been declared as Biosphere Reserves.

AGASTHIAYAMALAI BIOSPHERE RESERVE

State: Kerala

Area: 1,701sq. km.

Endemic Flora: Rudraksha tree, Black plums, Gaub tree, Wild dhaman

Endemic Fauna: Lion-tailed macaque, Slender loris, Great pied hornbill. The forest tracts of Neyyar, Peppara, Shendumey wildlife Sancturias and Achencoil, Thenmala, Konni, Punalur, Thiruvananthapuram Divisions and Agasthyavanam Specil Division are included in the this reserve. This reserve is likely to be extended to parts of Kanyakumari and Tirunelveli districts of Tamilnadu, the Mundanthurai Tiger Reserve and Kalakkadu wildlife sanctuary. Forest type includes thorn, moist deciduous and semi-evergreens. The area is rich in plant and animal diversity. This Biosphere Reserve harbours the most diverse eco-systems in Peninsular India. The forests, falling both in Tamil Nadu and Kerala, have many endemic species of plants unique to Peninsular India. As many as 35 of these plants are threatened or endangered species. Forests occur in the altitudinal range of less than 300 metres to more than 2,800 metres around Agasthyakudam.

Flora

So far, 2000 species of flowering plants have been reported. 30 new plant species are recorded from this region, about 100 endemic and 50 rare. A few examples are Aristolochia (Snake root), Cardiospermum (Faux persilo), Ceropegia (Taper vine), Dioscorea (Wild yam), Gloriosa (Glory lily), Rauvolfia (Serpentine wood) and Smilax (Laurel leaf greenbrier)

Fauna

Threatened animal species found in this reserve are tiger, lion-tailed macaque, great pied hornbill and slender loris.

Threats

The main threats are several settlements in the existing hydel and irrigation projects, cultivation of plantation crops and increase in the number of pilgrims to Agastyakudam area.

DIHANG – DIBANG BIOSPHERE RESERVE

State: Arunachal Pradesh
Area: 5,111 sq. km.
Endemic Flora: Cyathea(tree fern), Begonia, Lady's slipper orchid
Endemic Fauna: Red panda, Himalayan black bear, Green pit viper, Takin

The Dihang-Dibang Biosphere Reserve is situated in the districts of West Siang, Upper Siang and Dibang valley of Arunachel Pradesh. This reserve is almost totally forested, having villages and cultivated lands located on lower slopes and terrace edging the major river systems. The reserve is the last stronghold for many Himalayan species. A population of above 10,000 people who live in the reserve are primarily of the Adi, Buddhist and Mishmi tribes.

Flora

The vegetation types that occur in the Reserve are sub-tropical broad-leaved, sub-tropical pine, temperate broad-leaved, temperate conifer, sub-alphine woody shrub, alpine meadow (monton), bamboo brakes and grassland. The Reserve forms a part of the world's "Biodiversity Hot Spots". About 1500 flowering plants occur in this Biosphere Reserve. The rare orchid Vanda and 50 species of Rhododendron are found here. It is a shelter for saprophytes like Monotropa uniflora (Indian pipe) and Epipogium spp. (Widerbart) parasitic plants are found here. Some plants listed in "primitive" families are seen here and include Magnolia Compbellii (Charles raffill), Holboelli latifolia (Ribbonwood), and various species of rare and endangered species in the Reserve include Cyathea sp. (Rough tree fern) and Coptis teeta (Indian goldthread).

Fauna

About 45 species of insects, including moths and butterflies have been reported. There are about 195 species of birds, including, the pale-capped pigeon, a globally threatened species, others include Purple cochoa, Nepal cutia and pale blue flycatcher.

The Wedged billed wren-babbler, one of the rarest members of the Laughing thrush and babbler family (Timaliidae), has been found here. New species such as the water pipit, Japanese bush warbler, isabeline wheatear are reported here.

Mammals include leopard, snow leopard, golden cat, jungle cat, marbled cat and leopard cat. The critically endangered musk deer also lives at these elevations but it is confined to thick forest areas. Other mammals include gaur, serow, Himalayan black bear, sloth bear, Indian wild dog, red fox, deer, Assamese macaque, otter, squirrel and civet.

Threat

There are not many conservation problems except for poaching and collection of medicinal plants.

DIBRU-SAIKHOWA BIOSPHERE RESERVE

State: Assam

Area: 765 sq. km.

Endemic Flora: Rauvolfia (Sarpagandhi), Benteak, Livistona (orchid)

Endemic Fauna: White winged wood duck, Hollock-gibbon, Wild buffalo

This reserve is situated in the south bank of the river Brahmaputra in the extreme east of Assam state and includes Dibru- Saikhowa wildlife sanctuary and Dibru- Saikhowa National Park. Bio-geographically, the area exhibits the properties of both the Indian and the Malayan sub-regions. The annual rainfall ranges from 2300 mm to 3800 mm. There are around 38 villages in the buffer zone of the reserve. The forest types of the reserve comprises of semi-evergreen, deciduous, littoral and swamp forest and patches of wet evergreen forest. The entire reserve is flat terrain, situated on the flood plain of Brahmaputra.. There are a large number of perennial and seasonal channels namely Kolomi, Salbeel nala, Dadhia nala, Chabru nadi, Laikajan, Ananta nala, Hatighuli nala, Dimoruhola and Ajukhanala.

Flora

Major tree species of biosphere reserve are: Salix terasperma (India willow), Bishcofia javanica (Blume javanese bishopwood), Dillenia indica (Hondapara tree), Bombax ceibe (Red silk cotton tree), Lagerstromia parviflora (Landia), Anthoephalus cadamba (Indian seasde oak), Artocarpus chaplasa (Taungpienne), Mesua ferrea (Indian rose chestnut), Dalbergia sissoo (Sissoo) and Ficus spp. Commonly found orchids of the Reserve are Rhynocostylis retusa (Blume saccolabium blumei Lindley), and Pholidota articulate (Rattlesnake-tail orchid). Grassess such as Aurondo donax (Giant reed), Phragmities karka (Flute reed), Imperata cylindrica (Japanese blood grass) and Saccharum sp. are found in the Reserve.

Threatened and endangered medicinal plants of the Reserve are Rauvalfia serpentine (serpentine wood), Hydnocarpus kurizii (Chaulmoogra tree), Holarrhen antidysenterica (Conessi tree), Costus speciosus (Spiral ginger), Dioscorea alata (Winged yam)and Dioscorea bulbifor (Air potato vine).

Fauna

36 species of mammals are recorded from the Reserve. Of these, 12 belong to Schedule I of the Indian Wildlife Protection (Act) 1972. Royal Bengal tiger, leopard, clouded leopard, jungle cat, sloth bear, golden jackal, dole, Small Indian civet, small Asian mongoose, common mongoose, common otter, Malayan giant squirrel, Pallas' squirrel, Himalayan hoary bellied squirrel, common giant flying squirrel, Indian hare, pangolin, Himalayan mole, ground shrew, Gangetic dolphin, slow loris, capped langur, hoolock gibbon, Asian elephant, feral horses, wild boar, sambar, hog deer, barking deer and Asiatic water buffalo are the larger mammals found in the Reserve. There are two species of monitor lizards, eight species of turtles and eight species of snakes. The turtles such as Malayan box turtle, Asian leaf turtle, spotted pond turtle, brown roofed turtle, Assam roofed turtle, Indian tent turtle, Indian soft-shell turtle and narrow headed soft-shell turtle, besides 62 species of fish have been recorded.

About 350 resident as well as migratory birds have been recorded. These include great crested grebe, spot billed pelican, white bellied pelican, lesser adjutant stork, white winged duck, Bayer's pochard, greater spotted eagle, Bengal florican, pale capped pigeon, great pied hornbill, marsh babbler, Jordon's babbler, black breasted parrot bill, etc.

Threats

The annual floods cause a crisis for the wildlife. Grazing and siltation are also changing the habitat quality.

GREAT NICOBAR BIOSPHERE RESERVE

State: Andaman and Nicobar Islands

Area: 885 sq. km.

Endemic Flora: Screw pine, Nipa palm, Ceylon iron wood

Endemic Fauna: Crab eating macaque, Nicobar megapode, Giant robber crab,Nicobar serpent eagle

The Great Nicobar reserve with a total geographic area of 1044 sq.km. is the southern-most island of Andaman and Nicobar Archipelago and also the southern-most part of India. The area between the Alexandra River (West Coast) and Chengrappa Bay forms the Core zone-I and area in the southern part between Sahni and Anti Range of hills forms the Core zone-II. The core zone is kept absolutely undisturbed except for already existing settlements.

The Island presents varied natural panorama covered with virgin lush evergreen dense tropical forests extending from seacoast to the tip of the hills. This area is the home for the most endangered species, megapode as well as the edible-nest swiftlet (Collocalia fuciphaga). The area is the home of the Shomphens, one of the most primitive tribes of India

Flora

The Great Nicobar Reserve represents the topical rain forests. About 85 per cent of the forest in this Reserve is still in its virgin condition and rich in species content. Important species include 5 species of Ficus, 2 species of Terminalia (Gallnut), Pandanus tinctoria (Screw pine), Pinanga costata (Blume areca), Pterygota alata (Buddha's coconut) Ipomeoea spp. (Morning glory), Casuarina sp., (Beafwood) Nypa fruticans (Nipa palm), Albizia procera (White siris), Canarium euphyllum(Makok fan), Calophyllum spp (Lagarto caspi).

Syzygium cumini(Indian blackberry), Eleocarpus sphaericus (Rudhrakhsa tree), Manilkara littoralis (Sea mohwa), Rhizophora spp., (Red mangrove), Bruguiera spp., (Large-leaved orange mangrove), Cerips tagal (Tagal mangrove), bamboo and canes. The charsteristic Tree fern (Cyathea albosetacea) as well as the beautiful ornamental orchid, the Phalanopsis (Phalaenopsis speciosa)are confined to this southern most island.

Fauna

The unique fauna of this Reserve include Crab eating macaque (Macaca fascicularis), Salt water crocodile(Crocodylus porosus), Giant leather back turtle (Dermochelys coriacea), Malayan box turtle, Nicobar tree shrew, Nicobar megapode, reticulated python and the Giant robber crab (Birgus lactro). Other species are Andaman wild boar, palm civet, fruit bat, Nicobar pigeon, white bellied sea eagle, Nicobar serpent eagle, parakeets, Nicobar parakeets, water and monitor lizard.

Threats

The area around the reserve is inhabited by the aboriginal tribes of Shomphens and Nicobaris who are given rights under Section 65 of Wildlife (Protection) Act, 1972. These tribes hunt wild animals, particularly the Andaman wild boar and also collect other forest and wildlife produce. Hunting of Andaman wild boar is reported to have affected the population of this endemic and endangered species. Poachers from neighbouring countries frequently visit the reserve and the nearby islands mainly for collection of sea cucumber, nests of edible-nest swift let and for poaching crocodiles, turtles and other wildlife.

GULF OF MANNAR BIOSPHERE RESERVE

State: Tamil Nadu
Area: 10,500 sq. km.
Endemic Flora: Morning glory, Jatropha, Halophila grass
Endemic Fauna: Sea Cow, Sea Anemone, Sea fans

The Gulf of Mannar reserve is the first marine Biosphere Reserve established in India and is situated along the southern coast of Tamilnadu. The Biosphere Reserve includes the Gulf, the adjoining coasts and also the small islands dotting the gulf. The reserve also includes a Marine National Park.

Flora

About 160 species of algae have been recorded here of which some 30 species are edible seaweeds. The area is also rich in sea grasses which provide food for sea mammals, particularly the dugong. The mangrove vegetation of the islands consists of species of Rhizophora (Red mangrove), Avicennia (Black mangrove), Bruguieria (Large-leaved orange mangrove), Ceriops (Tagal mangrove) and Lumnitzera (Sandy mangrove). About 46 species of plants are endemic to Gulf of Mannar.

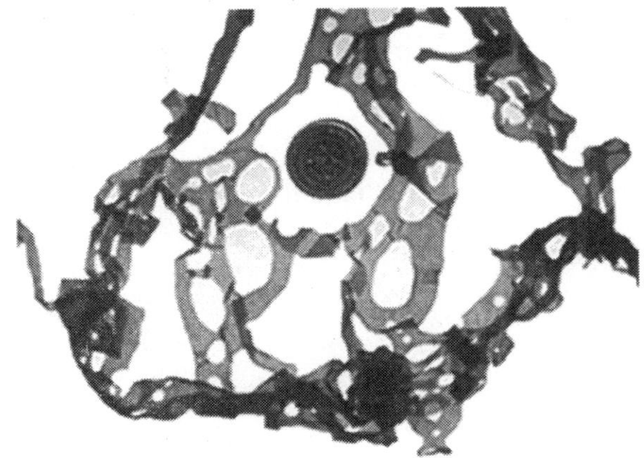

Fauna

The Gulf area has beautiful coral reefs that harbour a wide variety of marine vegetation and animals. Productive beds of pearl oysters, prawn species, edible bivalves, sea anemones, ascidarians and the sea cow (Dugong dugon) occur in the Reserve.Among the fauna, the invertebrates are represented by 280 species of sponges, 92 species of corals, 22 species of sea fans, 160 species of polychaetes, 35 species of prawns, 17 species of crabs, 7 species of lobsters, 17 species of cephalopods and 103 species of echinoderms.

Threats

Illegal coral mining for cement industries and indiscriminate collection of sea grass is the main threat to the reserve. 65 per cent of the existing coral reefs in the area are dead, mostly due to human interference.

KHANGCHENDZONGA BIOSPHERE RESERVE

State: Sikkim

Area: 2,619 sq. km.

Endemic Flora: Anemone, Uvaria, Sikkim Rhododendron, Sikkim Mahonia. Endemic Fauna: Tibetan sheep, Musk deer, Monal pheasant, Snow patridge. Khangchendzonga is bounded in the North by Khangchendzonga National Park and Lungnakla ridge, by the Teesta River in the East, and by reserve forests in the South and West Forest Divisions. This reserve is one of the high altitudes reserve of the Indian continent as Khangchendzonga Peak, the third highest peak in the world exists within the reserve. The reserve covers 70 per cent of the northern forest area and 30 per cent of the geographical area of the State, a total area of 2619 sq.km. It contains lofty, picturesque and beautiful peaks of heights ranging from 5825m and above. Glaciers and high altitude lakes constitute one of the world's highest ecosystems. The reserve is surrounded by a number of tiny villages. The population consists of Lepcha, Bhutia and Nepalese.

Flora

The forest types of the Reserve are sub-tropical broad leaved hill forest, Himalayan wet temperate forest, and temperate broad leaved forest, mixed coniferous forest, sub-alpine a forests and dry alpine forest. Dominant species of the temperate forests are Phalat, Shrub live oak, Ghoge champ, and Cinnamon. The mixed coniferous forests occur in higher altitude with fir, East Himalayan fir, Maple, Spruce, Juniper and Rhododendron. In the Alpine scrub and grasses, the common shrubs and herbs found are Primrose, Rhododendron, bottled gentian and stagger weed. There are many medicinal herbs in the Reserve like Aconite., Kutki root, Spikenard, Rhubarb and Ginseng.

Fauna

The high altitude alpine and plateau regions harbours rare and endangered species of animals. The snowleopard of the alpine land holds the position at the apex of the biological pyramid. Himalayan red panda, musk deer, nayan or the Tibetan sheep, bharal or blud sheep, Himalayan tahr, marco polo sheep, marmots and monkeys are a few animals are found here.

This reserve also harbours many species of birds. The pheasants include monal, trogopan and blood pheasant. The other species are Tibetan snow cock, Himalayan snow cock, snow partridge, Lammergier, forest eagle owl, Tibetan horned eagle, owl, eagles, falcons, hawks, snow and rock pigeon.

Threats

Landslides are one of the important factors causing hindrance to the conservation of bio-resources, resulting in the loss of soil as well as restrictions in the movement of wildlife during migration.

MANAS BIOSPHERE RESERVE

State: Assam
Area: 2,837 sq. km.
Endemic Flora: Catechu tree, Sissoo, White siris

Endemic Fauna: Pygmy hog, Golden lungur, Assam roofed turtle

Manas gets its name after the serpent Goddess 'Manasa' (Durga). The unique location of Manas at the confluence of the Indian, Ethiopian and Indo-Chinese realm along with hot and humid climate makes this reserve a treasure of immense diversity and endemism of the flora and fauna. It is a unique representation of tropical, humid "Bengal Rain Forests" in the Indo-Malaya realm. The reserve extends along the Himalayan forest hills to the north of Brahmaputra valley. The Manas River is the largest Himalayan tributary of the Brahmaputara River. Manas was described a World Heritage site by UNESCO in 1985.

Flora

Two major biomes are represented in Manas: the grassland biome (Savannah and Teri) and the forest biome (Bengal rain forests). A total of 543 plants species have been recorded from the core zone. Of these, 30 are Pteridophytes and Gymnosperms, 139 are monocots, and 37 are dicots.

Fauna

61 species of mammals, 327 birds, 2 reptiles, 7 amphibian and 54 species of fishes are recorded. Hispid hare, pigmy hog and golden langur are some of the rare species of animals found in the park, apart from tigers, elephants, rhinoceros, wild buffalo, wild boar, sambhar, swamp deer, and hog deer. During winter Manas is full of migratory birds like the riverchats, forktails, cormorants and ducks like the Ruddy Shell-Duck. Regular woodland birds like the Indian Hornbill. And the Pied Hornbill also found here.

The pygmy hog, hispid hare and golden langur are endemics among the animals. In fact the only viable population of the smallest and rarest wild suid, the pygmy hog, exists in Manas and nowhere else in the world. In 1988, the Assam roofed turtle, perhaps the least known of the species was found in Manas and listed under the IUCN Action Plan for Conservation. The Bodo communities are traditionally dependent on the forests for fodder, timber, firewood, thatch, wild vegetables and fruit, and fish. The other threats to the Reserve are soil erosion, weed infestation and trans-border management

NANDADEVI BIOSPHERE RESERVE

State: Uttaranchal

Area: 5,860 sq. km.

Endemic Flora: Salep Orchid, Silver weed, Fairy candelabra, Fairy Primrose. Endemic Fauna: Himalayan tahr, Brown bear, Koklas pheasant

The Nanda Devi reserve includes Nanda Devi National Park and Valley of Flowers National park in the core zone. Nanda Devi reserve represents the Himalayan zone of the bio-geographic zonation of India. Nanda Devi National Park was included in the list of World Heritage Site during 1991. It includes Nanda Devi and several other major peaks like Danagiri, Changbang, Trishul, etc. The reserve includes parts of Chamoli, Pithoragarh and Bagheswar districts in Uttaranchal. The main portion of the reserve falls in Chamoli district of Garhwal Himalayas. All the households depend entirely on the forests for fuel, fodder, timber and leaf litter for organic manure. Many plant species are used in traditional health care systems.

Flora

The major forest types of the Reserve are temperate forest, sub-alpine forest and alpine land. Botanical Survey of India has identified 800 species of plants. A few important species are: Potenlilla sp.,(Silver weed), Androsac sp.,(Fairy candelabra), primula sp., (Fairy primrose) Orchis latifolie (Salep orchid) and Rhododendron sp.(Satsuki azalea).

Fauna

The Biosphere Reserve has a rich fauna. The various surveys conducted by Zoological Survey of India and others have shown the presence of 18 mammals and nearly 200 birds in the area. Of these, seven mammals and eight birds are endangered species.

Mammals include Snow leopard, Black bear, Brown bear, Musk deer and Himalayan tahr. Endangered bird species include the monal pheasant, koklas pheasant, western tragopan, snow-cock, golden eagle, steppe eagle, black eagle and bearded vulture.

Cultural Heritage of the Area

The people of the Biosphere Reserve are very poor. Land holdings are very small and the literacy rate is very low due to the remoteness. They have their own culture, tradition and religious beliefs.

The principal occupation is agriculture and sheep rearing. Prior to 1962, Bhutias had a barter trade system with the Tibetans and were masters of Trans-Himalayan trade.

Threats

The major threats to the ecosystem are collection of endangered plants for medicinal use, forest fires, poaching and visits by pilgrims.

NILGIRI BIOSPHERE RESERVE

State: Karnataka, Kerala and Tamil Nadu
Area: 5,520 sq. km.

Endemic Flora: Vanda, Liparis, Bulbophyllum, Spiranthes, Thrixspermum
Endemic Fauna: Nilgiri tahr, Nilgiri langur, Lion – tailed macaque

The Nilgiri Biosphere Reserve was the first Biosphere Reserve in India. It is located in the Western Ghats and includes 2 of the 10 bio-geographical provinces of India. Wide ranges of ecosystems and species diversity are found in this region. Thus, it was a natural choice for the premier Biosphere Reserve of the country. The Nilgiri Biosphere Reserve falls under the biogeographic region of the Malabar rain forests. The Mudumalai Wildlife Sanctuary, Wyanaad Wildlife Sanctuary, Bandipur National Park, Nagarhole National Park, Mukurthi National Park and Silent Valley are the protected areas present within this reserve.

Vegetation

The forest types found in the Nilgiris are thorn scrub forest, dry deciduous forest, moist deciduous forest, wet evergreen forest, shoals, grasslands, marshes and swamps. Woody climbers and epiphytes are also found here.

Flora

The reserve is very rich in plant diversity. About 3,300 species of flowering plants can be seen here. Of the 3,300 species 132 are endemic to the Nilgiri Biosphere Reserve. The genus Baeolepis is exclusively endemic to the Nilgiris. Some of the plants entirely restricted to the Nilgiri Biosphere Reserve include species of Calacanthus (Carolina allspice), Baeolepis (Dogbanes), Arodina and Wagatea (False Thorn) etc. Of the 175 species of orchids found in the Nilgiri Biosphere Reserve, 8 are endemic to the Nilgiri Biosphere Reserve. These include endemic and endangered species of Vanda, Liparis (Lily-leaved), Bulbophyllum (Medusa's head orchid) Spiranthes (October ladies-tresses) and Thrixspermum (Chi-tou wind orchid).

Some of the common trees found here are sandal, Indian rosewood, jackfruit, jamun, ironwood, rhododendron, hill gooseberry, etc. Many varieties of orchids and the unique kurinji shrub that produces a blue coloured flower to which the Blue Mountains owe their name are also found here. Climbers like the black pepper and numerous herbs thrive here.

Fauna

The fauna of the Nilgiri Biosphere Reserve includes over 100 species of mammals, 350 species of birds, 80 species of reptiles and amphibians. 300 species of butterflies and innumerable invertebrates, 39 species of fish, 31 amphibians and 60 species of reptiles endemic to the Western Ghats also occur in the reserve. Fresh water fish such as Danio neilgheriensis (Pearl danio), Hypselobarbus dubuis and Puntius bovanicus are restricted to the reserve. The Nilgiri tahr, Nilgiri langur, slender loris, blackbuck, tiger, gaur, Indian elephant and marten are some of the animals found here.

People

Tribal groups like the Todas, Kotas, Irullas, Kurumbas, Paniyas, Adiyans, Edanadan Chettis, Cholanaickens, Allar, Malayan, etc. are native to the reserve. Except for Cholanaickens who live exclusively on food gathering, hunting and fishing, all the other tribal groups are involved in their traditional occupation of agriculture.

Threats

The Nilgiri Biosphere Reserve has been enduring human interference for a very long time through development projects such as hydroelectric power projects, agriculture, horticulture, intensive felling, monoculture, grazing, forest fires, development and construction activity and unplanned tourism have brought about substantial change in the ecology of the area. Environmental problems are noticed in different parts of the reserve.

Bibliography

Ajai Srivastava: *Taxonomy of Moths in India*, IBD Publication, Delhi, 2002.

Archana Singh and Kapila Shekhawat: *Soil Nutrient Management for Sustainable Agriculture*, Pointer Publication, Delhi, 2013.

D.R. Khanna, A.K. Chopra and G. Prasad: *Aquatic Biodiversity in India : The Present Scenario*, Daya Publication, Delhi, 2005.

Deepa Joshi: *Taxonomy of Microbiology*, Oxford Book Company, Delhi, 2012.

George H.M. Lawrence: *Taxonomy of Vascular Plants*, Scientific Publication, Delhi, 2012.

Jaspal Juneja: *Fundamental of Biodiversity Taxonomy*, Cyber Tech Publication, Delhi, 2010.

K. Muthuchelian: *Glimpses of Animal Biodiversity*, Daya Publication, Delhi, 2013.

Kumudranjan Naskar: *Aquatic and Semi-aquatic Plants of the Lower Ganga Delta: Its Taxonomy, Ecology and Economic Importance*, Daya Publication, Delhi, 1990.

Lingaraj Patro: *Aquatic Biodiversity*, Discovery Publishing House, Delhi, 2010.

M Sivadasan and Philip Mathew: *Biodiversity, Taxonomy and Conservation of Flowering Plants*, Mentor Books, Delhi, 1999.

M.P. Nayar, R.V. Varma and C.K. Peethambaran: *Biodiversity and Taxonomy*, Narendra Publishing House, Delhi, 2012.

M.P. Singh and S.G. Abbas: *Essentials of Plant Taxonomy and Ecology*, Daya Publication, Delhi, 2005.

Manas Malhotra and Susheela M. Das: *A Textbook of Taxonomy*, Wisdom Press, Delhi, 2012.

Pawan Kumar Bharti, Avnish Chauhan and H.A.H. Kaoud: *Aquatic Biodiversity and Pollution*, Discovery Publication, Delhi, 2013.

Philip Mathew and M. Sivadasan: *Diversity and Taxonomy of Tropical Flowering Plants: Prof. K.S. Manilal Commemoration Volume*, Mentor Books, Delhi, 1998.

R.K. Tandon and Prithipalsingh: *Biodiversity, Taxonomy and Ecology : Prof. K.M.M. Dakshini Festschrift*, Scientific Publication, Delhi, 1999.

R.M. Johri and Sneh Lata: *Taxonomy-III : Gamopetalae*, Sonali Publication, Delhi, 2005.

Rajeshwar Dayal: *Marine Biodiversity*, Surendra Publication, Delhi, 2011.

Rajiv K. Gupta: *Advancements in Invertebrate Taxonomy and Biodiversity*, Agrobios Publication, Delhi, 2011.

Ramesh Chandra Tripathi: *Biosystematics and Taxonomy*, University Book House, Delhi, 2005.

S. Kannaiyan and K. Venkataraman: *Marine Biodiversity in India*, Associated Publishing Company, Delhi, 2011.

Siddhartha Sarkar: *Accounting for Managers: Taxonomy of Financial Cost and Management Accounting*, Arise Publishers, Delhi, 2012.

Susheela M. Das: *A Textbook of Plant Taxonomy : Theory and Objectives*, Wisdom Press, Delhi, 2013.

T.N. Ananthakrishnan and K.G. Sivaramakrishnan: *Animal Biodiversity : Patterns and Processes*, Scientific Publication, Delhi, 2006.

T.N. Lakhanpal and K.G. Mukerji: *Taxonomy of the Indian Myxomycetes*, Bishen Singh Mahendra Pal Singh, Delhi, 2011.

Vaibhav Suri: *The Changing Landscape of Plant Sciences: Taxonomy and Ecology*, Cyber Tech Publication, Delhi, 2010.

Index